GRANULAR PHYSICS

The field of granular physics has burgeoned since its development in the late 1980s, when physicists first began to use statistical mechanics to study granular media. They are prototypical of complex systems, manifesting metastability, hysteresis, bistability and a range of other fascinating phenomena.

This book provides a wide-ranging account of developments in granular physics, and lays out the foundations of the statics and dynamics of granular physics. It covers a wide range of subfields, ranging from fluidisation to jamming, and these are modelled through a range of computer simulation and theoretical approaches. Written with an eye to pedagogy and completeness, this book will be a valuable asset for any researcher in this field.

In addition to Professor Mehta's detailed exposition of granular dynamics, the book contains contributions from Professor Sir Sam Edwards, jointly with Dr Raphael Blumenfeld, on the thermodynamics of granular matter; from Professor Isaac Goldhirsch on granular matter in the fluidised state; and Professor Philippe Claudin on granular statics.

ANITA MEHTA, a former Rhodes scholar, is currently a Radcliffe Fellow at Harvard University. She is well known for being one of the pioneers in granular physics, and is credited with the introduction of many new concepts in this field, in particular to do with the competition of slow and fast modes in granular dynamics.

GRANULAR PHYSICS

ANITA MEHTA

Harvard University

With contributions from

SIR SAM EDWARDS AND RAPHAEL BLUMENFELD
ISAAC GOLDHIRSCH
PHILIPPE CLAUDIN

CAMBRIDGE
UNIVERSITY PRESS

Shaftesbury Road, Cambridge CB2 8EA, United Kingdom

One Liberty Plaza, 20th Floor, New York, NY 10006, USA

477 Williamstown Road, Port Melbourne, VIC 3207, Australia

314–321, 3rd Floor, Plot 3, Splendor Forum, Jasola District Centre, New Delhi – 110025, India

103 Penang Road, #05–06/07, Visioncrest Commercial, Singapore 238467

Cambridge University Press is part of Cambridge University Press & Assessment,
a department of the University of Cambridge.

We share the University's mission to contribute to society through the pursuit of
education, learning and research at the highest international levels of excellence.

www.cambridge.org
Information on this title: www.cambridge.org/9780521660785

First published 2007

A catalogue record for this publication is available from the British Library

ISBN   978-0-521-66078-5   Hardback

Sables

Il n'est pas de désert si vaste
Que ne puisse traverser
Celui qui porte la musique des étoiles.
        Poem on the Paris Underground,
        attributed to Michel Le Saint

Sands

There is no desert so vast
that it cannot be traversed
by one who carries the music
of the stars.
        My translation

Contents

The colour plate section is situated between pages 62 & 63

Preface

This book was commissioned seven years ago, in Oxford, where I was an EPSRC Visiting Fellow at my alma mater, by Cambridge University Press. Its completion in Cambridge, Massachusetts, where I am a Radcliffe Fellow at Harvard University, owes a lot to the tranquillity of my initial and final conditions of work, where I am away from the regular pressures of my permanent position in India.

In the seven years since its conception, many things took priority over its writing, including, to a large extent, the research that has been presented in it. I feel this delay has been largely beneficial. In 1999, many of the developments that now seem obvious, that have now allowed granular media to be the focus of many conferences or multiple sessions at large meetings, were yet to happen. In particular, they changed the conception of the book itself, in my mind.

My initial idea, when I was approached to write a monograph on granular media, was to focus only on those areas where I had some understanding, or where I had myself been active. At that time, it was the so-called statistical mechanics of granular media, pioneered by Edwards, that held centre stage; people like myself were trying to make inroads into the dynamics of these fascinating systems. We focused in particular on what is now known as the jamming limit, which I thought even at the time had fascinating analogies to glasses. So little was known in the late nineties about powders – a feature that was at once attractive and challenging – that doing research on this field was really like stepping on the sand of a pristine beach, unaware of which step would lead to muddied waters, and which would land one on safe ground. I'd thought then of building a book around the new physics of these systems, referring people to traditional tomes on fluid dynamics and chemical engineering for everything else.

The seven years since then have seen a virtuous cycle – people have revisited old and seemingly known issues in the fluidised regime, and questioned the notion of the granular temperature, which had been set in stone by engineers. As always with physicists, people did not destroy an existing idea, but shed light on its fundamentals.

Now we know, for example, that although the kinetic energy of sand in the fluidised state does not yield a true thermodynamic temperature, it can nevertheless be useful in situations where the strict thermodynamics is less important than the use of a variable representing energy input. Additionally, people have embellished what were once only hypotheses; Edwards' compactivity, almost dismissed by many when he first seemed to get it out of thin air, has now been seen to be one of Sir Sam Edwards' many strokes of genius – it has been shown to have the *strict* characteristics of a thermodynamic temperature, despite its derivation from what was seen by many as a 'mere' analogy.

My original idea of focusing on only the dynamics of the jammed state is now simply not possible. What I have therefore done, to add to the modernity of the book, is to ask three distinguished colleagues, Profs. Sir Sam Edwards, Isaac Goldhirsch and Philippe Claudin, to contribute to it. The first of these, in collaboration with Prof. Rafi Blumenfeld, has contributed a chapter (Chapter 13) on his own ideas on the thermodynamics of granular matter, which has been complemented by a chapter (Chapter 14) on theoretical and experimental approaches to granular statics by Prof. Claudin. Prof. Goldhirsch (Chapter 12) has provided an excellent chapter which contains state-of-the-art references on granular media in the fluidised state. To all these colleagues, I owe my warmest thanks for their painstaking efforts, and the excellence of their results.

The plan of the book is as follows: Chapter 1 contains an introduction to many of the subfields that form the subject matter of the book. Chapter 2 contains an introduction to computer simulation approaches, while Chapter 3 expounds in detail on results that we have obtained on the structure of shaken granular material. Some of these results are still predictive and are virgin territory for enterprising experimentalists, while others have already been investigated thoroughly. Chapters 4 and 5 deal with cooperative phenomena in sand – focusing in turn on the dynamics of bridge formation and of the angle of repose – which are unique to such athermal systems. Chapter 6 sets out at length a way to probe the off-lattice and disordered nature of real sand, by setting forth the first of many approaches to model sand via random graphs. Chapters 7 and 8 discuss the shaking of a box of sand, the lattice-based formalism even extending to modelling grain shapes. Chapters 9, 10 and 11 contain very different approaches to the modelling of avalanches, that word from which it all began! – using in turn cellular automata, coupled-map lattice techniques, and the first of many approaches to coupled equations between surface and bulk in a sandpile. Since many of these subjects presented in different chapters are now veritable industries in the far enlarged scope of granular physics today, I make no apologies for presenting in some cases the original versions of current theories – this is done both in the interests of clarity, and because some of the most recent developments have yet to be fully verified in this continually evolving field.

Additionally, since these chapters contain largely my own work on the subject, I take responsibility for any errors, reserving the credit for my collaborators, who have been my constant sources of stimulation in my research. In particular it is to two of them, Dr. Gary Barker and Dr. Jean-Marc Luck, to whom I owe my unreserved thanks – without their active participation at various stages, this book would not have been possible.

It now only remains for me to thank the Editors of Cambridge University Press for their patience; the Service de Physique Theórique at CEA Saclay for allowing me the peace of mind to work on it on my frequent visits there; and of course the Radcliffe Institute of Advanced Study at Harvard University for gifting me the tranquillity of spirit and environment of intellectual stimulation which I so needed to finish this book.

1

Introduction

Sand in stasis or in motion – the image these words conjure up is one of lifelong familiarity and intuitive simplicity. Despite appearances, however, matter in the granular state combines some of the most complex aspects of known physical systems; to date, a detailed understanding of its behaviour remains elusive.

Granular media are neither completely solid-like nor completely liquid-like in their behaviour – they pack like solids, but flow like liquids. They can, like liquids, take the shape of their containing vessel, but unlike liquids, they can also adopt a variety of shapes when they are freestanding. This leads to the everyday phenomenon of the *angle of repose*, which is the angle that a sandpile makes with the horizontal. The angle of repose can take values between θ_r (the angle below which the sandpile is stationary) and θ_m (the angle above which avalanches spontaneously flow down the slope); in the intervening range of angles, the sandpile manifests *bistability*, in that it can either be at rest or have flowing down it. This avalanche flow is such that all the motion occurs in a relatively narrow boundary layer, so that granular flow is strongly non-Newtonian.

Sandpiles are not just disordered in their geometry – the shape and texture of the grains, on which physical parameters like friction and restitution depend, are also sources of disorder. These features, along with their amorphous packings, have important consequences for granular statics and dynamics. It is well known that sand must expand in order to flow or deform, since voids must become available for passing grains to flow through – this is the so-called phenomenon of Reynolds *dilatancy* [1], whose origin lies in the ability of powders to sustain voids. This also results in cooperative phenomena such as *bridge formation*, or its twin avatar, the propagation of *force chains*, both of which will be discussed comprehensively later in the book.

Granular Physics, ed. Anita Mehta. Published by Cambridge University Press. © A. Mehta 2007.

Since grains are typically massive, so that the ambient thermal energy kT is insufficient to impart to them kinetic energies of any significance, they do not undergo Brownian motion. Consequently, the phenomenon of thermal averaging does not occur, and hence bridges persist, once formed. This is unique to granular materials, since analogous structures would simply be thermally averaged away in gases or liquids. Bridge formation and kinetics are crucial to a proper description of the collective aspects of granular flow.

The *athermal* nature of granular media implies in turn that granular configurations cannot relax spontaneously in the absence of external perturbations. This leads typically to the generation of a large number of *metastable* configurations; it also results in *hysteresis*, since the sandpile carries forward a memory of its initial conditions. Bistability at the angle of repose is yet another consequence, since the manner in which the sandpile was formed determines whether avalanche motion will, or will not, occur at a given angle.

The above taken together, suggest that sandpiles show *complexity*; that is, the occurrence and relative stability of a large number of metastable configurational states govern their behaviour. Analogies between sandpiles and other complex systems, such as spin glasses, Josephson junction arrays, flux creep in superconductors and charge-density waves have been made: for example, de Gennes [2] has drawn analogies between vortex motion in superconductors and in sandpiles.

It should be mentioned at this point that granular matter has been studied extensively by engineers, and that it is beyond the scope of this book to provide a comprehensive review of all such contributions. In particular, there have been significant advances made in the study of the frictional properties of grains, for which the reader is referred to the book by Briscoe and Adams [3]. The regime of rapid flow in powders has also been extensively studied, and some of the relevant developments in an engineering context can be found in review articles by Savage [4].

1.1 Statistical mechanics framework, packing and the role of friction

As mentioned above, true thermal agitation in granular media takes place on an atomic rather than a particulate scale; therefore it is external vibration or shear that initiates and maintains the motion of grains. A characterisation of the relevant dynamic regimes was carried out in the pioneering work of Bagnold [5, 6]; he showed that, depending on the ratio of interparticle collision forces and interstitial viscous forces, a granular system could be in a macroviscous or grain-inertial regime. This ratio, subsequently named the Bagnold number N, was small ($N < 40$) for macroviscous flows (such as flows of slurries or mud where the viscosity of interstitial fluid predominates over grain inertia) and large ($N > 40$) for grain-inertial flows (such as granular flows in air, where fluid viscosity can be neglected in comparison with the effects of interparticle collisions).

Since this book is largely concerned with the flows of dry grains in air, it suffices to limit the discussion that follows to the grain-inertial regime; however, the nature of the externally applied shear needs to be specified. In the regime of rapid shear, a loosely packed granular system can be treated like a 'gas' of randomly colliding grains; 'kinetic theories' of grains based on a 'granular temperature' given by the root-mean-square of the fluctuating component of grain velocities [7] can be written down. These have been extensively studied via fluid-mechanical approaches [8]. However, such techniques are clearly inappropriate for situations when the applied shear is weak, and when the system under study consists of densely packed grains in slow, or no, motion with respect to one another. This regime of quasistatic flow needs new physical concepts, and it was to answer this need that Edwards [9] put forward a pathbreaking thermodynamics of granular media in the late 1980s. This was based on the observation that the volume occupied by a granular system (as measured by its packing fraction) is bounded, analogously to the energy in a thermal system.

Assemblies of grains normally pack in a disordered way; and the rigidity as well as the geometrical disorder of the packing are important determinants of granular flow. Although it has long been assumed without proof that the densest possible packing in three dimensions is the regular hexagonally close-packed structure with a volume fraction $\phi_{hcp} = 0.74$, the highest *available* packing for a disordered assembly such as a powder is closer to the random close-packed limit $\phi_{rcp} = 0.64$ in three dimensions [10]. The opposite limit of random loose packing, i.e. the least dense limit at which the powder is mechanically stable, is less clearly defined, but some experiments on sphere suspensions [11, 12] suggest values around $\phi_{rlp} = 0.52$.

Given the existence of these limits, Edwards [9] assumed that an analogy could be drawn between the volume V occupied by a powder and the energy of a thermal system. In addition, he put forward the concept of a new equivalent temperature for a powder; he called this the *compactivity* X, and defined it in terms of the configurational entropy S as $X = dV/dS$. The significance of the compactivity is that it is a measure of the disorder: when $X = 0$, the powder is constrained to be at its most compact, whereas the reverse holds for $X = \infty$. The importance of Edwards' formulation lies in the definition of this effective temperature, which is valid for powders at rest or in slow flow, unlike the previously defined granular temperature.

While the reader is referred elsewhere for further details of the statistical mechanics framework [9, 13, 14] and for a deeper explanation of the significance of the compactivity [15], it is pertinent here to mention Edwards' recent formulation of a pressure-related temperature, named by him as the 'angoricity'. Although still largely conceptual, this fills an important void in a theory of seminal importance in the physics of granular media.

The statistical mechanics framework of Edwards has been remarkably successful in various applications. It was used in its earliest form to examine the problem of segregation when a mixture of grains of two different sizes was shaken [16, 17]. An equivalent granular 'Hamiltonian' was written down and solved to increasing levels of sophistication. At the simplest level, the prediction of this model was total miscibility for large compactivities, and phase separation for lower compactivities. At a higher-order level of solution corresponding to the eight-vertex model of spins [18], the prediction for the ordered phase was more subtle: below a critical compactivity, segregation coexists with 'stacking', where some of the smaller grains nestle in the pores created by the larger ones. While it has so far not proved possible to carry out reliable three-dimensional investigations of granular packings at the particulate scale, experiments on concentrated suspensions for high Peclet number (where Brownian motion is greatly diminished) [19] support these predictions.

In our discussions so far, we have said little about the frictional forces that hold dry cohesionless powders together; the first attempt to formulate a macroscopic friction coefficient is attributed to Coulomb [20], who equated it to the tangent of the angle of repose, by defining it to be the ratio of shear and normal stresses on an inclined pile of sand. While the work of Bagnold [5, 6] made it clear that frictional force varied as the square of the shear rate for grain-inertial flow in the regime of rapid shear, it has long been recognised that the nature of the frictional forces in the quasistatic regime is complex; the frictional force between individual grains in a powder can take any value up to some threshold for motion to be initiated [21], so that considerations of global stability reveal little about the nature of microscopic stick–slip mechanisms [22, 23]. The proper microscopic formulation of intergrain friction remains an outstanding theoretical challenge.

1.2 Granular flow through wedges, channels and apertures

The flow of sand through hoppers [21] or through an hourglass [24] has been well studied, in particular to do with the dependence of the flow rate on the radius of the aperture, on the angle of the exit cone and on the grain size. Interest in this subject was rekindled by the experiments of Baxter *et al.* [25], who examined the flow of sand through a wedge-shaped hopper using X-ray subtraction techniques. They demonstrated that for large wedge angles, dilatancy waves formed and propagated upwards to the surface; their explanation was that these propagating regions were due to progressive bridge collapse. Thus, regions of low density trapped under bridges 'travel upwards' when they collapse due to the weight of oncoming material from the top of the hopper. This phenomenon is reversed for the case of small wedge angles, when waves propagate downwards and disappear altogether for totally smooth grains. Evesque [26] has also reported a related phenomenon in his observations of vibrated hourglasses; for large amplitudes of vibration, he observed that flow at the orifice was stopped. Naive reasoning would suggest that an increased flow might result as a consequence of the greater fluidisation of sand in the large-amplitude regime – the observation to the contrary confirms the well-known phenomenon of jamming [27, 28].

Theoretical approaches to this subject have been greatly restricted by their inability so far to deal with the fundamentally discrete and discontinuous aspects of granular flow through narrow channels. While existing kinetic theory approaches (see Chapter 12) can be adequate to cope with regions of the wedge where flow exists, they are inadequate for the regions where flow, if it exists, is quasistatic; an added complication from the theoretical point of view is that the transition between these two phases occurs discontinuously. Also, for narrow channels and orifices, the discreteness of the grains is very important and continuum approaches based on fluid mechanics are not really appropriate: despite this limitation, the continuum calculations of Hui and Haff [29] were able to reproduce experimentally observed features of granular flow in narrow channels, such as the formation of plugs. They predicted that for small inelastic grains, plug flow develops in the centre of the channel, with mobile grains restricted to boundary layers; for large elastic grains, on the other hand, plug flow does not occur at all, although the flow rate decreases near the centre. Caram and Hong [30] have carried out two-dimensional simulations of biased random walks on a triangular lattice based on the notion that the flow of grains through an orifice can be modelled as an upward random walk of voids; this yields a flavour of plug flow and bridge formation. Finally, Baxter and Behringer [31] have demonstrated the effects of particle orientation (see also Behringer and Baxter [32] for a fuller description); their cellular automaton (CA) model includes orientational interactions, whose results are in good agreement with their experiments on elongated grains. The results of both simulation and experiment indicate that elongated grains align themselves in the direction of flow, with the upper free surface exhibiting a series of complex shapes. More recent work on bridges [33] as well as on grain shapes [34, 35], will be discussed in detail in subsequent chapters.

1.3 Instabilities, convection and pattern formation in vibrated granular beds

The occurrence of convective instabilities in vibrated powders is among a class of familiar phenomena (see, for example [36]) that have been reexamined by several groups [37, 38]. When an initially flat pile of sand is vibrated vertically with an applied acceleration Γ such that $\Gamma > g$, the acceleration due to gravity, a spontaneous slope appears, which approaches the angle of repose θ; this is termed a convective instability, since it is then maintained by the flow down the slope, and convective feedback to the top. However, there is still considerable doubt about the mechanisms responsible for the spontaneous symmetry breaking associated with the sign of the slope. On the one hand, it seems very plausible that the presence of rogue horizontal vibrations (which are very difficult to eradicate totally) could be responsible for transients pushing up one side of the pile; the symmetry breaking thus achieved would lead to the resultant slope being maintained by convection in the steady state. Equally, a mechanism due to Faraday [39] has been invoked [38] to explain this, which relies on the notion that air flow in the vibrated pile is responsible for the initial perturbation of the grains and the consequent appearance of the 'spontaneous' slope. Finally, it is possible to draw analogies with the work of Batchelor [40] on fluidised beds, which suggests that one of the key quantities leading to instabilities in those systems is the gradient diffusivity of the grains, related to differences in their spatial concentration; however, for powders well below the fluidisation threshold, where interstitial fluid is expected to play a more minor role than in conventional fluid-mechanical systems, such analogies should be pursued with caution.

An associated problem is the extent to which the vibrated bed can indeed be regarded as fluidised in the sense required for the Faraday mechanism. While kinetic theory approaches suggest that a vibrated sandpile is more fluidised at the bottom than at the top [4], experiments [41] suggest the opposite; this scenario, i.e. that the free surface of a pile is more loosely packed than its base, is one that makes much more intuitive sense.

It is possible that the resolution of this controversy lies in the interpolation of granular temperatures discussed in [42]. In the regime of large vibration or when piles are loosely packed, grains can undergo a kind of Brownian motion in response to the driving force, so that the use of kinetic theories based on the concept of a conventional granular temperature is not inappropriate; it is then also conceivable that the extent of fluidisation is greatest at the base where the driving force is applied. On the other hand, for denser piles as used in the experiments of Evesque [41], providing the amplitude of vibration is not too large, the use of kinetic theory is limited, and the effective temperature is more likely to be the compactivity [15]; in such regimes, one would expect to see denser packings at the base which would then move like a plug in response to vibration, allowing for the greatest agitation to be felt at the free surface. The experiments of Zik and Stavans [43], where the authors measured the friction felt by a sphere immersed in a vibrated granular bed as a function of height from the base and applied acceleration, lend support to this scenario. They show that in a boundary layer at the bottom of the cell, the friction decreases rapidly with height, whereas it is nearly constant in the bulk; however, the size of this boundary layer decreases sharply with increasing acceleration, ranging from the system size at $\Gamma = 1$ to the sphere size at higher accelerations. They conclude that for large accelerations, grains are in a fluidised state, and respond as nearly Brownian particles; while for small accelerations and a denser packing, the presence of a systemwide boundary layer indicates strongly collective behaviour, with free particle motion restricted to the surface.

The phenomenon of convective instability has also been explored by computer simulations. Both Taguchi [44] and Gallas *et al.* [45] have employed granular dynamics schemes to simulate the formation of convective cells in two-dimensional vibrated granular beds containing a few hundred particles. These simulations are based on the molecular dynamics approach but they include parametrised interparticle interactions which model the effects of friction and the dissipation of energy during inelastic collisions. The form of this interaction, which allows a limited number of particle overlaps, precludes a direct quantitative comparison between the simulations and the behaviour of real granular materials.

However, it is clear that convection in a two-dimensional granular bed can be driven by a cyclic sinusoidal displacement imposed on the (hard) base of the simulation cell. In the steady state, a map of the mean particle velocity against position (in the frame of the container) shows two rolls which flow downwards next to the container walls and upwards in the centre. Although experiments have concentrated on the link between convection and heap formation, these simulations show the two phenomena as separate; a causal link between these two effects, if one exists, must be pursued in more realistic three-dimensional simulations. It is also clear that better models of the forces transmitted from the vibration source through grain contacts to the pile surface are necessary for the understanding of extended flow patterns in disordered granular systems. These issues will be further discussed later in this book.

For two-dimensional simulations containing a few hundred particles, the details of the driving force are paramount in determining the strength and the quality of the convective motion. Gallas *et al.* [45] show that there is a special (resonant) driving frequency for which convection is strongest and that the cellular pattern disappears if the vibration displacement amplitude is small. Taguchi [44] has shown that, for small vibration amplitude or large bed depth, convection is limited to an upper, fluidised layer while lower particles respond to the excitation, in large part, as a rigid body. The depth of the fluidised region increases with the vibration strength. Taguchi has identified the release of vertical stress during the vibrational part of the shake cycle as the origin of the convective motion. This occurs for acceleration amplitudes that are above a critical value ($\Gamma \approx 1$).

For larger accelerations yet, experiments report more and more complicated instabilities; Douady *et al.* [38] have reported period-doubling instabilities leading to the formation of spatially stationary patterns. Pak and Behringer [46] also observe these standing waves, and find in addition higher-order instabilities corresponding to travelling waves moving upward to the free surface. In some cases a bubbling effect is observed, where voids created at the bottom propagate upwards and burst at the free surface, indicating that the bed is fluidised. One of the most striking experimental observations is the oscillon, reported by Umbanhowar, Melo and Swinney [47–49]. While there is as yet an insufficient theoretical understanding of these difficult problems, it is clear [50] that the applied acceleration Γ, which has been used canonically as a control parameter for vibrated beds, is inadequate for their complete characterisation. This is corroborated by the experiments of Pak and Behringer [46], who point out that the higher-order instabilities they observe occur only for large amplitudes of vibration at a given value of the acceleration Γ. The previous use of Γ on its own was related to hypotheses [37] that a granular bed behaved like a single entity, e.g. an inelastic bouncing ball, in its response to vibration; while Γ is indeed the only control parameter for this system [51], the many-body aspects of a sandpile and its complicated response to different shear and vibratory regimes defy such oversimplification [52, 53]. We suggest, therefore, that competing regimes of amplitude and frequency should be explored for the proper investigation of pattern formation and instabilities in vibrated granular beds.

1.4 Size segregation in vibrated powders

Still keeping the convection connection, but in the context of segregation, we mention the work of Knight *et al.* [54] which identified convection processes as a cause of size segregation in vibrated powders. Size segregation phenomena, in which loosely packed aggregates of solid particles separate according to particle size when they are subjected to shaking, have widespread industrial and technological importance. For example, the food, pharmaceutical and ceramic industries include many processes such as the preparation of homogeneous particulate mixtures, for which shaking-induced size segregation is a concern. An assessment of the particulate mechanisms that underlie a segregation effect and of the qualities of the vibrations which constitute the driving forces is thus essential in these situations [55].

The convection-driven segregation proposed by Knight *et al.* [54] is clearly distinct from previously proposed segregation mechanisms (see below) which rely substantially on relative particle reorganisations. In a convection flow pattern all the particles, large and small, are carried upwards along the centre of each roll, but only particles which have sizes smaller than the width of the downward moving zone at the roll edges will continue in the flow and complete a convection cycle. Those particles that are larger than this critical size remain trapped on the top of a vibrating bed, and therefore segregation is observed. In the simplest case, such convection-driven segregation leads to a packing that is separated into two distinct fractions, respectively containing particles with sizes above and below critical. In the fully segregated state, there is a gradation of such phase separation: separate convection cells exist for each size fraction, with only a small amount of interference at their internal interface. The experiments of Knight *et al.* [54] show that such convective motion is driven by frictional interactions between the particles and the walls and disappears in its absence; they conclude also that convection is overwhelmingly responsible for size segregation in the regime of low-amplitude and high-frequency vibrations.

Size segregation is, however, frequently observed in vibrated particulate systems even when there is no apparent convective motion (see e.g. [56]). In the most significant process of this kind, collective particle motions cause large particles to rise, relative to smaller particles, through a vibrated bed. In a complementary process, that of interparticle percolation, vibrations assist the fall of small particles through a close-packed bed of larger particles. A large size discrepancy is not essential for these processes to proceed [57], and in many practical examples, it is the segregation of similarly sized particles that is most important. For these processes, it is often the excitation amplitude which is the appropriate control parameter.

Computer simulations have been instrumental in developing an understanding of these processes. The two-dimensional simulations of Rosato *et al.* [58] were designed to explain why Brazil nuts rose to the top, via a model that included sequential as well as nonsequential (cooperative) particle dynamics. They showed that, during a shaking process, the downward motion of large particles is impeded, since it is statistically unlikely that small particles will reorganise below them to create suitable voids. The large particles therefore rise with respect to the small ones, i.e. size segregation is observed.

In general, for a shaken bed containing a continuous distribution of particle sizes, a measure of the segregation is the weighted particle height,

$$s = \Sigma(R_i - R_o)z_i/(\langle z \rangle \Sigma(R_i - R_o)) - 1, \qquad (1.1)$$

where R_i is the size of the ith sphere at height z_i, R_o is the minimum sphere size and $\langle z \rangle$ is the mean height. This initially increases linearly with time [59] and, in the fully segregated state, fluctuates around a constant value; in this state there is a continuous gradation of particle sizes in the height profile.

Other simulations [58] follow the progress of a single impurity (tracer) particle that is initially located near the centre of a vibrated packing. For fixed vibration intensity, the mean vertical component, $\langle v \rangle$, of the tracer displacement per shake cycle varies continuously with the relative size, R, of the impurity such that $\langle v \rangle > 0$ when $R > 1$ and $\langle v \rangle < 0$ when $R < 1$. In three dimensions, there is a percolation discontinuity at small impurity sizes and $\langle v \rangle$ increases sublinearly for large impurity sizes. For $R \sim 1$, segregation is very slow and long simulation runs are necessary in order to measure accurately the segregation velocity of an isolated impurity. In this regime, the segregation takes place intermittently; that is, the impurity particle jumps sporadically, in between periods of inactivity. The process becomes continuous for larger relative sizes R. Another result from these simulations is that size segregation is retarded for shaking amplitudes which are smaller than some critical value [58].

The segregation results above must be considered carefully because they arise from nonequilibrium Monte Carlo simulations, for which dynamic results may depend on parameters such as the maximum step length and the termination criterion [60]. However, shaking simulations combining Monte Carlo deposition with nonsequential stabilisation which deploy a homogeneous introduction of free volume [61] as a response to shaking, lead to configurations of particles that are virtually independent of the simulation parameters [62]. Able to reproduce the qualitative features of segregation described above [63], their results [64] indicate that the competition between fast and slow dynamical modes determines the statistical geometry of the packing and therefore has a crucial influence on the mode of size segregation. Further details of this can be found in a subsequent chapter.

Convection and particle reorganisation mechanisms are clearly distinct, but they have some features in common which are essential in driving realistic segregation processes. Firstly, they both rely on nonsequential particle dynamics, so that the extent of the segregation (which depends, qualitatively speaking, on the competition between individual and collective dynamics) is dependent on the amplitude of the driving force. Secondly, both mechanisms rely on the complex coupling between a vibration source and a disordered granular structure, i.e. the fact that the driving forces are not transmitted to individual particles independently, but in a way that relies on many-body effects involving friction and restitution. The minimal ingredients for any convincing model of segregation thus must include nonsequential dynamics and complex force–grain couplings, to avoid unphysical results [63].

The above underlines the need for a precise specification of the driving forces if one is to build reliable models of shaking and any associated segregation behaviour. Thus, although the acceleration amplitude of the base is frequently chosen as the control parameter for a vibrated bed, in practice, details such as the extent and the location of free volume that is introduced into a packing at each dilation, and the contact forces at particle–wall collisions, may be required for an accurate analysis of segregation phenomena [50].

It has in fact been suggested [50] that convection-driven segregation dominates in the quasistatic regime of low-amplitude and high-frequency vibrations which result in free volume being introduced predominantly at the bottom of a granular bed. At larger amplitudes, free volume is introduced relatively evenly throughout the packing and particle reorganisations play a large part in the shaking response. In this case, convection rolls become unstable and the dominant mechanism of size segregation is the competition between independent-particle and collective rearrangements. If this picture is to be tested, it is clear that competing domains of amplitude and frequency need to be investigated experimentally; better control parameters than the acceleration Γ alone need to be found for a more accurate modelling of vibrated beds. This, along with further theoretical work, will be necessary for a more complete understanding of the phenomenon of size segregation in shaken sand.

1.5 Self-organised criticality – theoretical sandpiles?

The hypothesis of self-organised criticality (SOC) proposed by Bak, Tang and Wiesenfeld (BTW) [65] married the ideas of critical phenomena and self-organisation. It postulated that many large, multi-component and time-varying systems organise themselves into a special state, whose most striking feature is its invariance under temporal and spatial rescalings, so that no particular length or time scale stands out from any other.

A cellular automaton representation of a sandpile was constructed as an illustration of this concept; its 'grains' flowed down an incline in the direction dictated by gravity, provided that the local value of the slope exceeded some threshold. This was meant to represent, at its crudest level, the behaviour of a sandpile at its angle of repose, and statistics of the onset and duration of avalanches in the toy system were obtained. It was found within the context of this model that there were indeed no characteristic length or time scales, and that the power spectrum seemed to show $1/f$ behaviour; in other words, avalanches of all time and length scales were present, and uncorrelated one with the other, resulting in a set of independent events which gave rise to the observed flicker noise.

Analogies were then drawn between the sandpile at its angle of repose and a spin system at its critical temperature, with the angle of repose being an order parameter; at and above some critical value of this angle, avalanches of all lengths and times were to be expected, in a way befitting the onset of a second-order phase transition in a critical phenomenon [65]. The self-organised aspect came in via the ability of the sandpile to organise itself into this critical state: sand grains continued to accumulate till the critical angle of repose was reached, which was then maintained by avalanching.

Despite the theoretical appeal of SOC, its relevance to the dynamics of real sand is doubtful [66–70]. Before discussing more technical aspects, it is therefore pertinent to return to some facts about real sand.

Sandpiles are characterised not by a unique angle of repose θ_c, but by a range of angles of repose varying between θ_r (the angle of repose below which the sandpile is always stationary) and θ_m, the maximum angle of stability. Bistable behaviour is observed between θ_r and θ_m, in that the sandpile can either be at rest or in motion in this range [71]. The fact that angles of repose formed by pouring are very different [21] from those formed by draining underscores all of this; were θ_c a critical variable, relaxing from a supercritical state (draining) and building up from a subcritical state (pouring) would lead to the *same* angle of repose. Also, the fact that the angle of repose obtained in a sandpile depends on the conditions of formation, e.g. draining or pouring, shows that sandpiles exhibit hysteretic behaviour. This indicates already that a second-order phase transition as a function of the angle of repose is unlikely; that, if a phase transition exists, it is much more likely to be of first order [70].

The test of all these theories and counter-theories could be summed up in the following question: do sandpile avalanche statistics obtained experimentally show the predicted absence of characteristic length and time scales? The first experiments to answer this were carried out at the University of Chicago [72]; the average slope of a pile of sand was varied either by tilting the pile, or by randomly depositing particles on the top surface. Far from the predictions of SOC, what was observed was that avalanches of one particular size, separated by approximately regular intervals, dominated the flow.

The reason for this discrepancy was presumed to be that the sandpiles in the experiment were driven too hard, and that in order for SOC to be observed, one needed to drive the system very slowly relative to its relaxation rate [73]. An experiment which did just this was carried out by Held *et al.* [74]; sand was added to a pile one grain at a time in such a way that any resulting avalanche subsided before the next grain was dropped, so as to 'parallel more closely cellular-automaton models known to exhibit self-organised criticality' [74]. Their findings were as follows: for sandpiles built on plates with diameters below one and a half inches (3.8 cm), a broad distribution of avalanche sizes was detected, and a plot of weight against time showed similar fluctuations over one week to those over one hour. This scale-invariance was seen as clear evidence of self-organised criticality. By contrast, sandpiles built on three inch plates were characterised by the following behaviour: nearly all the mass flow of the sandpile occurred through large periodic avalanches, and therefore the scale-invariant characteristic of self-organised criticality was *not* observed on these larger piles. It was claimed as a result [74] that while SOC was observed in 'small' piles, there was a crossover to a quasiperiodic behaviour dominated by system-spanning avalanches for larger system sizes.

An explanation of this experiment was put forward by Mehta and Barker [67, 68], who subsequently quantified their explanation using cellular automaton sandpiles with evolving structural disorder [75]. In their model, avalanche motion on the surface of the pile as well as reorganisation in the interior occur as a result of deposition. Small avalanches result from local, and large from global, reorganisations: there is a boundary layer of constantly evolving disorder. The presence and size of this layer set up a natural length scale for the large avalanches and hence a preferred avalanche duration, in relation to which 'small' and 'large' events can be defined. This model will be discussed in detail later in the book.

In the experiments of Held *et al.* [74], it was observed that for the smaller sandpiles, all sand dropped at the top flowed out of the bottom; thus, particles are not stored in the boundary layer, avalanche flow predominates over cluster reorganisation, and no special length scale stands out, leading to the apparent observation of SOC. For the larger sandpiles, not all the particles deposited flowed out of the bottom, and large avalanches were seen to originate from 'below the surface' [74]; thus particles are stored in the boundary layer, whose periodic discharge leads to the characteristic large avalanches, and scale-invariance is lost [67, 68].

A related argument, put forward by the Chicago group, disputed even the limited claim of observing SOC in piles below some 'critical' size. They argued [70, 76] firstly that finite-size effects dominated for the smaller sandpile, whose size was insufficient for there to be a clear distinction between the minimum angle of repose θ_r and the maximum angle of stability θ_m. Secondly, they opined that the scaling behaviour predicted by SOC, if it exists, should be most manifest at large sizes and distances, whereas it was precisely at these distances that scaling disappeared in the results of Held *et al.* [74].

1.6 Cellular automaton models of sandpiles

Lattice-based sandpile models introduced [65] to illustrate the principle of SOC have since become widely used to study the flow of grains down a sloping surface. The discrete nature of lattice grain models, and in many cases their geometrical parallelism, are significant advantages for efficient computation; such lattice-based models, however, require considerable interpretation and analysis to be reliable indicators for the behaviour of real *irregular* granular systems. Most model sandpiles are concerned with the statistics for initiation and development of surface avalanches in driven systems, for comparison with experiment [72, 74]; for this they need to include the essential physical ingredients which would explain the observed predominance of large avalanches. To this end, the effects of grain inertia, structural disorder and damping have been included into simple lattice-based model sandpiles.

We first describe what is arguably the simplest such model which is nontrivial. Grains are unit squares, stacked in columns on a line of length L; their number in column i, $1 \geq i \geq L$, defines the column height z_i. New grains are added one at a time onto the tops of randomly chosen columns, at which point time is incremented by a unit: the model sandpile is then strictly a cellular automaton. If, after the addition of a grain, $z_i - z_{i+1} \geq 2$, then column i becomes unstable; two grains then slide from column i onto column $i + 1$. In turn this may make column $i + 1$ and/or column $i - 1$ unstable and a whole series of slides may ensue. The motion of several grains is called a model avalanche, which terminates when sliding leaves no further columns unstable. Column 1 rests against a hard wall so that no grains can slide onto it and column L borders an edge over which grains slide without trace. The number of grains n_x that exit column L as the result of adding a single grain is a convenient measure of avalanche size; however, other measures, such as the number of grain topplings, can also be used. In most practical implementations, local slopes $s_i = z_i - z_{i+1}$ are used, with $s_L = z_L$.

This sandpile model has been classified by Kadanoff *et al.* [77], as a one-dimensional local and limited model because, at each event, a limited number of grains (two) move locally, i.e. onto the neighbouring column. The model is fundamentally asymmetric because grains can only slide in one direction. In the steady state, which is independent of initial conditions, the mean slope of the pile fluctuates around a constant value; there are avalanches of many different sizes, with a mean size $\langle n_x \rangle = 1$. The total number of grains in the pile fluctuates only very slowly and the distribution function of avalanche sizes $x n_x$ varies smoothly and monotonically with avalanche size [77]. Kadanoff *et al.* [77] have shown that such distribution functions manifest a multifractal scaling. They have thoroughly examined many variants of this simple model, and conclude that all models obeying similar rules are subject to the same scaling, and therefore comprise a single universality class. This is constituted of several subclasses, where model sandpiles may be nonlocal (where grains can jump to distant neighbours) and/or unlimited (where unlimited numbers of grains topple after a deposition event).

In contrast, experimental observations of sandpiles do not show clear scaling [78]; the overwhelming consensus is that there is a preponderance of large avalanches in a characteristic size range [72, 74]. Sandpile models which include extra features such as disorder, nonconservation and grain inertia have been developed in order to explain this increased proportion of large avalanches and the associated absence of scale invariance.

As a fundamental departure from ordered sandpile automata, Mehta and Barker [75] introduced a model with evolving structural disorder. Here, surface dynamics are coupled to bulk structural rearrangements, leading to avalanche statistics with the appearance of characteristic time and length scales related to the surface–bulk couplings. Further details on this and related models will be found in succeeding chapters.

Another experimentally relevant model is that of Prado and Olami [79] whose sandpile cellular automaton leads to a special status for large avalanches. This fully ordered model is nonlocal and limited, with a toppling threshold which decreases with the number of topplings that have already occurred in an avalanche. Large avalanches are thus favoured, by this introduction of a 'snowball' effect which is a model of inertia. The resulting avalanche size distribution develops a peak at large sizes, which is manifest for sandpiles larger than a critical size. A drawback of this model is that the variations of sandpile mass are very large (sometimes as much as half the total mass of the pile) and very regular; their resultant time series resembles that of an oscillator much more than it does the irregular time series observed in sandpile experiments [72, 74]. Ding *et al.* [80] removed this unphysical regularity by including a stochastic element in the toppling threshold; this introduces a damping length which favours a characteristic size.

Lattice sandpile models in which grain motion is driven by height differences are conservative; that is, grain motions (apart from those at the boundaries) do not change the sum of the height differences. This feature is unrealistic – real sand grains dissipate their energy in frequent collisions across the surface of a pile. The role of nonconservative driving forces has been examined by Christensen *et al.* [81] and Socolar *et al.* [82], in their versions of sandpile models. They find that nonconservative driving forces do not automatically destroy scaling; they do, however, lead to nonuniversal exponents that depend on the degree of nonconservation. Barker and Mehta [22] have also developed a nonconservative coupled map lattice model of a reorganising sandpile which generates many large nonscaling avalanches. The observed departure from scaling is interpreted in terms of two key parameters, corresponding respectively to dilatancy and grain inertia.

The inclusion of realistic features of granular dynamics such as disorder, non-conservation and particle inertia thus leads to a breakdown of the scaling behaviour that appears in the simplest cellular automaton sandpile models. A formal correspondence between lattice grain models and continuum equations has so far not been established rigorously, despite their coincidence in a particular case or two [83]; this remains an important goal in the cellular automaton modelling of granular flows.

1.7 Theoretical studies of sandpile surfaces

Theoretical studies of sandpile surfaces have also been subject to a division similar to that mentioned in the previous section; namely those which have explored in great theoretical detail relatively simple models of generalised surfaces, and those which have concentrated on the modelling of increasingly complex features in their investigation of real sandpiles. Again, the motivations in each case are very different; in the first case, the aim is frequently the detailed study of theoretical concepts like SOC – for example, the identification of the crucial ingredients needed to observe scale invariance in a toy model. In the second case, the aim is typically the identification of the minimal physics needed to model real sandpiles. Sandpiles in the latter category are necessarily more complex, and resist the clear analytical solutions more accessible to the former case.

Well before the upsurge in interest in sandpiles, there were attempts to model evolving interfaces, such as those in colloidal aggregates and solidification fronts [84]. In all these models, the basic picture was of particle deposition on a surface; the growth of the interface in response by the rearrangement of local heights was modelled via Langevin equations, with noise representing the external perturbation. The seminal model in this series was due to Edwards and Wilkinson [85] (EW); the effects of surface tension were here represented by a diffusive term $\nabla^2 h$. Kardar *et al.* [86] added a term $(\nabla h)^2$ representing lateral growth to this, which was equivalent to using a form of the Burgers' equation [87]. The solution of this equation (widely known as the KPZ equation) has been an ongoing problem in theoretical physics; its critical exponents have been determined in some cases [86] by using dynamic renormalisation group approaches earlier applied [88] to the general form of the Burgers' equation. Among further variants of the KPZ equation to do with general growing interfaces has been one due to Maritan *et al.* [89] which comprises relativistic invariance under reparametrisation and leads to a crossover away from KPZ exponents in the long time limit, which the authors suggest is more relevant to the behaviour of growing interfaces.

There have also been attempts directed specifically at understanding sandpiles; these approaches, however, start from generalised considerations of symmetry rather than from specific physical considerations, and can in some sense be viewed as toy models of sandpiles. The first of these, due to Hwa and Kardar [90], started from symmetry conditions on a discrete sandpile model of the BTW variety; their system was open and anisotropic, with open boundaries at one end and closed boundaries at the other. A particular direction of transport being selected, the resultant absence of reflection symmetry along the direction of flow, and the presence of an inversion symmetry due to voids moving up as grains move down the pile, were incorporated into their lowest-order nonlinearity. Notably, the presence of grain conservation – with sand added being balanced on average by sand flowing out of the open system – excluded terms of the form h/τ, with τ being some characteristic relaxation time. The authors concluded that this conservation law was responsible for the absence of characteristic length and time scales, and the consequent presence of SOC.

It was pointed out by Grinstein and Lee [91] that this scale invariance, while characteristic of many noisy nonequilibrium systems with a conservation law governing their dynamics [92], was not uniquely a manifestation of SOC; such generic scale invariance had in fact been observed well before [93] in many other driven systems. In addition, the presence of temporal scale invariance in such systems does not always involve the concomitant presence of spatial scale invariance [92], in contradistinction to the predictions of SOC. More specifically, Grinstein and Lee [91] suggested that the joint inversion symmetry suggested by [90] was not a symmetry obeyed by generic dynamical rules for model sandpiles, whereas the invariance $h(x, t) \equiv h(x, t) + c$, corresponding to a uniform upward translation of the sandpile, was an important symmetry that had been overlooked by them. These authors [91] therefore had a different suggestion for the lowest order nonlinear term; this term, however, turned out to be asymptotically irrelevant so that the long-time behaviour of their equations [91] was diffusive, driven by the linear (EW) terms alone.

To recapitulate, all the above models were theoretical analyses of ordered systems based in one form or another on the BTW representation of the CA 'sandpile', and all of them manifested different incarnations of SOC. It was at this stage that attempts began to be made to represent more realistic sandpiles. The first such attempt was made within the framework of cellular automata when Toner [94] showed that the introduction of quenched disorder in Grinstein and Lee's equation [91] caused all traces of SOC to vanish. He explored the cases of weak (where only positional disorder was manifest) and strong (where there was additional randomness in grain sizes) disorder, and found purely diffusive behaviour in both cases [94].

All of the above approaches involved only one variable, the local surface height h. The dynamical coupling of moving grains and immobile clusters was introduced by Mehta *et al.* [42] – their phenomenological equations coupled the (global) dynamics of the angle of repose and the Bagnold angle [5, 6] representing the dilatancy of clusters. They also included an interpolation between different dynamical regimes via appropriate effective temperatures. This work gave rise to a more microscopic approach via equations which coupled the local surface height h to the local density of moving grains ρ, with noises representing the effect of external shaking and local packing [69, 95, 96]. These equations have been quite successful in modelling sandpile dynamics, with experiments finding their predicted surface roughening exponents [97]; they will be discussed later in the book.

2

Computer simulation approaches – an overview

Sand has many avatars – it can behave as a solid, liquid or gas, depending on external circumstances. This multiple identity is one of several reasons why the computer simulation of dry granular materials is difficult. Sand in the solid-like state responds to external stimuli on a very different timescale to sand in its liquidlike avatar – in contrast to most efficient computer simulation methods, which are typically tuned to one particular timescale such as a collision or relaxation time. Other features of sand which are difficult to simulate efficiently include complex, dissipative interparticle and particle–wall interactions, typically irregular grain shapes and strong hysteretic effects. Furthermore, the athermal nature of sand means that grains do not randomly sample all possible states ergodically – as a result, appropriate statistical averages can only be obtained by repeated (computationally demanding) simulations of a granular system.

For normal dry powders, interstitial fluid plays only a minor role – apart from exceptional cases when, say, there are small liquid pools at particle contacts which could seriously alter the pairwise nature of grain interactions. This is a clear distinction between granular systems and dense suspensions – for the former, interparticle interactions are restricted to short-ranged contact forces. In practice, the methods developed for granular simulations are quite similar to classical methods used to simulate simple liquids. Molecular dynamics and Monte Carlo methods have been adapted to model granular dynamics and powder shaking simulations, while more recent cellular automaton approaches (originally used in fluid dynamics) are by now widely used in the modelling of granular flow.

2.1 Granular structures – Monte Carlo approaches

A static powder may be considered as a random packing of its constituent grains. A particular configuration of grains is influenced in two ways by its method of

Granular Physics, ed. Anita Mehta. Published by Cambridge University Press. © A. Mehta 2007.

construction. Firstly, random dynamical fluctuations during shaking or pouring ensure that no two granular systems are identical. Secondly, the nature of the construction process – whether shaking, pouring or sedimenting – often leads to rather characteristic behaviour for structural descriptors such as particle contact numbers, bond angles or void volumes. These distributions are often indicative of a particular construction history, so that the static structure of a packing is history-dependent. The athermal nature of sand further implies that the structure determines transport properties, so that its dynamics is also history-dependent. Thus, from the point of view of computer simulations, ensembles of configurations built from independent realisations of the whole powder by a particular method can be used to evaluate representative material properties corresponding to it.

Random packing has been a subject of interest to physicists and mathematicians for a long time. Kepler formulated the most celebrated question on this subject: 'Can monodisperse spheres be arranged in a random way so that they occupy a fraction of the volume which exceeds the 74% occupied by the spheres in the densest regular packing?' The consensus so far is that the answer is no, and that in fact the maximum random close-packing fraction for monodisperse spheres is 64% [10]. This figure is widely accepted [98, 99], although some recent workers [100] have suggested that the definition is not mathematically precise.

We will shortly discuss some simulation methods for generating random packings of three-dimensional powders, since two-dimensional random structures are not really representative of granular materials [101]. All the packings we consider are constructed, for computational convenience, from non-cohesive hard spheres, which are a reasonable representation of real grains. The simulation of irregular grain shapes incurs computational complexity and does not markedly affect the gross structural descriptors of a packing.[1] Also, attractive forces are usually only relevant for very small particles and lead to relatively open (less dense) structures [102]. By contrast, the packings we consider are gravitationally stable, so that grains within them occupy positions that are local potential energy minima under gravity. This means, operationally, that each grain is in contact with at least three others and its centre lies above a triangle defined by theirs.

Simulations of random packing can be classified as sequential or nonsequential. Sequential simulations, where grains are added one at a time, are divided into site search and site deposition models. In site search models, the list of available sites is continually updated as particles are added; new particles are added, one at a time, at any one of these sites chosen according to a predetermined rule. In the generalised Eden model, e.g. [103], all possible sites have equal a-priori

---

[1] The issue of grain shape is, however, crucial to dynamics in the jammed state, and will be the subject of a subsequent chapter.

probability for occupation, while in the Bennett model (originally established for particles in a central force field) [104] incoming particles always occupy the site with lowest potential energy. All such models lead to packings in which the mean coordination number of particles is 6.0 in three dimensions; this corresponds to a given particle being stabilised by three grains above and three grains below it. The volume fractions corresponding to different schemes can, however, be different; for the Eden and Bennett models, they are respectively 0.57 and 0.6. This is a clear indication of the absence of a simple (one-to-one) relationship between the volume fraction and the coordination.

In sequential deposition models, incoming particles follow noninteracting trajectories, which are terminated irreversibly when a local potential energy minimum is reached; this in turn implies that the dynamics is influenced by the configurations of previously deposited particles. In the simplest case, these trajectories are ballistic until the surface is reached; spheres then roll without slipping, down the path of steepest descent, into a local potential energy minimum in contact with three supporting spheres. This process has been studied extensively, e.g. [103, 105]. For monodisperse spheres, there are boundary layers of quasi-ordered configurations extending for approximately five sphere diameters above the base and below the surface. Away from them, the mean coordination number of the packing is 6.0 corresponding to three-particle stabilisation, while the corresponding packing fraction is 0.5815 ± 0.0005 [106]. These values are not altered substantially by introducing a small amount ($\sim 5\%$) of polydispersity.

Extensive manipulations of a powder, such as stirring, shaking and pouring, lead, however, to particle trajectories which are fundamentally nonsequential: any one trajectory cannot be computed without simultaneously computing many others. In general therefore, sequentially constructed packings are not representative of realistic granular structures. To generate the latter, it is essential that simulations contain collective restructuring, so that particles reorganise at the same time as deposition occurs. The resulting granular configurations reflect the essentially cooperative nature of the process, containing bridges [33] and a wide variety of void shapes and sizes, none of which occur in sequentially deposited structures. Since bridges are stable arrangements in which at least two grains depend on each other for their stability, they cannot be formed by sequential dynamics; they are, on the other hand, a natural consequence of the cooperative resettling of closely neighbouring grains. These and related issues will be discussed in subsequent chapters.

Nonsequential (non-Abelian) construction of random packings can of course be done in many different ways. A particular way could be the simulation of shaking, which we will discuss at length later. We stress here that the result of a nonsequential process depends not only on the particular prescription used, but also on the choice of the initial conditions; i.e., the structure of a nonsequential deposit depends non-trivially on the initial grain configurations. This history dependence is a reflection of the very real hysteresis in granular media, and is thus a very physical feature of nonsequential simulations; by contrast, sequential deposits do *not* depend on their process histories, and sequential dynamics remain Abelian.

Computer simulations of nonsequential random close packing are most easily initiated from expanded sequential close packings [62], from other well characterised sequential configurations such as the RSA configurations [107], or from perturbed ordered configurations [108]. In general, initial configurations of this kind can be parametrised by a single parameter (such as an expansion factor or an initial packing fraction) which can be used as a control parameter for the final (nonsequential) packing.

Soppe [109] examined the Monte Carlo compression of random ballistic deposits via a scheme without an explicit stabilisation mechanism, so that the resulting structures are not really representative of granular materials. The packing fraction obtained is $\phi = 0.60$, even in the presence of a small amount of polydispersity. Jodrey and Tory [107] produced dense nonsequential sphere packings with $\phi = 0.64$, by using an isotropic and deterministic compression method. Although their final configurations are unrealistic because they contain non-contacting spheres, their final packing density increases with decreasing compression, a feature which has been observed in more realistic simulations. Mehta and Barker [61, 62] have made extensive investigations of nonsequential hard sphere packings using a hybrid simulation method that includes both Monte Carlo and nonsequential random close packing phases, with a well defined control parameter. A wide variety of stable, nonsequential packings, with volume fractions in the range $0.55 < \phi < 0.60$, have been obtained, which have many features in common with real granular materials. Simulation results, including pair distribution functions, connectivities and pore sizes, show that in general, less dense initial configurations lead to looser, less ordered packings which have rougher surfaces. These results will be detailed in the following chapter. Also, Nolan and Kavanagh [108] have performed nonsequential random close packing simulations for hard spheres, using an extension of a compressed gas method. This technique produces stable structures, which contain finite concentrations of bridges, with volume fractions in the range $0.51 < \phi < 0.64$. Again, denser initial configurations lead to denser final packings with more short-range order.

The above authors [62, 108] show that stable nonsequential hard sphere packings have coordination numbers in the range $4.5 < z < 6.0$: in fact lower values ($z \sim 4.5$) are typical of a nonsequential process. The contrast with the fixed value, $z = 6.0$, obtained from random sequential stabilisations can be understood as follows. The stabilisation of each sphere in a sequential process leads to the formation of three bonds; hence, it leads to an increase of the network coordination by six for each added sphere. In contrast, the addition of, say, two bridged spheres causes the formation of five new bonds, causing an increase in coordination by five per added sphere. More complex nonsequential structures have lower coordinations: the deviation of the mean coordination from $z = 6.0$ is thus a reflection of the *cooperative* nature of grain stabilisations. These issues, and their relevance to friction, will be dealt with in succeeding chapters.

2.2 Granular flow – molecular dynamics approaches

Granular flows run the gamut between rapid (e.g. hopper or chute flows) and very slow (e.g. mudslides). The former are characterised by instantaneous and energetic binary collisions; in the latter, grains move slowly and collectively, while grain collisions have finite durations and are not generally decomposable into ordered binary sequences. The fluid-like behaviour of a granular assembly thus covers the range from a dense gas to a viscous liquid, a dynamical range which is too large to be modelled efficiently by a single technique. Granular dynamics simulations are therefore usually tailored to one of two regimes corresponding to the above limiting cases: the *grain-inertia* regime in which instantaneous and inelastic two-particle collisions dominate the motion, and the *quasistatic* regime where particles interact collectively. Simulations in the grain-inertia regime contain grains with high kinetic energies and are most efficient at moderate particle densities $\phi \sim 0.3 - 0.4$. In the quasistatic regime, the specification of particle contact forces is the most important component of computer simulations tailored to model the collective behaviour of densely packed particles with $\phi \sim 0.55$.

In the grain-inertia regime, granular motion is reminiscent of molecular motion in a dense gas; the implementaton of granular dynamics simulations follows standard methods for molecular dynamics simulations of rough hard spheres [110]. Particle trajectories are traced by the solution of Newton's equations of motion, combined with the repeated application of a binary collision operator. Hard-particle simulations use a flexible list structure to identify the next instantaneous binary collision (these are often referred to as 'event-driven simulations' [111, 112]) , while soft-particle methods employ an iterative solution with a time step $\Delta t \sim 10^{-3} - 10^{-5}$ seconds. An important distinction between molecular and granular dynamics arises because intergrain collisions (unlike intermolecular ones) are inelastic; this is typically incorporated by including a single coefficient of restitution into the collision operator, although real collisions need a greater complexity of description.

Walton [113, 114] has performed some of the earliest and most significant granular dynamics simulations in the grain-inertia regime. For steady-state shear with fixed friction and restitution, he finds that the granular temperature (the random component of the kinetic energy) and the effective viscosity increase approximately linearly with the applied strain rate. Among his many nontrivial results are the appearance of boundary layers and anisotropic velocity distributions, in the presence of highly dissipative interactions and extremal (low or high) particle densities.

Campbell [115, 194] has also studied steady-state rapid shear flow of rough particles in both two and three dimensions. In two-dimensional chute flow simulations, Campbell and Brennan [115] identify a high-temperature, low-density zone at the chute base; they note that two-dimensional simulation results depend strongly on system parameters such as friction coefficients, in addition to exhibiting fluctuations characteristic of small thermodynamic systems. Their three-dimensional shear flow simulations [194] with standard Lees–Edwards boundary conditions, indicate that the stress tensor is symmetric, with the the the only nonzero off-diagonal element being the in-plane component of the shear. Typically the stress tensor is large for high ($\phi \sim 0.5$) and low ($\phi < 0.05$) components of the volume fraction, with a minimum in between.

Cellular automata have also been used to study specific problems in the grain inertia regime. Baxter and Behringer [31] have used a cellular automaton to simulate the flow of irregular particles from a wedge-shaped hopper. Their particle states are described by discrete velocity and orientation variables on a two-dimensional triangular lattice; these evolve in separate collision and propagation steps at each timestep. The outcome of a collision is determined by the minimisation of a local energy function, subject to particle conservation [116]. Instantaneous snapshots of their configurations clearly resemble their experimental photographs of grass seeds flowing from a narrow perspex wedge; many features of the flow, including long-range orientation correlations, the appearance of stagnant regions of steady-state flow and a (relatively) time- and depth-independent discharge rate, are, remarkably, also reproduced. Other phenomena in granular flow, such as steady Couette flow, flow down a vertical channel and particle size segregation, have also been modelled using cellular automata [117, 118]. All of this suggests that despite their simplicity, lattice-based models (if used judiciously) can provide powerful probes of granular flow over a range of dynamic regimes: their development to include higher dimensionalities, anisotropy, disorder and grain coordinations and shapes [34, 35] are valuable areas of ongoing research.

Turning now to quasistatic flow, one finds again a vast range of granular dynamics simulations. A majority of these simulations have been performed with customised versions of the TRUBAL computer code originally developed by Cundall [119] for soil science applications. The crucial features of such simulations are: (1) the efficient evaluation of intergrain forces and (2) the simultaneous solution of many coupled equations of motion for interacting grains. The accurate computation of forces is paramount in the quasistatic regime, and consequently, time steps of simulations are much smaller than the typical duration of an interaction. Although the specification of interparticle forces is an essential part of such simulations, a first-principles understanding of grain interactions is absent. Typically, hard-particle simulations require adequate representations of static friction and tangential restitution, while Hertzian deformations and associated restoring forces are used for soft particles. The distinct element method pioneered by Cundall and Strack [119] uses a parametrised approximation to the contact force, composed of parallel viscous and harmonic elements.

Further details of the grain-inertia regime and the quasistatic regime can be found in the chapters by Goldhirsch and Claudin respectively. In between these two extremes, there are few molecular dynamics approaches which are both efficient as well as accurate; however, the simulations carried out by Silbert *et al.* [120] are a notable exception, and we will return to them in subsequent chapters.

We end with a few general remarks. For granular flow to be sustained, grains must be coupled to an external energy source; this can be provided by gravity (for instance during the discharge of grains from a hopper) or through interactions between grains and moving boundaries (during shear, for example). The mechanical energy supplied is then dissipated by grain–grain collisions in the bulk. The proper modelling of intergrain interactions and external energy sources has already been mentioned; however, the modelling of wall effects is also crucial, given the nonequilibrium nature of real granular media, despite its introduction of extra complications like particle–boundary couplings and finite size effects.

2.3 Simulations of shaken sand – some general remarks

It is appropriate to make a few remarks on simulations of shaken sand, in view of the special role that vibration plays in granular excitation. The details of the typically incoherent driving force, such as the strength of its harmonic components and/or their couplings to individual particle motions, are usually unknown, so it is a task of some subtlety to model them. The qualitative picture is that while grains are frozen into particular configurations in the absence of any external stimulus, the application of mechanical energy in the form of, say, shaking introduces periods of release; during these, grains can rearrange, and the powder 'jumps' between different macroscopic configurations. Since granular media must expand [1] in order to flow or deform, volume expansion is an essential component of these periods of release. The dynamic response of the powder consists of these particulate rearrangements – a response that has both transient and steady-state components – which lead in turn to fluctuating grain configurations. A path, depending on the dynamics of individual grains as well as on the quality of the driving force, is thus traced by a shaken powder in configurational phase space.

The complicated behaviour of shaken powders originates in the fact that a typically dense, disordered and many-component system is subjected to ill-defined and complex driving forces. In fact, driving forces that seem qualitatively similar may result in rather different behaviour, sometimes enhancing mixing and at other times causing size segregation. Gentle shaking or 'tapping' may be used as a means of powder compaction, while stronger vibrations can lead to fluidisation. The aim of a good simulation of shaking is to probe, in terms of individual and collective dynamics, the different microscopic responses which underlie the macroscopic response of a variously vibrated powder.

The nonsequential packing mechanisms described above can of course be used to model aspects of shaking-induced particle rearrangements. However, since these are compaction-dominated, i.e. they generically cause the packing fraction to increase, they cannot model the diversity of granular responses to shaking. In order to do so, they need to be coupled to a modelling of the effects of the driving force, such as the periodic insertion of free volume during the dilation cycle, or an inherent stochasticity which prevents the build-up of static steady states. These effects are heavily reliant on the coupling of the external energy source to the granular medium, the details of which are generally unknown. Current shaking simulations avoid microscopic models of shaking as a result, concentrating instead on general characteristics such as the incoherence of the driving force; they take a macroscopic, noise-based view of the complex coupling of grains to the external driving force.

Most shaking processes take place in a series of regimes which traverse the spectrum from grain inertia to quasistatic. For example, grains may experience widely varying particle densities ($\phi \approx 0.3 - 0.6$) as well as wide-ranging dynamics, during a shake cycle. Shaking is thus difficult to simulate with continuously tuned granular dynamics; successful molecular dynamics simulations of shaking over all dynamical regimes have yet to be performed. Some early and purely sequential simulations of hard sphere shaking [105] combined a sequential close packing scheme with a search for a global minimum of the potential energy. Configurations were built by adding grains one by one at sites of minimum potential energy chosen at random; the resulting packings, which remain fully sequential, have volume fractions $\phi \sim 0.6$, which are greater than those for unshaken configurations but still significantly below the random close packing value $\phi = 0.64$. More recently, Rosato *et al.* [58] introduced a two-dimensional Monte Carlo method to study shaking-induced size segregation. Their method includes important nonsequential features but does not include a criterion for stability; hence it cannot be used directly to follow changes in volume fraction or particle coordinations induced by the applied vibrations.

Stability criteria were introduced in a model with nonsequential reorganisation by Duke *et al.* [121] to study the steady relaxation of the slope of a two-dimensional pile of hard particles under vertical vibrations. These simulations indicated that collective particle motions over extended length scales were important dynamical features of the dynamics. Barker and Mehta [62] next developed a three-dimensional, nonsequential model of a vibrated granular bed. In these models [62, 121] the periodic driving force leads to clearly defined periods of dilation for the powder assembly, punctuated by static granular configurations. The uniaxial driving force is coupled homogeneously to the powder, so that free volume is introduced uniformly. During dilation, grain motion is dominated by the effects of the gravitational field and hard-core interactions between the grains themselves as well as with the container base. This scheme approximates precise particle trajectories by using a low-temperature Monte Carlo method supplemented by a nonsequential random close packing algorithm. The compromise inherent in this scheme involves losing the details of the granular dynamics while efficiently reproducing static structures over a wide range of dynamical regimes. We will discuss the details of this scheme, as well as its results, in succeeding chapters.

3

Structure of vibrated powders – numerical results

A microscopic description which focuses on the essential discreteness of grains is a fundamental part of understanding many fascinating features of a granular medium. Its response to vibration can be understood in terms of the competition between individual-particle and cooperative dynamical mechanisms [17, 122] intrinsic to the grains in a disordered assembly. We present below computer simulation results for some features of granular structure which depend only weakly on details of grain size or material, e.g. the finite range for the packing fraction or shaking-induced size segregation.

3.1 Details of simulation algorithm

In this section, we describe the details of the simulations first reported in [61, 62]. Simulations are performed for a bed of monodisperse hard spheres above a hard base at $z = 0$. The granular bed is periodic, with a repeat distance of L sphere diameters in two perpendicular directions x and y in the plane. Each primary simulation cell contains N spheres. A unidirectional gravitational field acts downwards, i.e. along the negative z-direction. Initially, spheres are placed in the cell using a sequential random close-packing procedure; the packing is then subjected to a series of non-sequential N-particle reorganisations. Each reorganisation is performed in three parts: first, a vertical expansion or dilation, second, a Monte Carlo consolidation, and finally a nonsequential close-packing procedure. We call each full reorganisation a shake cycle or, simply, a shake. The duration of model shaking processes as well as the lengths of other time intervals are measured in units of the shake cycle.

The first part of the shake cycle is a uniform vertical expansion of the sphere packing, accompanied by random horizontal shifts of sphere positions. Sphere i at height z_i is raised to a new height $z_i' = (1 + \epsilon)z_i$. For each sphere, new lateral

Granular Physics, ed. Anita Mehta. Published by Cambridge University Press. © A. Mehta 2007.

coordinates are assigned, according to the transformation $x' = x + \xi_x$, $y' = y + \xi_y$, providing they do not lead to an overlapping sphere configuration; here ξ_x and ξ_y are Gaussian random variables with zero mean and variance ϵ^2. The expansion introduces a free volume of size ϵ between the spheres, and facilitates their cooperative rearrangement during the next two parts of the shake cycle.[1] We use the free volume ϵ as a measure of the intensity of vibration, on the assumption that particles are freer to move as the intensity of vibration increases.

In the second phase, compression occurs by a series of random sphere displacements according to a very low-temperature hard-sphere Monte Carlo algorithm. A trial position for sphere i is given by $\mathbf{r}'_i = \mathbf{r}_i + \mathbf{a}d$, where $\mathbf{a}$ is a random vector with components $-1 \leq a_x, a_y, a_z \leq 1$ and d defines the size of a neighbourhood for the spheres. A move is accepted if it reduces the height of sphere i without causing any overlaps; in fact all successful moves reduce the overall potential energy of the system. This process continues until the efficiency with which moves are accepted, measured by batch sampling, falls below a threshold value e. Here d and e are free parameters which are chosen to optimise the simulation; our results given below are essentially independent of this choice [62].

In the third and final phase, the assembly is stabilised by a nonsequential packing method: spheres are chosen in order of increasing height and then allowed to roll and fall into stable positions. At this point, spheres can roll over or rest on any other spheres, including those which are yet to be stabilised; such touching particles can continue to move together until no further rolling/falling is possible. This mimics a cooperative dynamics, allowing for the formation of bridges – in a way that would be impossible to do via a purely sequential process.

The outcome of a shake cycle can be variously viewed as the replacement of one stable close-packed configuration by another, or as a cluster reorganisation scheme. Viewed in the latter way, expansion challenges the integrity of the clusters while the Monte Carlo compression reintegrates disbanded clusters and/or creates new ones. Finally, the stabilisation phase repositions particles in the clusters just created. In practice, during the expansion phase of the nth shake cycle, the mean volume fraction of the assembly falls from ϕ_{n-1} to $\phi_{n-1}/(1 + \epsilon)$; in the Monte Carlo recompression phase, the volume fraction steadily increases to $\phi_n \cong \phi_{n-1}$, while in the stabilisation phase, it remains approximately constant. In contrast, the mean coordination number is reduced from c_{n-1} to zero in the expansion phase of the nth shake, remains zero throughout the Monte Carlo compression and finally, during stabilisation, increases steadily to $c_n \cong c_{n-1}$.

The continuous evolution of particle positions and velocities which occurs during a real shaking process is replaced in this model by a time-ordered and discrete

---

[1] It is important to emphasise that this expansion is virtual; we seek to introduce a free volume rather than to model a physical expansion.

set of static N-particle configurations. Each such set consists of nonsequentially reorganised close packings which are obtained after integral numbers of completed shake cycles, starting from a given initial configuration, and may be labelled by three parameters e, d and ϵ. Descriptors, such as volume fraction and mean coordination number, have been evaluated for each such set; in order to minimise surface effects, particles have been chosen in the centre of a given packing such that in every case, more than fifty per cent of the spheres in the simulation cell are included.

Repeated nonsequential reorganisation leads to vibrational steady states, i.e. to packings which are macroscopically insensitive to further vibrations. Such steady states are independent of initial conditions, depending only on the shaking intensity. The properties described below have been obtained by taking averages from sets of m consecutive configurations in the steady-state shaking regime with $m \sim 50$. The simulations typically contain $N \sim 1300$ and $L = 8$ particle diameters; the mean depth of the packing is approximately 20 particle diameters. For fixed bed depths, it has been shown [62] that serious size dependence is absent for $L \geq 8$.

We remark in closing that Monte Carlo consolidation is structurally the most influential, and computationally the most intensive, part of each shake cycle. The duration of this phase, which can be measured in terms of the number of Monte Carlo steps per particle, can be increased either by decreasing e (the terminating efficiency of the Monte Carlo sequence) or by decreasing d (the maximum size of each Monte Carlo step). It has been shown elsewhere [62] that for long Monte Carlo consolidations i.e. for sufficiently small values of e, measured volume fractions are independent of, while measured coordination numbers are only weakly dependent on, a given choice of d.

3.2 The structure of shaken sand – some simulation results

In this section, we focus on some of the most important descriptors of granular structure. Overwhelmingly the most important of these is the (steady-state) volume fraction ϕ, whose variation with vibration intensity ϵ is shown in Fig. 3.1. For large intensities (e.g. $\epsilon > 1$) the volume fraction fluctuates weakly around $\phi \approx 0.55$; it however rises sharply as the intensity is reduced and for very low intensities (e.g. $\epsilon \leq 0.2$), takes on values that are *inaccessible* via purely sequential dynamics, as the powder is packed to beyond 0.58.[2] This is a first indication, to be buttressed later on, that collective dynamics is essential for really compact granular packings.

Figure 3.2 shows the variation of the (steady-state) mean coordination number c with the intensity of vibration ϵ; this too is largely constant after $c \sim 4.48$, and decreases as ϵ increases for $\epsilon \leq 0.25$. It is important to note that the entire range

---

[2] It is well known that the threshold for randomly and sequentially deposited packings in three dimensions is 0.58.

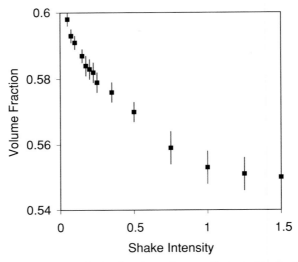

Fig. 3.1 The steady state volume fraction of monodisperse hard spheres plotted against the shaking intensity.

inhabited by the coordination number for these nonsequentially generated packings is *substantially below the c* = 6 *value for a sequential deposit*; this is a *clear* indication of the presence of bridges [33]. More recent molecular dynamics simulations [120] show that the presence of friction leads to coordination numbers in the range shown in Fig. 3.2, which suggests the following coherent physical picture: spheres slip individually (sequentially) in the absence of friction, so that $c = 6$, while friction allows bridges to be stably sustainable, leading to lower coordination numbers $c \sim 4.4 - 4.6$.

Also shown in the inset of Fig. 3.2 are the mean fractions $P(n)$ (for $n = 3 - 9$) of n-fold coordinated spheres submitted to vibrations at two different intensities ($\epsilon = 0.05$ and $\epsilon = 0.5$). Although most spheres touch four or five of their neighbours, larger values of ϵ lead to lower coordinations: the shift of the histogram peak to lower values suggests that *bridges are more space-wasting for high-intensity vibrations*. On the other hand, an examination of Figs. 3.1 and 3.2 shows that for $0.25 \leq \epsilon \leq 1$, the volume fraction ϕ steadily decreases with ϵ while the mean coordination number c remains constant. All of these facts taken together suggest that while the density of bridges is independent of ϵ over this range, bridge *shapes* become more eccentric (and therefore more wasteful of space) for higher vibrational intensities.

If we recall that ϕ remains below 0.58 for three-dimensional packings which are sequentially generated, while it rises sharply above this value for nonsequential (bridge-containing) packings (Fig. 3.1), we seem to be in the presence of an apparent paradox: why is it that high packing densities are achievable with structures which are, after all, characterised by lower coordination numbers? The resolution of this

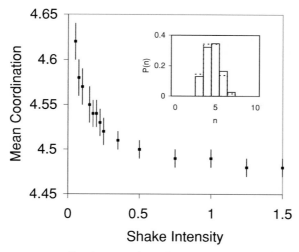

Fig. 3.2 The mean coordination number of monodisperse hard spheres plotted against the shaking intensity. The inset shows the mean fractions $P(n)$ of spheres which are n-fold coordinated in the steady-state regime of the shaking process which has shaking intensity $\epsilon = 0.05$ (full lines) and $\epsilon = 0.5$ (broken lines).

is as follows: *cooperative (nonsequential) motion of grains is able to shave down voids in a way that sequential dynamics disallows.* Thus, voids in sequentially generated packings around 0.58 are typically frozen in, i.e. not removable by further vibrations; bridges in nonsequential packings at $\phi \sim 0.58$ can, on the other hand, relax collectively (for low ϵ) to minimise void spaces between them, so that higher packing densities are reachable (Fig. 3.1). This, already hinted at in the inset of Fig. 3.2, will be borne out below.

We first study the evolution of the contact network of a stable configurations of spheres; this is the network formed by joining the centres of all pairs of touching spheres. Figure 3.3 shows the qualitative effects of low- and high-intensity shaking on a sphere assembly. In Fig. 3.3b, we see the contact network of the sphere assembly shown in Fig. 3.3a; Fig. 3.3c shows that the network deforms only very slightly after a low-intensity shake, while Fig. 3.3d shows that shaking at high intensities completely destroys the original network, with new bonds replacing older ones. This information is shown more quantitatively via the evolution of the autocorrelation function (Fig. 3.4). This is defined by the following prescription: for each sphere i at time t, define an $(N-1)$-dimensional vector, $\mathbf{b}_i(t)$, such that the jth element of $\mathbf{b}_i(t)$ is unity if sphere i is touching sphere j at time t, and zero otherwise. The average autocorrelation function so defined, $z(t) = \langle \mathbf{b}_i(t').\mathbf{b}_i(t+t')/|\mathbf{b}_i(t')||\mathbf{b}_i(t+t')|\rangle$ at two shaking intensities $\epsilon = 0.05$ and $\epsilon = 0.5$, is plotted as a function of time in Fig. 3.4. Its evolution shows in essence the rate at which a given particle changes its neighbours; we notice that at low vibration intensities, particles tend to have

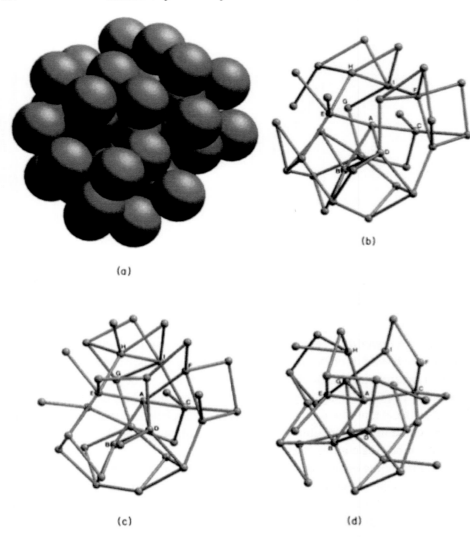

(a)

(b)

(c)

(d)

Fig. 3.3 (a) A three-dimensional cluster of spheres generated by shaking at $\epsilon =$ 0.05. (b) The corresponding contact network to the cluster in (a). Small balls represent the centres of the spheres, while bonds represent the contacts between them. The centres of spheres B, C, D and E in contact with the central sphere A have been coloured red. (See the version in the colour plate section.) (c) The contact network after cluster (a) is further shaken with $\epsilon = 0.05$; notice the similar topology to (b). (d) The contact network after cluster (a) is shaken with $\epsilon = 0.5$; notice the completely changed topology of this network relative to (b).

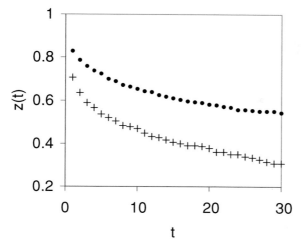

Fig. 3.4 The autocorrelation function, $z(t)$, of the contact network plotted against the number of shake cycles, t, for monodisperse hard spheres in the steady-state regime. The plots correspond to shaking intensities $\epsilon = 0.05 (\bullet)$ and $\epsilon = 0.5 (+)$.

long-lived neighbours relative to the high-intensity case, when contact correlations disappear rapidly, and fresh neighbours continue to replace existing ones. This behaviour is evidently consistent with the snapshot observations of consecutive contact network configurations [62] shown in Fig. 3.3. We can safely conclude from this evidence that low-intensity shaking causes the coherent motion of neighbouring particles, while high-intensity shaking leads to a series of different topologies, of essentially unrelated configurations.

The volume fraction ϕ and the mean coordination number z are the simplest structural descriptors for random-close-packed solids. However, there is not a simple one-to-one relationship between them, and hence the specification of both ϕ and z as independent quantities is insufficient to enumerate adequately many of the structurally dependent properties of random packings. The study of higher-order descriptors such as correlation functions is thus necessary for a holistic understanding of granular structure.

We first examine the role of position correlations. Pair distribution functions $h(r)$ for interparticle distances r in the lateral (horizontal) plane and $g(z)$ for interparticle distances along the vertical (longitudinal) direction z are illustrated in Fig. 3.5 for vibration intensities $\epsilon = 0.05$ and $\epsilon = 0.5$. These averages are evaluated over all non-overlapping sphere pairs and over ~ 25 shake cycles in the steady state. In both directions the structure is similar to that expected for dense hard sphere fluids; both functions indicate the presence of a second shell of neighbours at a separation of approximately two particle diameters. Such short-range order is most

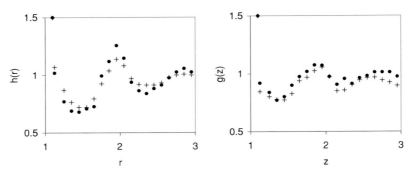

Fig. 3.5 The pair distribution functions of particle positions, $h(r)$ and $g(z)$ for monodisperse hard spheres in the steady-state regime, plotted against horizontal displacement r and vertical displacement z. The plots correspond to shaking intensities $\epsilon = 0.05(\bullet)$ and $\epsilon = 0.5(+)$. The peak heights, which are not shown, are $h(1) = 6.35$ (6.20) and $g(1) = 4.40$ (4.25) for $\epsilon = 0.05$ (0.5).

pronounced in the lateral plane, and diminishes significantly as the intensity of vibration is increased. This dependence on vibration intensities, while still present for correlations $g(z)$ along the direction of vibration z, is less pronounced. This is reasonable: while vibration induces ordering via cluster formation in all directions, clusters formed along the direction of vibration are more easily destroyed than those in a lateral direction, which have a more coherent response. Also, we can safely conclude that short-range order diminishes with increasing vibration intensity in all spatial directions (as would happen, for example, if temperature were increased in a fluid).

The study of displacement correlations (later termed 'dynamical heterogeneities' in the context of glasses [123]) in vibrated powders proves [62] to be even more revealing, especially in the context of compaction in the jamming limit. Such correlations show the extent to which grains move in a correlated fashion, giving rise to so-called 'dynamical clusters' [62]. If the lth grain undergoes a displacement $\Delta \mathbf{r}_l$ (where $\Delta \mathbf{r}_l = \Delta x_l \mathbf{i} + \Delta y_l \mathbf{j} + \Delta z_l \mathbf{k}$ with $\mathbf{i}, \mathbf{j}, \mathbf{k}$ taken to be unit vectors in the x, y and z directions respectively), the full correlation function $\langle \Delta \mathbf{r}_l \Delta \mathbf{r}_m \rangle$ defines the displacement correlation for all grain pairs l and m. We focus for concreteness on only the z component of the displacements, remarking that for vibration intensities in the range that we have studied, the mean size of vertical displacements $\langle |\Delta z_l| \rangle$ during a shake cycle is a monotonically increasing function of the intensity. We now consider grain pairs along the longitudinal (z) and transverse $(x - y$ plane) directions and define the corresponding correlations $H(r)$ and $G(z)$ as:

$$H(r) = \langle \Delta z_l \Delta z_m \delta(|t_{lm}| - r)\Theta(|z_{lm}| - 1/2)\rangle/\langle|\Delta z_l|\rangle^2 \qquad (3.1)$$

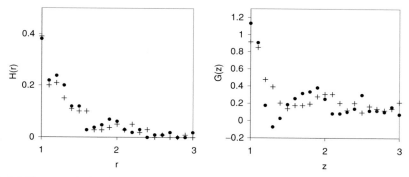

Fig. 3.6 The correlation functions, $H(r)$ and $G(z)$, for the vertical displacements of spheres during a single cycle of the steady-state shaking process, plotted respectively against horizontal displacement r and vertical displacement z. The plots correspond to shaking intensities $\epsilon = 0.05(\bullet)$ and $\epsilon = 0.5(+)$.

and

$$G(z) = \langle \Delta z_l \Delta z_m \delta(|z_{lm}| - z)\Theta(|t_{lm}| - 1/2)\rangle/\langle|\Delta z_l|\rangle^2 \qquad (3.2)$$

with

$$t_{lm}^2 = (x_l - x_m)^2 + (y_l - y_m)^2, \; z_{lm} = z_l - z_m, \qquad (3.3)$$

where $\Theta(x)$ is the Heaviside step function. These definitions ensure[3] that the averages run over all displacements of (non-overlapping) sphere pairs in the horizontal $(x - y)$ plane and vertical (z) direction respectively; they are evaluated over ~ 25 shake cycles in the steady state.

In Fig. 3.6, transverse and longitudinal correlation functions for vertical grain displacements have been plotted for vibration intensities $\epsilon = 0.05$ and $\epsilon = 0.5$. We consider first longitudinal displacement correlations measured in the transverse plane; weak to begin with ($H(r)$ decreases rapidly to zero with increasing r), they show only a small decrease in magnitude as the shaking intensity is increased. The data allow for a qualitative estimate for the horizontal range over which spheres move coherently in the vertical direction; they thus provide a measure of typical dynamical clusters in the plane transverse to vibration. The picture is of weakly correlated transverse blobs of grains bobbing up and down coherently during vertical vibrations; as they are vibrated more violently, the range of this coherent motion decreases.

The picture changes quite dramatically when we examine the *longitudinal* correlation functions for *longitudinal* grain displacements $G(z)$. Grain motions along the direction of vibration must evidently be far more correlated in that direction as a result of excluded volume effects (whereas one could almost visualise the

---

[3] Units are in sphere diameters.

longitudinal motions of neighbouring grains in the plane *transverse to vibration* as being quasi-independent). Figure 3.6 illustrates this; first, the correlations shown in $G(z)$ are visibly stronger than those shown in $H(r)$, i.e. $G(z)$ has a large first peak and at large displacements it decreases more slowly than $H(r)$. Second, $G(z)$ depends strongly on vibration intensity. Third, and most remarkably, it was one of the first [62] indicators of *grain anticorrelations* near jamming, which have been the subject of much recent work [34, 35, 124]; thus, for low vibrational intensities, $G(z)$ has a distinct (negative) minimum at approximately $z = 1.3$ sphere diameters. Such separations are typical of vertical grain separations in shallow bridges; we can thus infer that *grain anticorrelations at these separations are consistent with the slow collapse of (shallow) bridges*. (Also, and interestingly, the *transverse* correlation function of *transverse* grain displacement is negative at small separations for low-intensity vibrations, consistent with spheres moving closer together via anticorrelated displacements in the transverse plane [62].)

The most obvious conclusion from the above is that the range of coherent, correlated motion decreases as granular media are vibrated more intensely. However, the analysis of the longitudinal component of longitudinal grain displacements (as well as its analogue in the transverse plane) shows that *low vibrational intensities engender strong anticorrelations in grain displacements, consistent with bridge collapse*. We use this fact to create a consistent picture of granular compaction in the jamming limit as follows.

As vibration intensities decrease, grain contact networks deform rather than break (Fig. 3.3) with grains tending to retain their neighbours (Fig. 3.4).[4] In the jamming limit (defined as the limit when grains are 'locked' into place, with no whole voids available for a grain to move into), the only mechanism left to increase the volume fraction further is the elimination of partial voids. Bridge collapse, which occurs via the anticorrelated motion of grains surrounding voids (Fig. 3.6), has the effect of 'squeezing' out such partial voids. This bears out our assertion made earlier regarding the resolution of the apparent paradox presented in Fig. 3.1, where packings containing bridges, and relaxing via a cooperative dynamics, are able to achieve far higher densities than their sequential analogues. Such cooperative relaxations predominate in the region above $\phi = 0.58$ (the limit for sequentially generated packings) in Fig. 3.1, so that the only way to attain high packing densities in practice (for example, when one is trying to get the last few coffee granules to fit into a jar) is to tap very gently indeed!

Another second order descriptor which contains valuable additional information about granular structures is the distribution of the orientations for particle–particle

---

[4] For high vibration intensities, bridges form and break stochastically, and grain neighbourhoods change constantly; there is thus no long-lived coherent process that allows for the systematic minimisation of voids in this case.

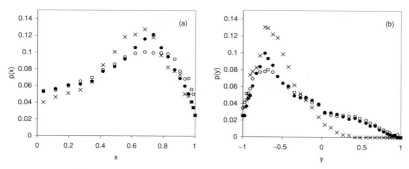

Fig. 3.7 The distribution of orientations for particle–particle contacts in random packings of monosize spheres. Crosses correspond to sequentially constructed packings while open (closed) circles correspond to (nonsequentially) vibrated packings for $\epsilon = 0.05$ (1.0). (a) Here $x = \cos(\zeta)$ is the cosine of the angle between a particle–particle contact vector and the z-axis. (b) Here $y = \cos(\psi)$ is the cosine of the angle between the z-axis and the contact vectors which form the stabilising contacts of each particle.

contacts: this underlies most of the tensor properties of granular materials [125]. We study this orientational distribution in terms of two angles: first, the angle ζ between the z-axis and a line drawn between the centres of contacting particles, with $0 \le \zeta \le \pi/2$. (For example, a pair of grains aligned along the x-axis will have $\zeta = \pi/2$.) The second angle ψ, such that $0 \le \psi \le \pi$, is the angle between the z-axis and a line from the centre of each particle to the centre of each one of its three stabilising contacts. These stabilising contacts are the three special neighbours of a grain which define the local potential energy minimum in which it rests, in a stable close packing. (Thus for example, a grain supported by a grain above and two below itself, will have $\psi \ge \pi/2$ and $\cos(\psi) \le 0$ for its two lower neighbours, and $\psi \le \pi/2$ and $\cos(\psi) \ge 0$ for its upper neighbour.)

In Fig. 3.7, we plot the orientational distributions for ζ and ψ for configurations taken from the stable close-packed phase of steady states under shaking at $\epsilon = 0.05$ (open circles), $\epsilon = 1.0$ (closed circles) respectively, with crosses denoting the data for purely sequential dynamics. For the contact angle distribution, notice that the sequential data are most anisotropic; the data for nonsequential shaking with $\epsilon = 1.0$ are less so, while those for $\epsilon = 0.05$ are the most isotropic. This is easily explained: for sequential dynamics, particles essentially stick to positions where they are stable, so that configuration shapes are in general more jagged, leading to a greater anisotropy in the distribution for ζ. With nonsequential dynamics, groups of particles can reorganise so that sharp configurations are smoothened; this effect is enhanced as the vibration intensity is lowered, and cooperative relaxations predominate. Turning now to the distribution for the stabilising angle ψ, we note that there are many more data points for $\cos(\psi) \ge 0$ in both nonsequential cases as

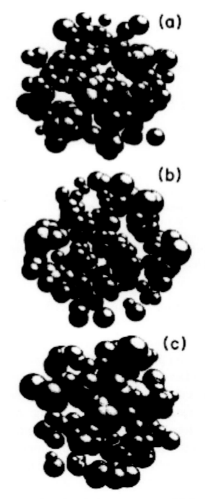

Fig. 3.8 Sections of the overlapping hole structures which are topologically complementary to the structures formed by the spheres. The shaking intensity is (a) $\epsilon = 0.05$, (b) $\epsilon = 0.5$, (c) $\epsilon = 1.5$.

compared to the sequential one. Realising that $\cos(\psi) \geq 0$ (and $\psi \leq \pi/2$) corresponds to *upward stabilisations*, i.e. stabilisations by particles whose centres are at higher z, we see that these are strong indicators of bridges; since we have premised our nonsequential restructuring on its generation of bridges, this leads to a satisfying inner consistency. Interestingly, our data indicate that the *number* of bridges (upward stabilisations) observed in shaken packings is not strongly dependent on ϵ (something which was already hinted at in Fig. 3.2); what happens is that bridge *shapes* are 'shaved down' to become more space-saving as the granular assembly undergoes predominantly cooperative relaxations at low intensities.

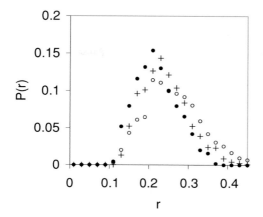

Fig. 3.9 Distribution functions $F(r)$ for the radii r of the overlapping holes presented in Fig. 3.8. Closed circles correspond to $\epsilon = 0.05$, $+$ to $\epsilon = 0.5$, and open circles to $\epsilon = 1.5$.

We have concentrated above on properties and correlations of spheres in a random close-packed structure; equally fundamental (and intimately related) problems concern the nature of the continuous network of empty space [62], consisting of pores, necks and voids, which complement the physical structure. In order to investigate the pore space of shaken packings, we have constructed the complex structures formed from overlapping holes, where we define a hole as a partial void [62]. For a close-packed bed of spheres, the overlapping holes are another species of spheres, each of which touches four packed spheres; holes may overlap each other but cannot intersect any of the packed spheres. For a monodisperse close packing, the maximum hole size is approximately the same as the sphere size (i.e. it is nearly the size of a complete void) and the minimum hole diameter is 0.224 times that of the spheres (corresponding to the hole at the centre of a regular tetrahedron formed from four spheres). Figure 3.8 shows a small section of overlapping hole structures for vibrated packings with $\epsilon = 0.05, 0.5$ and 1.5; note that low-intensity shaking leads to large numbers of isolated holes, whereas high intensities generate clearly defined strings of connected, overlapping holes. This suggests that transport of grains occurs *percolatively* when the hole strings are connected, an idea which has been exploited in the context of avalanches [126] and more recently in the context of dynamical arrest in glasses [127].

Ancillary data on the corresponding distribution functions for hole radii (Fig. 3.9) indicate that low-intensity shaking is an efficient method of removing larger holes from the overlapping hole structure (and, therefore, a method for removing large voids from a packing) without producing a regular structure. This reinforces our earlier comments about efficient compaction in the low-intensity regime: we conclude from all the above that *prolonged vibration at low intensities leads to the removal of voids by grains moving cooperatively 'inwards' (i.e. with displacement anticorrelations), with the gradual collapse of long-lived bridges.* Finally, the robustness of our results has been proved in recent simulations that have used our algorithm to probe similar quantities of interest in a vertically tapped system, with very similar results on key quantities [128].

3.3 Vibrated powders: transient response

In the previous section, we have shown the variation of key structural descriptors with shaking intensity for vibrated powders in the steady state of shaking. These results support the existence of two distinct relaxation mechanisms for granular media – *fast* dynamics, which involve the motions of independent (decoupled) grains and *slow* dynamics, which involve collective (coupled) motions of grains.

Here we develop this picture further by considering the transitions between steady states. We have chosen our reference (steady-state) configurations to be those appropriate to a shaking intensity of $\epsilon = 2.0$ – in this state, the spheres are relatively loosely packed, with $\phi \sim 0.55$. The vibrations applied to this packing are chosen to be such that $\epsilon \leq 2.0$, which drive the system to steady states of denser packing. Our motivation in doing this is to model the familiar observation [129] that dry granular material which is poured into a container can be consolidated by tapping. The connection with this scenario can be made if one recognises that the reference configuration is generated by the algorithmic analogue of *pouring* – a single large-intensity nonsequential reorganisation – while the low-intensity and low-frequency shakes that are applied to it are the algorithmic analogue of *tapping*.

This reference configuration is [130] subjected to algorithmic shaking, with intensities in the range $0.05 \leq \epsilon \leq 0.75$. The granular bed consists of a periodic arrangement of cells; the primary cell has a square cross-section of side 64 sphere diameters in the $x - y$ plane, with a depth of 20 sphere diameters in the z direction, along which vibrations are applied; this primary cell is then repeated in the x and y directions. Volume fractions in the close-packed phase of the shaking cycle are measured from the central portion of the bed (in order to minimise surface effects), and time is measured in units of the shaking cycle. The transient response of the volume fraction for shaking at five different intensities is shown in Fig. 3.10; each data set is an average over at least eight independent simulations. Also plotted in each case are two nonlinear least-squares fitted functions. Broken curves show the best single exponential fit $\phi(t) = a_0 - a_1 \exp(-a_3 t)$, while the full lines show the best fit with a sum of two exponentials $\phi(t) = a_0 - a_1 \exp(-a_3 t) - a_4 \exp(-a_5 t)$.

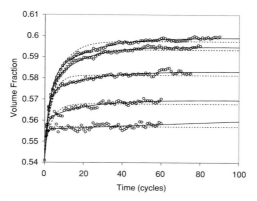

Fig. 3.10 The variation of the volume fraction with time in computer simulations of shaken granular deposits. The five data sets correspond, from top to bottom, to shaking intensity $\epsilon = 0.05, 0.1, 0.25, 0.5$ and 0.75. Broken curves show the best single exponential fits and full curves show the best two-exponential fits.

Before making comparisons between the data and the fits, we point out the purpose of the latter: rather than looking to see what is the 'best' fit for granular relaxation, we are trying to establish that *more than one* dynamical mechanism is responsible for it.[5] The simulation results show smooth, monotonic variations of volume fraction from the poured steady-state value to the shaken steady-state value. We note that the transient response of the volume fraction reflects a transition between steady states corresponding to different densities, i.e. to different magnitudes of trapped void space. If compaction were the result of a single vibration-driven process then we would expect this void space to decay with a single relaxation time – however, the poor fit achieved using the best single exponential relaxation is very noticeable. The improved fit, using a sum of two exponentials, indicates that the above expectation is unrealistic, and that the granular medium has a more complex response – similar improvements can be achieved by fitting a single stretched exponential, or indeed a logarithm [72]. Furthermore, for each value of ϵ, the two time constants obtained from the double exponential fit are very different. For $\epsilon = 0.05, 0.1, 0.25$ and 0.5, the two relaxation times are about 3 and 20 cycles respectively, but for $\epsilon = 0.75$ these times are respectively 1 and 50 cycles; also, the relevant fitted coefficients (a_1, a_4) in each case are of comparable magnitudes. From this we can infer three things: firstly, that the timescales are *well separated* (one slow and one fast), secondly, that the corresponding relaxation times depend, as one would expect, on the external vibration, and thirdly, that neither one can be ignored in comparison with the other. Once again, we emphasise that this is a coarse-grained version of the

---

[5] In fact our data can also be reasonably well fitted to a logarithm [106], which is the standard experimental fit to granular relaxation [72].

truth – in reality, there are many fast and slow timescales in a vibrated powder, and by separating them into two sets, we are only drawing attention to the coexistence of fast and slow dynamics in this system.

Evidently, in these computer simulations, shaking-induced particle reorganisations are only subject to geometrical constraints – for real powders, consolidation is far more complicated, and includes other factors such as cohesive forces and particle fragmentations. The success of our simulation technique lies in its ability to isolate fundamental geometrical constraints from other extraneous effects – the results provide a valuable benchmark for evaluating more realistic situations in experiment and industry [129].

Next, we focus on the phenomenon of self-diffusion [130]. In each cycle of a finite-amplitude and low-frequency shaking process, a granular bed has periods of both quasistatic (low kinetic energy) and volatile (high kinetic energy) behaviour. In the quasistatic regime, particles are either static or move together in tandem, but in the volatile regime, they are mobile, and apt to lose information concerning their relative positions. This loss of information can be considered a diffusive process as follows – since the position of a given particle, measured at the same phase point of consecutive shake cycles, will be slightly displaced, a sequence of these finite displacements can be visualised as a three-dimensional random walk. Since these comparisons concern the relative displacements of the same particle, we can regard the entire process as self-diffusion [43] due to shaking.

We have measured the average displacements of approximately two hundred spheres over thirty cycles in the steady-state regime for $0.05 \leq \epsilon \leq 1.0$. In each case, we observe a linear increase of the squared displacement with time. In Fig. 3.11, we have plotted, as a function of ϵ, the gradients D_T and D_Z which are obtained from least-squares fits of $\langle (\Delta x)^2 + (\Delta y)^2 \rangle$ against $2t$ and $\langle (\Delta z)^2 \rangle$ against t respectively. Here Δx and Δy are two orthogonal displacements in the plane transverse to, while Δz is the displacement parallel to, the direction of shaking. For all intensities, $D_Z > D_T$ (as expected for vertical shaking under gravity), since the diffusive motion of a particle in the direction of shaking will always be greater than that in the transverse plane.

The results in Fig. 3.11 indicate the existence of two different diffusive regimes: there is a fast regime for $\epsilon > 0.2$, where D_T and D_Z are linearly dependent on ϵ, and a slower regime at smaller shaking intensities. This picture is in qualitative agreement with experimental observations of self-diffusion in vibrated beds of granular material [43], which have been interpreted via a hydrodynamic approach that is considered to be appropriate for rapid flows and large voidage. Our simulations span both slow and rapid flows as well as large and small volume fraction; the resulting interpretation of the observed self-diffusion thus encompasses both the so-called hydrodynamic and viscous regimes.

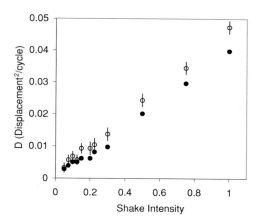

Fig. 3.11 The effective diffusion coefficients, D_Z(open circles) and D_T(full circles), against the shaking intensity ϵ. For clarity only one set of error bars is included.

The processes underlying self-diffusion under vibration are complex. In each shake cycle, spheres spend some time subject to a direct fluctuating force arising from effective collisions between pairs of moving particles; in addition, they also spend some time following deterministic trajectories (including rolling and falling) on a complicated potential energy surface. However, this energy surface changes from one shake cycle to another so that it, too, can be considered to fluctuate. Thus the random displacements of the spheres during one shake cycle result from a combination of different fluctuations – phenomenologically, this corresponds to 'hopping' between potential wells, where both the hopping times and the energy landscape are complex functions of the shaking intensity [66].

The results in Fig. 3.11 can be interpreted in terms of two distinct hopping processes. The major contribution to particle displacements for steady-state shaking with $\epsilon > 0.2$ occurs during the expanded and volatile regime. At these shaking intensities, the free volume available per particle is sufficient to destroy a large number of particle clusters, so that many particles spend a certain time (whose duration is proportional to the shaking intensity concerned) in random motion before they form new clusters. For steady-state shaking with $\epsilon \leq 0.2$, on the other hand, local clusters remain largely intact for the whole of the shake cycle; particle displacements are usually the result of deterministic motion inside their (slightly deformed) local environments.[6] These two different mechanisms underlie the crossover between the fast and slow self-diffusion observed in the results presented in Fig. 3.11.

---

[6] The size of the cluster deformations is not strongly dependent on ϵ for $\epsilon \leq 0.2$ [130].

Additionally, we note that the so-called hydrodynamic regime studied by [43] corresponds to expanded configurations where the volume of grains equals the volume of voids, so that existing voids are large enough to occupy entire grains. This corresponds [130] in our simulations to shaking intensities of $\epsilon = 0.2$; it is reasonable to expect, therefore, that a change of hopping behaviour occurs for $\epsilon < 0.2$, when grains have only partial voids to move through. Flow becomes cooperative and slower, leading to the slower diffusion observed below this crossover intensity.

3.4 Is there spontaneous crystallisation in granular media?

It is commonly assumed in the literature that the highest density that is spontaneously attainable by granular media under external perturbation is the so-called 'random close packing' (RCP) threshold, usually associated with the experiments of Bernal [10], where its numerical value was found to be 0.64 in three spatial dimensions. The physical interpretation of this threshold is that it represents the highest density at which the powder is randomly packed. However, a little bit of thought will show that there is no reason for an externally stimulated powder to restrict itself to a thoroughly disordered state, at its state of highest packing. On the contrary, given the predominance of nucleation phenomena in high density disordered systems such as colloids, one might in fact expect that granular media *should* crystallise spontaneously to states of higher density than the RCP threshold. These results below (first published in [131–133]) serve to illustrate this claim.

In general, when a powder in a loose-packed state (with $\phi \sim 0.54$, say) is shaken at a fixed intensity, steady-state values of the packing fraction are attained after short or long transients, depending on the value of the shaking intensity (see e.g. Fig. 3.10). However, our findings are [131, 132] that, at least within a range of shaking intensities, the powder can undergo a *first-order transition* to a more ordered and close-packed state. Again, our findings indicate that such spontaneous crystallisation does not occur outside this range (at least for the simulation times we have chosen), although we cannot rule out the possibility that longer times at lower shaking intensities might engender it.

The details of the simulations can be found in [131, 132]. Figures 3.12a–3.12c show the variation of the packing fraction with time t measured in shake cycles, as the spheres are shaken at amplitudes[7] $A = 0.05, 0.5$ and 1.2 respectively. For $A = 0.5$ (Fig. 3.12b), a sharp rise in packing fraction to about $\phi = 0.68$ at $t \sim 900$ occurs, which does *not* happen in the other two cases, for times of observation of up to $t \sim 2.10^5$ cycles. For large shaking amplitudes

---

[7] The amplitude of shaking A is also parametrised in terms of free volume, as the intensity ϵ was.

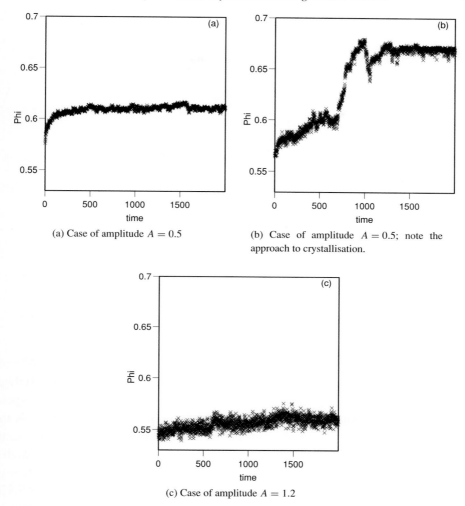

(a) Case of amplitude $A = 0.5$

(b) Case of amplitude $A = 0.5$; note the approach to crystallisation.

(c) Case of amplitude $A = 1.2$

Fig. 3.12 Plots of packing fraction ϕ vs time t

(Fig. 3.12c) the dynamics seems akin to that of fluidisation, while for very small amplitudes (Fig. 3.12a), the powder appears to be stuck in 'supercooled' configrations.

Our interpretation of this crystallisation is in terms of a nucleation scenario. When grains are already sufficiently well packed that available free volume is in short supply, the transition to crystallinity can only occur if, on the one hand, existing order is maintained under shaking, while at the same time, grains in 'non-ideal' positions are given enough free volume under shaking to move to their ideal positions. Thus, perfect packing is realisable in granular media only via an optimisation process: shaking intensities must be large enough to give grains enough free volume

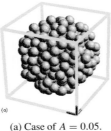

(a) Case of $A = 0.05$.

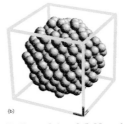

(b) Case of $A = 0.5$. Note the
near-crystalline ordering.

Fig. 3.13 An example of typical clusters obtained after 2000 timesteps. To see the full colour version, please refer to the colour plates.

to move (to avoid the supercooled scenario of Fig. 3.12a) into a state of lower potential energy and not so large as to destroy the ordered structures around them (as happens in Fig. 3.12c). However, the specific range of amplitudes where spontaneous crystallisation occurs is clearly dependent on the observation time; as the latter increases, the probability of a nucleation event enabling a supercooled assembly to jump to a state of near-crystalline packing increases correspondingly, so that the lower limit of the range of amplitudes where crystallisation occurs *decreases*.

The structures obtained at the end of the dynamics represented by Fig. 3.12a and 3.12b are extremely different, and add weight to the scenario of spontaneous crystallisation we propose. In Figs. 3.13a–3.13b, clusters of approximately 300 spheres which represent the corresponding structures are shown. It is evident that the two structures are fundamentally different, with the structure obtained after the 'jump' of Fig. 3.12b being much more ordered. Finally, while some theoretical support for this scenario has been found in the context of ongoing work [134], more detailed simulations and experiments [133] are clearly necessary to establish it beyond reasonable doubt.

3.5 Some results on shaking-induced size segregation

As mentioned earlier, size segregation phenomena concern the situation when solid particle mixtures separate according to particle size [56]. These include percolation, fractionation, and the preferential rejection of large particles during pile formation with a distribution of particle sizes [135]. These processes are dominated by individual particle dynamics and are most effective at separating particles that have large size disparities (i.e. size ratios $\gg 1$).

We deal here, however, with shaking-induced segregation, which is the dominant segregation process during many real granular materials handling operations. This causes large particles to rise through a shaken bed of smaller particles, while assisting smaller particles to fall through a shaken bed of larger ones. For this mechanism, large size ratios are *not* essential, and one of its main applications concerns in fact the separation of nearly similarly sized particles. Such segregation is generated largely by collective dynamics, with, often, the excitation intensity playing the role of an appropriate control parameter. Illustrations of this process [136] abound, one of the most celebrated being the segregation of Brazil nuts [58].

Two-dimensional simulations described in [58] have allowed for a clearer interpretation of the underlying mechanism for the 'Brazil nuts' phenomenon. They suggest that when grains under shaking are redeposited onto a substrate, smaller grains move collectively to fill voids under large grains, thus impeding their downward motion; these correlations are at the basis of the observed size segregation. In three-dimensional simulations [64] of a shaken bed of spheres with a continuous distribution of sphere sizes, more quantitative studies have been done. A measure of segregation is the weighted particle height

$$ s = \Sigma (R_i - R_o) z_i / (\langle z \rangle \, \Sigma (R_i - R_o)) - 1, \qquad (3.4) $$

where R_i is the size of the ith sphere at height z_i, R_o the minimum sphere size and $\langle z \rangle$ the mean height.

Figure 3.14 shows the increase of s with time for a set of shaken spheres. The spheres have sizes distributed uniformly between R_o and $1.5 R_o$ and the six datasets illustrate the segregation caused six different shaking intensities. Clearly, larger shaking intensities lead to a faster segregation; note also that the segregation rate appears linear, a feature that has been confirmed by using partially segregated initial packings [64].

In Fig. 3.15 the initial segregation rate $|ds/dt|_{t=0}$ obtained from the data in Fig. 3.14 is plotted against the dimensionless shaking amplitude $A/2R_o$; there is evidence of a single segregation regime in which the segregation rate increases monotonically with the shake amplitude, and for $A/2R_o \le 2$, it is found that $|ds/dt|_{t=0} \sim A^{0.45}$ [64].

Figure 3.16 shows results of simulations performed with a single (tracer) particle initially located near the centre of the container. The results show that the mean vertical component of its velocity $\langle v \rangle$ varies continuously with the relative size ratio

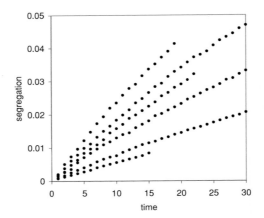

Fig. 3.14 A measure of the segregation *s* plotted against time for a shaken bed of polydisperse particles. The data sets correspond to $A/2R_o = 0.1, 0.2, 0.5, 0.75.1.0, 2.0$, moving from the bottom up. The spheres have sizes which are uniformly distributed in $[R_o, 1.5R_o]$.

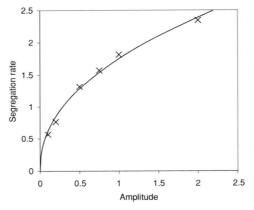

Fig. 3.15 The initial segregation rate plotted against amplitude for the data sets displayed in Fig. 3.14. The continuous curve represents a segregation rate which scales as $A^{0.45}$.

of the impurity $r = R_i/R_o$ for a fixed shaking intensity [64]. Figure 3.16 shows that for $r < 1$, $\langle v \rangle$ is negative, showing that smaller particles move downwards; while for $r > 1$, $\langle v \rangle$ is positive, showing the upward segregation of larger particles, as in the Brazil nut effect. There is a percolation discontinuity at small impurity sizes and a slow increase in the segregation velocity (below linear) for larger impurity sizes. These simulations also show that the segregation is retarded for shaking amplitudes smaller than a critical amplitude $A_c = 0.6R_o$. For driving forces below

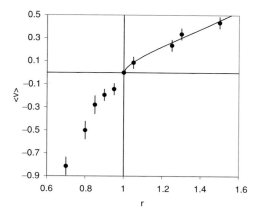

Fig. 3.16 Plot of the mean vertical component of the tracer velocity $\langle v \rangle$ against the relative impurity size r for an isolated tracer in a monodisperse bed. The continuous curve represents a scaling form $\langle v \rangle \sim (r^3 - 1)^{0.6}$.

this threshold amplitude, steady tracer motion is undetectable [64]. Devillard [137] has suggested a scaling form for the tracer segregation rate

$$v \sim (r^d - 1)^\alpha f((A - A_c)(r^d - 1)^z) \tag{3.5}$$

and for $d = 2$, has given effective exponents $\alpha = 0.6$, $z = 0.1$ with $A_c = 0.3$. This scaling is only approximate and breaks down for large impurity sizes in two dimensions. A better estimate in three dimensions can be obtained from Fig. 3.16, where the scaling curve $\langle v \rangle \sim (r^3 - 1)^{0.6}$ is plotted. Devillard's scaling form (3.5) can therefore be extended to three dimensions with $\alpha + z = 0.6$ for $r \leq 2$ [64].

Before passing onto some more general issues, we mention in passing that in addition to simple size effects, there are also effects due, e.g. to relative density [57] and orientation [106]. In the latter case, large anisotropic particles align themselves with the direction of shaking, before rising through the bed. The reader is referred to Refs. [55, 57, 106] for more details.

We sum up some of the crucial phenomena [50, 64, 106] related to the 'Brazil nut' phenomenon of size segregation by shaking. Here, granular media are dilated during the 'upward' part of a shake cycle: as the grains settle, smaller grains fall freely, while larger grains need large voids to fall into. Since large voids can only be generated by the collective motion of many small grains, a statistically unlikely process, large grains are seen to 'rise' through a shaken bed. There is no size threshold below which the segregation ceases, but with decreasing size ratios between the smaller and larger particles, the process can be intermittent; for larger size ratios, the process is continuous. In order to model these features correctly, care must be

taken to include simultaneous and collective grain dynamics, and complex couplings between the grains and the driving force. The omission of these features lead
to unrealistic stationary configurations in simulations [135], which lead in their
turn to unphysical features such as the prediction of spurious size thresholds for
segregation. These issues are extensively discussed in Ref. [63].

There are of course other possible mechanisms which could drive shaking-
induced size segregation. Of these, the convective mechanism is particularly significant; it was investigated by Knight *et al.* [54] and observed [138] using magnetic
resonance imaging. They suggest that under conditions of low amplitude and high
frequency, convection is the dominant mechanism for inducing size segregation.
Here, both large and small particles are borne upwards along the middle of a convection cell, but while small particles move downwards in a thin convection zone
at the cell edges, particles larger than the zone width remain trapped on top. In this
case, the packing becomes separated into two distinct fractions, whereas in the earlier 'Brazil nut' scenario (driven by particle reorganisations), there is a continuous
gradation of particle sizes. Note that friction at the container walls is essential for
the initiation of a convective pattern in granular material [54].

Although the convection mechanism implies mass flow, and therefore no explicit
mechanism to distinguish between different particle properties, the separation proceeds with a rate that depends on particle size, flow of interstitial fluid and particle
density [139]. The competition between these parameters could result in the so-
called 'reverse Brazil nut' effect – that is, large, light particles that actually sink to
the bottom of a shaken mixture of grains [140]. This behaviour is, however, still
the subject of some debate [141]. What is clear is that, at least in the convective
regime, changes in density can be used to reverse the direction of separation. A final
question concerns dynamic regimes: when does convection-induced size segregation predominate over that induced by particle reorganisations? A tentative answer
[50] is that the convective mechanism predominates for low amplitudes and high
frequencies of vibration; larger amplitudes render the convective rolls unstable and
particle reorganisations become the dominant inducers of segregation. However,
detailed investigations of the competing domains of amplitude and frequency need
to be undertaken before convincing answers are found to this important question.
This discussion also underlines the inappropriateness of using the the acceleration
$\Gamma \equiv A\omega^2$ as a control parameter for shaking – while this is totally appropriate for a
one-particle system [51, 52], it is deeply unsuitable for the modelling of many-body
effects that are crucial to granular media.

Vertical shaking is, of course, not the only process that generates size segregation;
for instance, pouring a particulate mixture to make a conical pile can also cause
segregation of particle sizes such that large particles are found predominantly at the
base of the pile. The degree of segregation depends on deposition rate, and other

particulate properties [142]. In certain cases, stratification can also arise as the pile grows [143].

Other parameters of importance include attractive interactions such as cohesion, which can significantly change the segregation response [144]. Using a rotating cylinder greatly increases the range of size segregation phenomena [145], with patterns which include both radial and axial segregation. In order to explain stratification, great use has been made of noninvasive observations, such as NMR imaging [146], which suggest that different mechanisms such as percolation, convection, avalanching and diffusion are always in competition during driven granular flows. The authors of Ref. [146] hypothesise that stratification could occur when rapid mixing, e.g. resulting from a convective mechanism, is restricted by some of these microscopic mechanisms; since diffusion is generally inefficient in granular media, the 'edges' of flow bands remain, resulting in stratification. These issues have been probed using weak fluidisation by Conway *et al.* [147], who have identified a series of well-defined instabilities in granular flows. These are observed in a classical Couette geometry and arise from the development of vortices, at intermediate scales, which are similar to the primary Taylor instability for fluids. Transitions are controlled by the shear stress and a characteristic hierarchy of banded patterns is observed. The vortices observed by Conway *et al.* [147] are accompanied by novel segregation transitions and also spawn additional vortices that modify the scale of the kinetic interactions.

4

Collective structures in sand – the phenomenon of bridging

4.1 Introduction

From a general introduction to phenomena in vibrated sand, we focus now on (granular) matter in the jammed state, which has become a focus of interest for physicists in recent years. Glasses [148–150] manifest jamming, in addition to densely packed granular media [27, 28, 151]; while the mechanisms of jamming in each case show strong similarities, the ineffectiveness of temperature as a dynamical motor in granular media leads to vastly more surprising effects.

A direct consequence of such athermal behaviour in sandpiles is the stable formation of cooperative structures such as bridges [33], or indeed the very existence [21] of an angle of repose [23]; neither would be possible in the presence of Brownian motion. In this chapter and the next, we show that the dynamics of bridge formation is a typical result of collective dynamics, as is the relaxation of the angle of repose; competition between density fluctuations and external driving forces can, on the other hand, result in sandpile collapse. The chapters which follow these, with their focus on jamming, unify aspects of glasses and granular media; one of them uses random graphs to illustrate competitive and cooperative effects in granular compaction [152, 153], while the other makes clear how asymmetric grains can orient themselves suitably so as to waste less space, as the jamming limit is approached [34, 35].

4.2 On bridges in sandpiles – an overarching scenario

The athermal nature of granular media results in the following fact: all granular dynamics is the result of external stimuli. These result in grains competing with each other to fall under gravity to a point of stability; when, instead, the process is one of cooperation so that two or more grains fall together to rest on the substrate

Granular Physics, ed. Anita Mehta. Published by Cambridge University Press. © A. Mehta 2007.

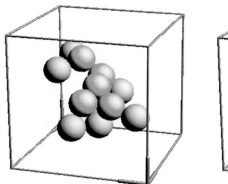 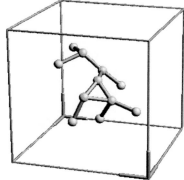

Fig. 4.1 A five particle *complex bridge*, with six base particles (left), and the corresponding contact network (right). Thus $n = 5$ and $n_b = 6 < 5 + 2$. A colour version of this figure appears in the colour plates section.

with mutual support, bridges [33] are formed. These can be stable for arbitrarily long times, since the Brownian motion that would dissolve them away in a liquid is absent in sandpiles – grains are simply too large for the ambient temperature to have any effect. As a result, bridges can affect the ensuing dynamics of the sandpile; a major mechanism of compaction is the gradual collapse of long-lived bridges in weakly vibrated granular media, resulting in the disappearance of the voids that were earlier enclosed [61, 62, 130]. Bridges are also responsible for jamming in granular processes, for example, as grains flow out of a hopper [21].

We first define a bridge in more quantitative terms. Consider a stable packing of hard spheres under gravity, in three dimensions. Each particle typically rests on three others which stabilise it, in the sense that downward motion is impeded. *A bridge is a configuration of particles in which the three-point stability conditions of two or more particles are linked; that is, two or more particles are mutually stabilised.* Bridges thus cannot be formed sequentially, but are ubiquitous in generic powders. While it is impossible to determine bridge distributions uniquely from a distribution of particle positions, it is possible via sophisticated algorithms to obtain the most likely positions of bridges in a given scenario [33].

We now distinguish between *linear* and *complex* bridges via a comparison of Figs. 4.1 and 4.2. Figure 4.1 illustrates a *complex* bridge, i.e., a mutually stabilised cluster of five particles (shown in green), where the stability is provided by six stable base particles (shown in blue). Of course the whole is embedded in a stable network of grains within the sandpile. Also shown is the network of contacts for the particles in the bridge: we see clearly that three of the particles each have two mutual stabilisations. Figure 4.2 illustrates a seven particle linear bridge with nine base particles. This is an example of a *linear* bridge. The contact network shows

Bridges in sand

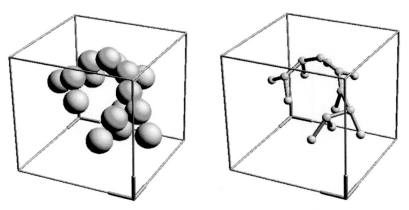

Fig. 4.2 A seven particle *linear bridge* with nine base particles (left), and the corresponding contact network (right). Thus $n = 7$ and $n_b = 9 = 7 + 2$. See the version of this figure in the colour plates section.

that this bridge has a simpler topology than that in Figure 4.1. Here, all of the mutually stabilised particles are in sequence, as in a string. A linear bridge made of n particles therefore always rests on $n_b = n + 2$ base particles. For a complex bridge of size n, the number of base particles is reduced ($n_b < n + 2$), because of the presence of *loops* in their contact networks.

An important point to note is that bridges can only be formed sustainably in the presence of friction; the mutual stabilisations needed would be unstable otherwise! Although the hybrid Monte Carlo simulations described in earlier chapters do not contain friction explicitly, the configurations generated by them include its effects because of their cooperative nature [61, 62, 130]; in particular, the coordination numbers of typical configurations lie in a range consistent with the presence of friction [120, 154].

4.3 Some technical details

The nonsequential restructuring algorithm [61, 62, 130] described in earlier chapters, whose main modelling ingredients involve *stochastic* grain displacements and *collective* relaxation, was used to generate and examine bridge structures. We review its main ingredients here, which occur in three distinct stages.

(1) The granular assembly is dilated in a vertical direction (with free volume being introduced homogeneously throughout the system), and each particle is given a random horizontal displacement; this models the dilation phase of a vibrated granular medium.
(2) The assembly is compressed in a uniaxial external field representing gravity, using a low-temperature Monte Carlo process.
(3) Individual spheres in the assembly are stabilised using a steepest descent 'drop and roll' dynamics to find local potential energy minima.

Steps (2) and (3) model the quench phase of the vibration, where particles relax to locally stable positions in the presence of gravity. Crucially, during the third phase, the spheres are able to roll in contact with others; *mutual stabilisations* are thus allowed to arise, mimicking collective effects. The final configuration has a well-defined contact network where each sphere is supported by a uniquely defined set of three other spheres.

The simulation method recalled above builds a sequence of static packings. Each new packing is built from its predecessor by a random process and the sequence achieves a steady state, where structural descriptors such as the mean packing fraction and the mean coordination number fluctuate about well-defined mean values. The steady-state mean volume fraction Φ typically evolves to values in the range $\Phi \sim 0.55 - 0.61$, depending on the shaking amplitude; the mean coordination number is always $Z \approx 4.6 \pm 0.1$. We recall that for frictionless (isostatic) packings in d dimensions, $Z = 2d$ ($= 6$ for $d = 3$) [155], while for frictional packings the *minimal* coordination number is $Z = d + 1$ ($= 4$ for $d = 3$) [154]; the configurations described here [33] thus clearly correspond to those generated in the presence of friction. This is confirmed by the results of molecular dynamics simulations of sphere packings in the limit of high friction, which yield a mean coordination number slightly above 4.5 [120].

In the configurations analysed here, segregation is avoided by choosing monodisperse particles: a rough base prevents ordering. A large number of restructuring cycles is needed to reach the steady state for a given shaking amplitude: about 100 stable configurations (picked every 100 cycles in order to avoid correlation effects) are analysed, corresponding to $\Phi = 0.56$ and $\Phi = 0.58$. From these configurations, and following specific prescriptions, a very particular algorithm identifies bridges as clusters of mutually stabilised particles [33].

Figure 4.3 illustrates two characteristic descriptors of bridges used in this work. The *main axis* of a bridge is defined using triangulation of its base particles as follows: triangles are constructed by choosing all possible connected triplets of base particles, and the vector sum of their normals is defined to be the direction of the *main axis* of the bridge. The orientation angle Θ is defined as the angle between the main axis and the z-axis. The *base extension b* is defined as the radius of gyration of the base particles about the z-axis; note that this is distinct from the radius of gyration about the main axis of the bridge.

4.4 Bridge sizes and diameters: when does a bridge span a hole?

In the following, we present statistics for both linear and complex bridges. While we recognise that bridge formation is a collective dynamical process, we adopt an ergodic viewpoint [15] here. Inspired by polymer theory [156], we visualise a linear bridge as a random chain which grows as a continuous curve, i.e. 'sequentially' in

Fig. 4.3 Definition of the angle Θ and the base extension b of a bridge. The main axis makes an angle Θ with the z-axis; the base extension b is the projection of the radius of gyration of the bridge on the x-y plane. A colour version of this figure may be found in the colour plates section.

terms of its arc length s. (For complex bridges, this simplification is not possible in general – a direct consequence of their branched structure.) This replacement of what is in reality a *collective phenomenon in time by a random walk in space* is somewhat analogous to the 'tube model' of linear polymers [156]: both are simple but efficient *effective* pictures of very complex problems.

We first address the question of the length distribution of linear bridges. We define the length distribution f_n as the probability that a linear bridge consists of exactly n spheres. We make the simplest and the most natural assumption that a bridge of size n remains linear with some probability $p < 1$ if an $(n+1)$th sphere is 'added' to it: this leads naturally to an exponential distribution with respect to n.

$$f_n = (1 - p)p^n. \tag{4.1}$$

The exponential distribution above can also be derived by means of a continuum approach. Here, a linear bridge is viewed as a continuous random curve or 'string', parametrised by the arc length s from one of its endpoints. We assume also that such a bridge disappears at a constant rate α per unit length, either by changing from linear to complex or by collapsing. The survival probability $S(s)$ of a linear bridge up to length s thus obeys the rate equation $\dot{S} = -\alpha S$ and falls off exponentially, according to $S(s) = \exp(-\alpha s)$. Consequently, the probability distribution of the length s of linear bridges reads $f(s) = -\dot{S}(s) = \alpha \exp(-\alpha s)$, a continuum analogue of (4.1).

This is in good accord with the results of independent simulations, which exhibit an exponential decay of linear bridges of the form (4.1), with $\alpha \approx 0.99$ [33], which is clearly seen until $n \approx 12$. Around $n \approx 8$, complex bridges begin to predominate; these have size distributions which show a power-law decay:

$$f_n \sim n^{-\tau} \tag{4.2}$$

with $\tau \approx 2$ [33].

The diameter R_n of linear and complex bridges of size n, which is such that R_n^2 is the mean squared end-to-end distance, has also been measured. Simulation data on diameters and size distributions [33] indicate that linear bridges in three dimensions start off as *planar self-avoiding walks*, which eventually collapse onto each other because of vibrational effects; on the other hand, complex bridges look like *3d percolation clusters*.

Another issue of interest is the jamming potential of a bridge. A measure of this, in the case of a linear bridge, is its base extension b (see Fig. 4.3); this is the horizontal projection of the 'span' of the bridge. Again, simulation results [33] indicate that *three-dimensional bridges of a given length have a fairly characteristic horizontal extension*, making it relatively easy to predict whether or not they would 'jam' a given hole.

In order to compare simulations with experiment [157–159], we plot in Fig. 4.4 the logarithm of the probability distribution of base extensions $p(b)$ against the (normalised) base extension $b/\langle b \rangle$. This figure emphasises the exponential tail of the distribution function, and also shows that bridges with small base extensions are unfavoured. We note that this long tail is characteristic of three-dimensional experiments on force chains in granular media [159–161]. The sharp drop at the origin, as well as the long tail in Fig. 4.4, are observed in normal force distributions obtained via molecular dynamics simulations of particle packings [162], in the limit of strong deformations; they are also remarkably similar to measurements of normal force distributions of the 'isotropically compressed' force chain experiments of Majmudar and Behringer [160]. Realising that the measured forces propagate through chains of particles, we use this similarity to suggest that *bridges are really just long-lived force chains*, which have survived despite strong deformations. We

Bridges in sand

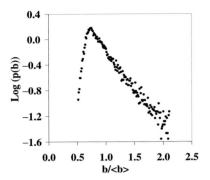

Fig. 4.4 Distribution of base extensions of bridges, for $\Phi = 0.58$. The logarithm of the normalised probability distribution is plotted as a function of the normalised variable $b/\langle b \rangle$, where $\langle b \rangle$ is the mean extension of bridge bases.

suggest also that with the current availability of 3D visualisation techniques such as NMR [163], bridge configurations might be an easily measurable and effective tool to probe inhomogeneities in shaken sand.

4.5 Turning over at the top; how linear bridges form domes

Recall that a linear bridge is modelled as a continuous curve, parametrised by its arc length s. We here focus on its most important degree of freedom, the tilt with respect to the horizontal; the azimuthal degree of freedom is neglected. Accordingly, we define the local or *link* angle $\theta(s)$ between the direction of the tangent to the bridge at point s and the horizontal, and the *mean angle* made by the bridge from its origin up to point s, also with the horizontal:

$$\Theta(s) = \frac{1}{s} \int_0^s \theta(u)\, du. \tag{4.3}$$

The local angle $\theta(s)$ so defined may be either positive or negative; it can even change sign along the random curve which represents a linear bridge. Of course, the orientation angle Θ measured in numerical simulations [33] is positive by construction, being defined as the angle between the main bridge axis and the z-axis (see Fig. 4.3). (Note that by simple geometry, this 'zenith angle' made by the *bridge axis* with the *vertical* equals the mean angle made by the *basal plane* of the bridge with the *horizontal*.)

Results of simulations [33] show that the mean angle $\Theta(s)$ typically becomes smaller and smaller as the length s of the bridge increases. Small linear bridges are almost never flat [33]; as they get longer, assuming that they still stay linear, they get 'weighed down', arching over as at the mouth of a hopper [21]. Thus, in addition to our earlier claim that long linear bridges are rare, we claim further here that (if and) when they exist, they typically have flat bases, becoming 'domes'.

We use these insights to write down equations to investigate the angular distribution of linear bridges. These couple the evolution of the local angle $\theta(s)$ with local density fluctuations $\phi(s)$ at point s (with $'$ denoting a derivative with respect to s):

$$\theta' = -a\theta - b\phi^2 + \Delta_1\eta_1(s), \tag{4.4}$$
$$\phi' = -c\phi + \Delta_2\eta_2(s). \tag{4.5}$$

The effects of vibration on each of θ and ϕ are represented by two independent white noises $\eta_1(s)$, $\eta_2(s)$, such that

$$\langle \eta_i(s)\eta_j(s') \rangle = 2\,\delta_{ij}\,\delta(s - s'), \tag{4.6}$$

whereas the parameters $a, \ldots, \Delta_2$ are assumed to be constant.

The phenomenology behind the above equations is the following: the evolution of $\theta(s)$ is caused, in our effective picture, by the *sequential* addition of particles to the bridge at its ends. The fluctuations of local density ϕ at a point s are caused by *collective* particle motion [95]. The first terms on the right-hand side of (4.4) and (4.5) say that neither θ nor ϕ is allowed to be arbitrarily large. Their coupling via the second term in (4.4) arises as follows: if there are density fluctuations ϕ^2 of large magnitude at the tip of a bridge, these will, to a first approximation, 'weigh the bridge down', i.e., decrease the angle θ locally.

Reasoning as above, we therefore anticipate that for low-intensity vibrations and stable bridges, both density fluctuations $\phi(s)$ and link angles $\theta(s)$ will be small. Accordingly, we linearise (4.4), obtaining thus an Ornstein–Uhlenbeck [164] equation:

$$\theta' = -a\theta + \Delta_1\eta_1(s). \tag{4.7}$$

Let us make the additional assumption that the initial angle θ_0, i.e., that observed for very small bridges, is itself Gaussian with variance $\sigma_0^2 = \langle \theta_0^2 \rangle$. The angle $\theta(s)$ is then a Gaussian process with zero mean for any value of the length s. Its correlation function can be easily evaluated to be [164]:

$$\langle \theta(s)\theta(s') \rangle = \sigma_{eq}^2\, e^{-a|s-s'|} + \left(\sigma_0^2 - \sigma_{eq}^2\right)e^{-a(s+s')}. \tag{4.8}$$

It follows from this that the variance of the link angle is:

$$\langle \theta^2 \rangle(s) = \sigma_{eq}^2 + \left(\sigma_0^2 - \sigma_{eq}^2\right)e^{-2as}, \tag{4.9}$$

We see from the above that orientation correlations decay with a characteristic length given by $\xi = 1/a$; also, in the limit of an infinite bridge, the variance $\langle \theta^2 \rangle$ relaxes to $\sigma_{eq}^2 = \Delta_1^2/a$ [33]. Thus, as the chain gets longer, the variance of the link angle relaxes from its initial value of σ_0^2 (i.e. that for the initial link) to σ_{eq}^2 for infinitely long chains. These results are very similar to those of the correlations of the 'isotropically compressed' force chains of Majmudar and Behringer [160].

Given the above, it can be shown that the mean angle $\Theta(s)$ will also have a Gaussian distribution. Its variance can be derived by inserting (4.8) into (4.3):

$$\langle \Theta^2 \rangle(s) = 2\sigma_{\mathrm{eq}}^2 \frac{as - 1 + \mathrm{e}^{-as}}{a^2 s^2} + (\sigma_0^2 - \sigma_{\mathrm{eq}}^2) \frac{(1 - \mathrm{e}^{-as})^2}{a^2 s^2}. \tag{4.10}$$

The asymptotic result,

$$\langle \Theta^2 \rangle(s) \approx \frac{2\sigma_{\mathrm{eq}}^2}{as} \approx \frac{2\Delta_1^2}{a^2 s}, \tag{4.11}$$

confirms our earlier statement that *the longest bridges form domes*, i.e. they have bases that are almost flat. Each such bridge can be viewed as consisting of a large number $as = s/\xi \gg 1$ of independent 'blobs' of length ξ; this result suggests, yet again, strong analogies between linear bridges and linear polymers [156].

The result (4.11) has another interpretation. As $\Theta(s)$ is small with high probability for a very long bridge, its extension in the vertical direction reads approximately

$$Z = z(s) - z(0) \approx s\,\Theta(s), \tag{4.12}$$

so that $\langle Z^2 \rangle \approx s^2 \langle \Theta^2 \rangle(s) \approx 2(\Delta_1/a)^2 s$. Switching back to a discrete picture of an n-link chain, we have

$$Z_n \sim n^{1/2}. \tag{4.13}$$

The vertical extension of a linear bridge is thus found to grow with the usual random-walk exponent $1/2$, in agreement with experiments on two-dimensional bridges [157]. Our observations on horizontal extensions of three-dimensional bridges have yielded [33] a nontrivial exponent $\nu_{\mathrm{lin}} \approx 0.66$. Putting all of this together, our results predict that *long linear bridges are domelike; also, they are vertically diffusive but horizontally superdiffusive*. Evidently, jamming in a three-dimensional hopper would be caused by the planar projection of such a *dome*.

We now compare the results of this simple theory with data on bridge structures obtained from *independent* numerical simulations of shaken hard sphere packings

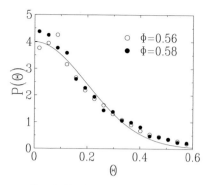

Fig. 4.5 Plot of the normalised distribution of the mean angle Θ (in radians) of linear bridges of size $n = 4$, for both volume fractions. The $\sin \Theta$ Jacobian has been duly divided out, explaining thus the larger statistical errors at small angles. Full line: common fit to (half) a Gaussian law.

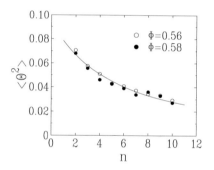

Fig. 4.6 Plot of the variance of the mean angle of a linear bridge, against size n, for both volume fractions. Full line: common fit to the first (stationary) term of (4.10), yielding $\sigma_{eq}^2 = 0.093$ and $a = 0.55$. The 'transient' effects of the second term of (4.10) are invisible with the present accuracy.

[61, 62, 130]. Figure 4.5 confirms that the mean angle is Gaussian to a good approximation, while Fig. 4.6 shows the measured size dependence of the variance $\langle \Theta^2 \rangle (s)$. The numerical data are found to agree well with a common fit to the first (stationary) term of (4.10) – the 'transient' effects of the second term of (4.10) are too small to be significant at our present accuracy. We thus conclude that this simple theory [33] captures the principal structural features of linear bridges.

4.6 Discussion

We end this chapter with the following remarks. First, more subtle effects, including the effects of transients via the second term of (4.10), and the dependence of the parameters σ_{eq}^2 and a on the packing fraction Φ, are deserving of further investigation. Second, we might expect that with increasing density Φ, branched structures would become more and more common; linear bridge formation, with its 'sequential' progressive attachment of independent blobs, would then become more and more rare. Our theory should therefore cease to hold at a limit packing fraction Φ_{lim}, which is qualitatively reminiscent of the single-particle relaxation threshold density [152, 153] (see Chapter 6). Finally, our investigations suggest that *long-lived bridges are natural indicators of sustained inhomogeneities in granular systems*; their most accurate realisation in real systems can be found in the so-called 'isotropically compressed' force chains of the Behringer group [160], whose normal force distributions and weak angular correlations are in good agreement with our results above.

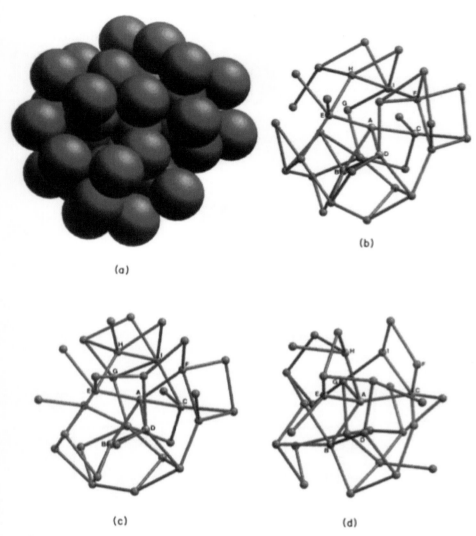

Fig. 3.3 (a) A three-dimensional cluster of spheres generated by shaking at $\epsilon =$ 0.05. (b) The corresponding contact network to the cluster in (a). Small balls represent the centres of the spheres, while bonds represent the contacts between them. The centres of spheres B, C, D and E in contact with the central sphere A have been coloured red. (c) The contact network after cluster (a) is further shaken with $\epsilon = 0.05$; notice the similar topology to (b). (d) The contact network after cluster (a) is shaken with $\epsilon = 0.5$; notice the completely changed topology of this network relative to (b).

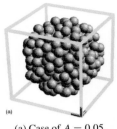

(a)

(a) Case of $A = 0.05$.

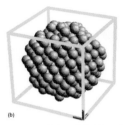

(b)

(b) Case of $A = 0.5$. Note the
near-crystalline ordering.

Fig. 3.13 An example of typical clusters obtained after 2000 timesteps.

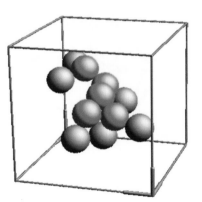 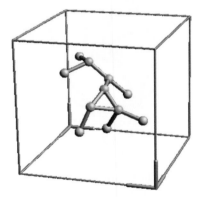

Fig. 4.1 A five particle *complex bridge*, with six base particles (left), and the corresponding contact network (right). Thus $n = 5$ and $n_b = 6 < 5 + 2$.

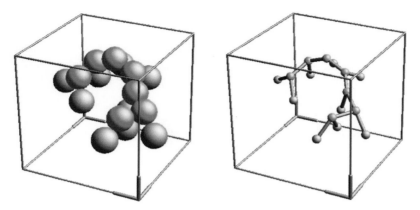

Fig. 4.2 A seven particle *linear bridge* with nine base particles (left), and the corresponding contact network (right). Thus $n = 7$ and $n_b = 9 = 7 + 2$.

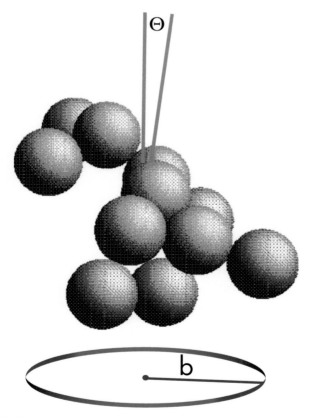

Fig. 4.3 Definition of the angle Θ and the base extension b of a bridge. The main axis makes an angle Θ with the z-axis; the base extension b is the projection of the radius of gyration of the bridge on the x-y plane.

5

On angles of repose: bistability and collapse

The phenomenon of the angle of repose is unique to granular media, and a direct consequence of their athermal nature; this manifests itself in the fact that, typically, the faces of a sandpile are inclined at a finite angle to the horizontal. The angle of repose θ_R can, in practice, take a range of values before spontaneous flow occurs as a result of the sandpile becoming unstable to further deposition; the limiting value of this angle before such avalanching occurs is known as the *maximal angle of stability* θ_m [21].

Also, as a result of their athermal nature, sandpiles are strongly hysteretic; this results in *bistability* at the angle of repose [126, 165, 166], such that a sandpile can either be stable or in motion at any angle θ such that $\theta_R < \theta < \theta_m$. However, despite the above, it is possible for a sandpile to undergo spontaneous collapse to the horizontal; this is, in general, a rare event. We propose a theoretical explanation [23] below for both bistability at, and collapse through, the angle of repose via the coupling of fast and slow relaxational modes in a sandpile [69].

5.1 Coupled nonlinear equations: dilatancy vs the angle of repose

Our basic picture is that fluctuations of local density are the collective excitations responsible for stabilising the angle of repose, and for giving it its characteristic width,

$$\delta\theta_B = \theta_m - \theta_R, \tag{5.1}$$

known as the Bagnold angle [6]. Such density fluctuations may arise from, for instance, shape effects [34, 35] or friction [21, 154]; they are the manifestation in our model of *Reynolds dilatancy* [1].

Granular Physics, ed. Anita Mehta. Published by Cambridge University Press. © A. Mehta 2007.

The dynamics of the angle of repose $\theta(t)$ and of the density fluctuations $\phi(t)$ are described [23] by the following stochastic equations, which couple their time derivatives $\dot{\theta}$ and $\dot{\phi}$:

$$\dot{\theta} = -a\theta + b\phi^2 + \Delta_1 \eta_1(t), \tag{5.2}$$

$$\dot{\phi} = -c\phi + \Delta_2 \eta_2(t). \tag{5.3}$$

The parameters $a, b, c, \Delta_1, \Delta_2$ are phenomenological constants, while $\eta_1(t), \eta_2(t)$ are two independent white noises such that

$$\langle \eta_i(t)\eta_j(t') \rangle = 2 \delta_{ij} \delta(t - t'). \tag{5.4}$$

The first terms in (5.2) and (5.3) suggest that neither the angle of repose nor the dilatancy is allowed to be arbitrarily large for a stable system. The second term in (5.2) affirms that dilatancy underlies the phenomenon of the angle of repose; in the absence of noise, density fluctuations *constitute* this angle. The term proportional to ϕ^2 is written on symmetry grounds, since the the magnitude (rather than the sign) of density fluctuations should determine the width of the angle of repose. The noise in (5.2) represents external vibration, while that in (5.3) embodies the slow granular temperature, otherwise known as Edwards' compactivity [15], being related to purely density-driven effects. We note that these equations bear more than a passing resemblance to those in the previous chapter on orientational statistics of bridges: the underlying reason for this similarity is the idea [23, 33] that bridges form by initially aligning themselves at the angle of repose in a sandpile.

Examining the above equations, we quickly distinguish two regimes. When the material is weakly dilatant ($c \gg a$), so that density fluctuations decay quickly to zero (and hence can be neglected), the angle of repose $\theta(t)$ relaxes *exponentially fast* to an equilibrium state, whose variance

$$\theta_{eq}^2 = \frac{\Delta_1^2}{a} \tag{5.5}$$

is just the zero-dilatancy variance of θ. The opposite limit, where $c \ll a$, and density fluctuations are long-lived, will be our regime of interest here. When, additionally, Δ_1 is small, the angle of repose has a *slow* dynamics reflective of the slowly evolving density fluctuations. These conditions can be written more precisely as

$$\gamma \ll 1, \quad \epsilon \ll 1, \tag{5.6}$$

in terms of two dimensionless parameters (see (5.13)):

$$\gamma = \frac{c}{a}, \quad \epsilon = \frac{ac^2\Delta_1^2}{b^2\Delta_2^4} = \frac{\theta_{eq}^2}{\theta_R^2}. \tag{5.7}$$

The parameter γ, which sets the separation of the fast and slow timescales, is an inverse measure of *dilatancy* in the granular medium; small values of this imply a granular medium that is 'stiff' to deformation, resulting from the persistence of density fluctuations. The parameter ϵ measures the ratio of fluctuations about the (zero-dilatancy) angle of repose to its full value in the presence of density fluctuations: from this we can already infer that it is a *measure of the ratio of the external vibrations to density-driven effects*, which are explicitly contained in the ratio (Δ_1^2/Δ_2^4). Realising that *external vibrations and density/compactivity respectively drive fast and slow dynamical processes in a granular system*, we see that a quantity which measures their ratio has all the characteristics of an effective temperature [69] in the slow dynamical regime of interest to us here. This temperature-like aspect will become much more vivid subsequently, when we discuss the issue of sandpile collapse.

To recapitulate: the regime (5.6) that we will discuss below is characterised as *low-temperature and strongly dilatant*, governed as it is by the *slow dynamics* of density fluctuations.

5.2 Bistability within $\delta\theta_B$: how dilatancy 'fattens' the angle of repose

Suppose that a sandpile is created in regime (5.6) with very large initial values for the angle θ_0 and dilatancy ϕ_0. In the initial transient stages, the noises have negligible effect and the decay is governed by the deterministic parts of (5.2) and (5.3):

$$\theta(t) = (\theta_0 - \theta_m)e^{-at} + \theta_m e^{-2ct}, \tag{5.8}$$

$$\phi(t) = \phi_0 e^{-ct}, \tag{5.9}$$

with

$$\theta_m \approx \frac{b\,\phi_0^2}{a}. \tag{5.10}$$

Thus, density fluctuations $\phi(t)$ relax exponentially, while the trajectory $\theta(t)$ has two separate modes of relaxation. First, there is a fast (inertial) decay in $\theta(t) \approx \theta_0\,e^{-at}$, until $\theta(t)$ is of the order of θ_m; this is followed by a slow (collective) decay in $\theta(t) \approx \theta_m\,e^{-2ct}$. When $\phi(t)$ and $\theta(t)$ are small enough [i.e., $\phi(t) \sim \phi_{eq}$ and $\theta(t) \sim \theta_R$, cf. (5.11) and (5.13)] for the noises to have an appreciable effect, the above analysis is no longer valid. The system then reaches the equilibrium state of the full nonlinear stochastic process represented by (5.2) and (5.3), a full analytical solution of which is presented in [23].

In order to get a feeling for the more qualitative features of the equilibrium state, we note first that the equilibrium variance of $\phi(t)$ is:

$$\phi_{eq}^2 = \frac{\Delta_2^2}{c}. \tag{5.11}$$

We see next that, to a good approximation, the angle θ adapts instantaneously to the dynamics of $\phi(t)$ in regime (5.6):

$$\theta(t) \approx \frac{b\,\phi(t)^2}{a}. \tag{5.12}$$

The two above statements together imply that the distribution of the angle $\theta(t)$ is approximately that of the square of a Gaussian variable. The *typically observed* angle of repose θ_R is the time-averaged value

$$\theta_R = \langle\theta\rangle_{\text{eq}} = \frac{b\,\phi_{\text{eq}}^2}{a} = \frac{b\Delta_2^2}{ac}. \tag{5.13}$$

Equation (5.12) then reads

$$\theta(t) \approx \theta_R \frac{\phi(t)^2}{\phi_{\text{eq}}^2}. \tag{5.14}$$

Equation (5.14) entirely explains the physics behind the multivalued and history-dependent nature of the angle of repose [70, 72]. Its instantaneous value depends directly on the instantaneous value of the dilatancy; its maximal (stable) value θ_m is noise-independent [cf. (5.10)] and depends only on the maximal value of dilatancy that a given material can sustain stably [21]. Sandpiles constructed above this will first decay quickly to it; they will then decay more slowly to a 'typical' angle of repose θ_R. The ratio of these angles is given by

$$\frac{\theta_m}{\theta_R} = \frac{\phi_0^2}{\phi_{\text{eq}}^2}, \tag{5.15}$$

so that $\theta_m \gg \theta_R$ for $\phi_0 \gg \phi_{\text{eq}}$. Within the Bagnold angle $\delta\theta_B$ (i.e. for sandpile inclinations which lie in the range $\theta_R < \theta < \theta_m$), this simple theory also demonstrates the presence of *bistability*. Thus, sandpiles submitted to low noise are stable in this range of angles (at least for long times $\sim 1/c$); on the other hand, sandpiles submitted to high noise (such that the effects of dilatancy become negligible in (5.2)) continue to decay rapidly in this range of angles, becoming nearly horizontal at short times $\sim 1/a$.

Our conclusions are that bistability at the angle of repose is a natural consequence of applied noise (tilt [126, 165] or vibration) in granular systems. For sandpile inclinations θ within the range $\delta\theta_B$, sandpile history is all-important: depending on this, a sandpile can either be at rest or in motion at the *same* angle of repose.

5.3 When sandpiles collapse: rare events, activated processes and the topology of rough landscapes

When sandpiles are subjected to low noise for a sufficiently long time, they can collapse [69], such that the angle $\theta(t)$ vanishes. Such an event is expected to be very rare in the regime (5.6); in fact it occurs only if the noise $\eta_1(t)$ in (5.2) is sufficiently negative for sufficiently long to compensate for the strictly positive term $b\phi^2$. It can be shown [23] that the equilibrium probability for θ to be negative, $\Pi = \mathrm{Prob}(\theta < 0)$, scales throughout regime (5.6) as:

$$\Pi \approx \frac{(2\epsilon)^{1/4}}{\Gamma(1/4)} \mathcal{F}(\zeta), \quad \zeta = \frac{\gamma}{\epsilon^{1/2}} = \frac{b\Delta_2^2}{a^{3/2}\Delta_1}. \tag{5.16}$$

The scaling function $\mathcal{F}(\zeta)$ decays [23] monotonically from $\mathcal{F}(0) = 1$ to $\mathcal{F}(\infty) = 0$; to find out when the angle of repose first crosses zero, we should explore the latter limit, i.e. the regime $\zeta \gg 1$. Here, the equilibrium probability of collapse vanishes exponentially fast:

$$\Pi \sim \exp\left(-\frac{3}{2}\left(\frac{\gamma^2}{\epsilon}\right)^{1/3}\right). \tag{5.17}$$

The above suggests that sandpile collapse is an *activated* process, with a *competition* between 'temperature' ϵ and 'barrier height' γ^2. Collapse events occur at Poissonian times, with an exponentially large characteristic time given by an Arrhenius law:

$$\tau \sim 1/\Pi \sim \exp\left(\frac{3}{2}\left(\frac{\gamma^2}{\epsilon}\right)^{1/3}\right). \tag{5.18}$$

The stretched exponential with a fractional power of the usual 'barrier-height-to-temperature ratio' γ^2/ϵ is suggestive of glassy dynamics [149, 150]; it also reinforces the idea that sandpile collapse is a *rare event*.

While the reader is referred to a longer paper [23] for the derivation of the stretched exponential, the physics behind it is readily understood by means of an exact analogy with the problem of random trapping [167], which we outline below.

Consider a Brownian particle in one dimension, diffusing (with diffusion constant D) among a concentration c of Poissonian traps. Once a trap is reached, the particle ceases to exist, so that its survival probability $S(t)$ is also the probability that it has not encountered a trap until time t. Assuming a uniform distribution of starting points, the fall-off of this probability can be estimated by first computing the probability of finding a large region of length L without traps, and then weighting this with the probability that a Brownian particle survives within it for a long time t:

$$S(t) \sim \int_0^\infty \exp\left(-cL - \frac{\pi^2 Dt}{L^2}\right) \, dL. \qquad (5.19)$$

The first exponential factor $\exp(-cL)$ is the probability that a region of length L is free of traps, whereas the second exponential factor is the asymptotic survival probability of a Brownian particle in such a region, $\exp(-Dq^2 t)$. The integral is dominated by a saddle-point at $L \approx \left(\dfrac{2\pi^2 Dt}{c}\right)^{1/3}$, whence we recover the well-known estimate

$$S(t) \sim \exp\left(-\frac{3}{2}\left(2\pi^2 c^2 Dt\right)^{1/3}\right). \qquad (5.20)$$

Notice the similarity in the forms of (5.17) and (5.20); it turns out that the steps in their derivations are identical [23], and form the basis of an exact analogy. In turn the analogy allows us to formulate an *optimisation-based* approach to sandpile collapse, which makes for a much more intuitive grasp of its physics.

Accordingly, let us visualise the angle θ as an 'exciton' whose 'energy levels' are determined by the magnitude of θ. It diffuses with temperature ϵ in a frozen landscape of ϕ (dilatancy) barriers of typical energy γ. Only if it succeeds in finding an unusually low barrier can it escape via (5.17), to reach its ground state ($\theta = 0$) – this of course corresponds to sandpile collapse. Taking the analogy a step further, we visualise the exciton as 'flying' at a 'height' θ, surrounded by ϕ-peaks of typical 'height' γ in a rough landscape. Flying too low would cause the θ exciton to hit a ϕ barrier fast, while flying too high would cause the exciton to miss the odd low barrier. It turns out [23] that flying at $\theta \sim \epsilon^{1/3}$ allows the exciton to escape via (5.17) (cf. the arguments leading to $L \sim t^{1/3}$ above). Translating back to the scenario of sandpile angles, the above arguments imply the following: angles of repose that are too low are unsustainable for any length of time, given dilatancy effects, while angles that are too large will resist collapse. Thus *optimal angles for sandpile collapse are found to scale as $\theta \sim \epsilon^{1/3}$*; sandpiles with these inclinations show a finite, if small, tendency to collapse via (5.17).

Clearly, the frequency of collapse will depend on the topology of the ϕ-landscape; the form (5.17) was valid for a landscape with Gaussian roughness [23]. What if the landscape is much rougher or smoother than this? To answer this question, we look at two opposite extremes of non-Gaussianness.

First, let us assume that density fluctuations are peaked around zero; typical barriers are low, and the ϕ-landscape is much flatter than Gaussian. The exciton's escape probability ought now to be greatly increased. This is in fact the case [23];

it can be shown that in the $\gamma \to 0$ limit, the collapse probability scales as $\epsilon^{1/4}$. Switching back to the language of sandpiles, this limit corresponds to a nearly *non-dilatant material*; it results in a 'liquid-like' scenario of *frequent collapse*, where a finite angle of repose is hard to sustain under any circumstances.

In the opposite limit of an extremely rough energy landscape, where large values of ϕ are more frequent than in the Gaussian distribution, one might expect the escape probability of the θ exciton to be greatly reduced. If, for example, the jaggedness of the landscape is such that $|\phi(t)|$ is always larger than some threshold ϕ_{th}, the stretched exponential in (5.17) reverts (in the $\epsilon \ll 1$ regime considered) to an Arrhenius law in its usual form:

$$\Pi \sim \exp\left(-\frac{(\phi_{\text{th}}/\phi_{\text{eq}})^4}{2\epsilon}\right). \tag{5.21}$$

In the language of sandpiles, this limit corresponds to *strongly dilatant* material; here, as one might expect, sandpile collapse is even *more strongly inhibited* than in (5.17). Wet sand, for example, is strongly dilatant; its angles of repose can be far steeper than usual, and still resist collapse.

5.4 Discussion

The essence of our theory above is that dilatancy is responsible for the existence of the angle of repose in a sandpile. We claim further that bistability at the angle of repose results from the difference between out-of-equilibrium and equilibrated dilatancies. We are also able to provide an analytical confirmation of the following everyday observation: *weakly dilatant sandpiles collapse easily, while strongly dilatant ones bounce back.*

5.5 Another take on bistability

As mentioned above, the angle of repose of a sandpile, θ_r, is the typical inclination of the free surface of a stationary pile. It is well known [21] that sandpiles exhibit bistable behaviour at and around this angle; this corresponds to a range of values for the measured angle of repose which varies as a function of different configurational histories. It is conventional to define this range in terms of another angle, θ_m, called the *angle of maximal stability*; this is the minimum value of the angle of the sandpile at which avalanching is inevitable. Clearly, $\theta_m > \theta_r$ and the range $\delta\theta_B \equiv \theta_m - \theta_R$ (defined in Eq. (5.1) as the Bagnold angle) corresponds to a range of angles between the free sandpile surface and the horizontal such that *either* a stable stationary state *or* avalanching can result depending on how the sandpile was produced.

In this portion of the chapter, we show that such bistable behaviour is obtained when a model sandpile with time-varying disorder is tilted [168]. The resulting findings on the correlation between avalanche shapes, and the angle of tilt of the underlying sandpile surfaces, match recent experimental results [166]; additionally, a theoretical explanation for these results is provided in terms of concepts of directed percolation.

The model sandpile used here is a two-dimensional version of an earlier (one-dimensional) disordered and non-abelian sandpile [121]. 'Grains' are rectangular blocks with dimensions $1 \times 1 \times \alpha$ which are embedded in two dimensions: they are placed on the sites i, j of a square lattice of size L with $1 \leq i, j \leq L$. A grain within column i, j may rest on either its square (1×1) face or its rectangular ($1 \times \alpha$) face. We denote these two states pictorially by $-$ or $|$ because they contribute respectively $\Delta z = \alpha$, 1 to the total columnar height $z(i, j)$.

Grains are deposited on the sandpile with a given probability of landing in the $-$ or the $|$ orientation. The square face down $(-)$ configuration of grains is considered to be more stable and this implies that in general, and certainly well away from the surface, grains contribute $\Delta z = \alpha$ to the column height. However, incoming grains, as well as all other grains in the same column, can 'flip' to the other orientation with probabilities:

$$P(- \rightarrow |) = \exp(-d/d_-),$$
$$P(| \rightarrow -) = \exp(-d/d_|), \tag{5.22}$$

where d_-, $d_|$ are scale heights. This 'flip' embodies the elementary excitation involved in the collective dynamics of clusters since, typically, clusters reorganise by grain reorientation. The depth dependence reflects the fact that surface deposition is more likely to cause cluster reorganisation near the surface than deep inside the sandpile. After deposition and possible reorganisation, each column has a local slope $s(i, j)$ given by:

$$s(i, j) = z(i, j) - \frac{1}{2}(z(i + 1, j) + z(i, j + 1)). \tag{5.23}$$

If $s(i, j) > s_c$, where s_c is the critical slope threshold for grains to topple, then the two uppermost grains fall from column i, j onto its neighbours [77] or, when i or $j = L$, exit the system. This process could lead to further instabilities and hence avalanching.

This model is, despite its simplicity, capable of manifesting great complexity and diversity of behaviour. We will use it (with minor modifications) in succeeding chapters to investigate subjects as diverse as surface roughening and the effect of granular shape. Here we use it to investigate the effect of tilt on the angle of repose, a topic that has been the subject of experimental investigations [166].

We first define the angle of repose in the context of this cellular automaton model. It can be easily seen [168] that the macroscopic slope $\tan \theta_r$ measured in experiments

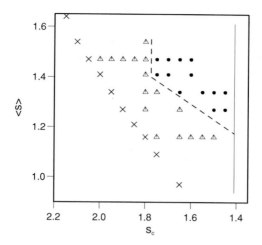

Fig. 5.1 A stability diagram for two-dimensional sandpiles with $L = 32$, $\alpha = 0.7$, $d_- = 2$ and $d_| = 20$. The measured mean slope $\langle s \rangle$ is plotted against the critical slope s_c, which should be interpreted as an inverse tilt (see text). The crosses (x) represent the values of $\langle s \rangle$ attained in the steady state when the sandpile is started with the corresponding values of s_c. The full line represents the spontaneous flow threshold, at which avalanching continues forever until the sandpile is emptied.The triangular and circular symbols correspond to the $\langle s \rangle$ that results when the pile (built at a smaller angle or larger s_c) is tilted to the corresponding s_c. The symbols correspond to triangular, predominantly downhill ($\triangle$) or uphill ($\bullet$) avalanches; the broken line separates the two regions.

is given by the mean slope $\langle s \rangle = (\Sigma_{i,j} z(i, j)/L^2(L + 1))$ of the cellular automaton sandpile. Next, we reflect on the effect of tilt: clearly, the greater the (positive) tilt angle made by the base of a sandpile with the horizontal, the more unstable will be the pile to avalanching. *Evidently, therefore the change in global slope engendered by tilting the sandpile affects the stability of local slopes such that those that were previously stable will now be unstable to avalanching.* In effect, therefore, tilting the pile leads to a *decrease* in the critical slope s_c; by reversing the logic, therefore, we can model the effect of a (positive) macroscopic tilt of the sandpile, by a decrease of its critical slope threshold.

We now use the above insights to look at the effect of tilt on various observables in a sandpile. Clearly, since the critical slope s_c is the threshold for permissible local slopes (i.e. those which can be sustained without avalanching), it is a strong determinant of the allowable granular configurations in the sandpile, and in par-ticular the relative populations of the ordered ($-$) and disordered ($|$) states. Fixing other parameters, we first look at the effect of tilt on the angle of repose. This is shown in Fig. 5.1 in a plot of $\langle s \rangle$ against s_c, as a line of crosses. The results indicate that $\langle s \rangle$, i.e. $\tan \theta_r$, decreases proportionately with s_c. With the interpretation (see above) that decreasing s_c corresponds to an increasing angle of macroscopic tilt, this indicates that *large* tilt angles (low s_c) should result in *lower* angles of repose θ_r, as might intuitively be expected.

Next, we follow experiment [166] in examining the topology resulting from a sudden tilt of the sandpile. We mimic this sudden tilt by reducing the critical slope of a sandpile constructed at a particular s_c to some $s_c' < s_c$. A direct result of this is that erstwhile stable slopes become unstable, avalanching occurs, and the sandpile stabilises to a new mean slope $\langle s \rangle$. Of course, when the angle of tilt is so large (i.e. the critical slope is so small) that spontaneous flow occurs continuously, we get the situation shown by the full line in Fig. 5.1; this sets in for critical slopes $s_c' \leq 2\alpha$, since such thresholds make even ordered stackings of flat ('−') grains unstable.

In experiments [166], distinctions have been made between so-called 'triangular' (where, overall, grains *below* a given grain are destabilised by its motion) and 'uphill' avalanches (where, overall, grains *above* a given grain are destabilised by its motion). These are the different kinds of avalanche 'footprints' generated when a sandpile is tilted through different angles and then submitted to additional deposition [166]. In the simulations under discussion [165, 168], such avalanche footprints have been extensively analysed. The triangular and circular symbols in Fig. 5.1 correspond respectively to numerical observations [165] of triangular and uphill avalanches. We illustrate this with an example: when a sandpile built with, say, $s_c = 2.05$ is tilted so that $s_c' \sim 1.75$,[1] generated by further deposition are, on average, triangular in shape (Fig. 5.2a). Beyond this value of s_c', uphill avalanches result (Fig. 5.2b). Thus, as in experiment [166], smaller tilt angles result in triangular avalanches, whereas larger tilt angles result in uphill avalanches. We will content ourselves with this agreement for the moment, noting that it will enable us to give a more theoretical basis for the experimental observations, an issue to which we will shortly return.

We reflect on the difference between the principal symbols on the busy diagram that is Fig. 5.1. The crosses denote angles of repose (mean slopes $\langle s \rangle$) obtained in the steady state when the sandpile is constructed with the corresponding value of tilt (critical slope threshold s_c). The triangular and circular symbols represent phenomena which are essentially nonequilibrium in character; they denote the values of $\langle s \rangle$ obtained when the sandpile is *tilted to* the corresponding s_c from some lower angle. These differences are crucial to the understanding of the *bistable and hysteretic* behaviour manifested by this simple model.

Consider thus a typical value of critical slope, say $s_c = 1.85$, in Fig. 5.1. The steady state mean slope for a sandpile constructed with this critical slope is given

---

[1] From this point on we use s_c' to refer to the *tilt* angle with the unprimed version referring to the *steady-state* angle.

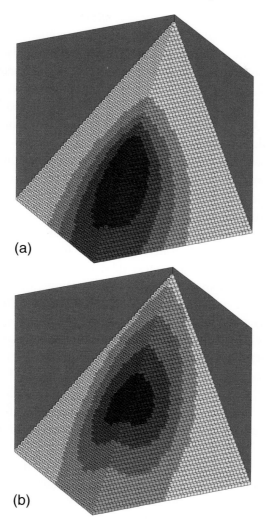

(a)

(b)

Fig. 5.2 Contour plots showing the frequency with which avalanches initiated near the centre of a sandpile cover its different regions. The piles have $L = 50$, $\alpha = 0.7$, $d_- = 2$, $d_| = 20$, $s_c = 2.05$. In (a) $s_c' = 1.95$; a 'triangular' avalanche results. In (b) $s_c' = 1.75$; an 'uphill' avalanche results. The contours correspond to 20, 40, 60 and 80 per cent coverage, and the images are built from ≈ 3500 test events.

by $\langle s \rangle = 1.21$. On the other hand, if a sandpile is constructed with a smaller tilt angle (corresponding in this case [168] to $s_c = 2.05$), and subsequently tilted to the angle corresponding to the *same* critical slope $s_c' = 1.85$, the corresponding angle of repose is *larger*, corresponding in this case to $\langle s \rangle = 1.39$. This is a clear indication of *hysteresis*, since it shows that *different angles of repose are attained for the same final tilt angle, depending on the prior history of the pile*. Next, consider a typical value of angle of repose, say that corresponding to $\langle s \rangle = 1.4$. We observe that at this angle, the sandpile can be *stationary* (i.e. in the steady state), with $s_c \sim 1.85$ or *avalanching* (note that all the triangular and circular symbols correspond to avalanching at that angle), which is of course proof of *bistable* behaviour.

Hysteresis also manifests itself in the distributions of local slopes for tilted piles (Figs. 5.3a–5.3c). Figure 5.3a shows the smooth distribution of local slopes obtained when a pile is constructed in the steady state with $s_c = 1.85$. Figure 5.3b shows the much more discontinuous values of local slopes obtained when, after construction with $s_c = 2.05$, the sandpile is tilted to the same angle, viz. $s'_c = 1.85$. Of course, this demonstrates clear hysteresis; that is, the configuration of a sandpile at a given macroscopic angle of tilt is determined by its dynamical history. Also, if we realise that jagged distributions of local slopes imply a rougher mean surface, we see that tilt induces rougher surfaces; if we now recognise also that the 'roughest' surface sustains the most dilatancy, we see that (by the logic of the previous section) this should build up larger angles of repose. Satisfyingly, the example of the previous paragraph, where the freshly tilted sandpile had a larger angle of repose ($\langle s \rangle = 1.39$) than the steady-state one ($\langle s \rangle = 1.21$) built at $s_c = 1.85$, agrees with this (see Fig. 5.1). Figure 5.3c is the result of even larger tilt ($s'_c = 1.75$); note the extremely jagged distributions of local slopes obtained in this case. This suggests that *the larger the magnitude of the sudden tilt, the rougher the surface, and the larger the angle of repose*; this is in agreement with the data of Fig. 5.1. The two very different viewpoints [23, 168] upon which this reasoning is based have an appealing self-consistency, which adds a robustness to its conclusions.

These differences in the local slope distributions before and after tilting, (Figs. 5.3a–5.3c) have implications for surface roughness in sandpiles [83, 96]. As we have seen, the quasi-continuous distribution of local slopes (Fig. 5.3a) of steady-state sandpiles results in a more continuous, locally 'smoother', surface with a large number of disordered grain configurations. On the other hand, a more discontinuous, locally 'rougher', surface results when there are very few permissible local slopes (Fig. 5.3c) for the case of strong tilt. We will see in later chapters, however, that the disordered surfaces of weakly tilted sandpiles are *globally rougher* [62], despite being *locally smoother*. In contrast, the ordered surfaces of sandpiles submitted to large tilt are *globally smoother*, despite being locally rougher, with a discrete distribution of local slopes. This preliminary indication of *anomalous roughening* [169] in sandpile automata can be explained, tentatively, in terms of local slope variations: When disorder is present, the sandpile surface can 'wander' over length scales corresponding to many columns by making small changes in local slope, thus adding to the surface width without compromising stability. However, in ordered sandpiles, the allowable changes in local slopes are comparatively large

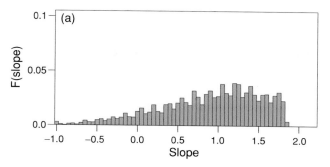

(a) A steady state pile with critical slope $s_c = 1.85$.

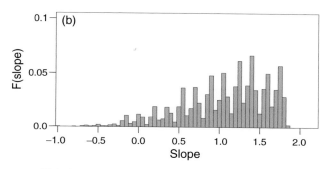

(b) A pile subjected to small tilt; $s_c = 2.05$ and $s'_c = 1.85$.

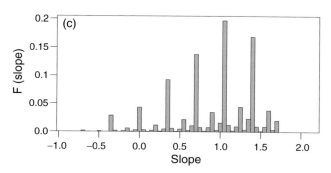

(c) A pile subjected to large tilt; $s_c = 2.05$ and $s'_c = 1.75$.

Fig. 5.3 Distribution functions for local slopes in disordered sandpiles with $L = 50$, $\alpha = 0.7$, $d_- = 2$ and $d_| = 20$.

and the width cannot increase indefinitely without making the sandpile unstable. The observation of an abrupt decrease in the surface width of the sandpile at the dotted line separating triangular and uphill avalanches in Fig. 5.1 supports this argument. In a later chapter, the issue of anomalous roughening in sandpile surfaces, whereby its roughening depends on the length scale at which it is probed [96], will be probed in greater depth.

The observation that steady-state sandpiles have a smooth distribution of local slopes, while tilted sandpiles manifest a *clustering* of values of local slopes is easily explained at a more microscopic level. The upper layers of a steady-state sandpile, when subjected to sudden tilt, are 'avalanched' away, exposing bulk grains which form the new surface. These tend to be in the − orientation, with $\Delta z = \alpha$; hence slope differences tend to be clustered around multiples of $\alpha/2$. (This is in clear contrast to the situation at the surface of steady-state or 'equilibrated' sandpiles, where | and − orientations are stochastically generated due to deposition and rearrangement.) The effect of tilting the sandpile through larger angles is to bring grains to the surface which were ever deeper in the bulk; these tend to be more and more ordered as a function of their depth [168], leading to increasingly discrete distributions of local slopes.

We next use the agreement of this model with experiment [166] to explain the apparition of triangular and uphill avalanches. Small tilts, as seen above (Fig. 5.3b) result in only minor changes in the continuous distribution of local slopes. Typically, an avalanche generated by deposition in these circumstances will only traverse a few sites, until a local slope much smaller than the threshold for toppling s_c is encountered. The resulting avalanches, somewhat mystifyingly referred to as 'triangular' in the experiment of Ref. [166], consist of instabilities which propagate downwards most of the time. For sandpiles tilted through large angles, on the other hand, simulations suggest [168] that only a few discrete values of local slopes are permissible (Fig. 5.3c); the toppling of one grain very likely disturbs the stability of grains uphill of its initiation site, causing the propagation of large avalanches [168], referred to in Ref. [166] as 'uphill'.

In Fig. 5.4, two different observables which can help to distinguish between triangular and uphill avalanches are plotted as a function of s_c'. The crosses (corresponding to the label 'Upper Fraction') show the average fraction of the avalanche which is uphill from its point of initiation; this is a measure of how 'uphill' an avalanche really is. Of course, most grains topple grains *below* themselves, so that the centre of mass of every avalanche is *below* the initiation point. The dots (corresponding to the label 'Distance') denote the average distance of the 'centre of mass' of the avalanche below the point of initiation; thus the more uphill the avalanche, the closer the centre of mass will be to the point of initiation. In Fig. 5.4, we see that there is a sharp transition for *both* indicators, at $s_c' \approx 1.8$; this is in fact the transition between triangular and uphill avalanches, as will be explained below. For low tilt angles (high s_c'), avalanche centres are well below the point of initiation, with small fractions of the avalanche being uphill. We infer therefore that for $s_c' \geq 1.8$, the avalanches propagate largely downward, and correspond to those referred to as 'triangular' in experiments [166]. For high tilt angles (low s_c'), the avalanche centres are much closer to the initiation point, and

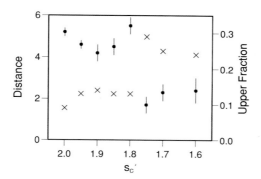

Fig. 5.4 The mean distance (•) between the centre of mass of the avalanche and its point of initiation, and the average fraction of avalanche sites uphill from the initiation point (x), plotted against the degree of tilt for a model sandpile with parameters $L = 50$, $\alpha = 0.7$, $d_- = 2$, $d_| = 20$ and $s_c = 2.05$.

substantial fractions of them are uphill from their initiation point; we infer that for $s_c' \ll 1.8$, they are largely of the kind referred to as 'uphill' in the experiments [166]. Similar data, for other values of the initial base angle s_c, were used to construct the 'phase diagram' in Fig. 5.1 distinguishing the two avalanche morphologies.

It is easy to see why no (discrete) uphill avalanches are observed for sandpiles prepared with initially large tilt ($s_c \approx 1.9$ in Fig. 5.1); further large tilts simply bring the sandpile to its maximal angle of stability with respect to the horizontal, θ_m. Denoted by the full line in Fig. 5.1, this is the angle at which a stationary sandpile becomes unstable; spontaneous flow sets in, with continuum resulting from the destabilisation of *all* sites.

Lastly, we provide a theoretical framework [126, 165, 168] for the onset of uphill avalanching in terms of directed percolation. We ask the question: what should be the condition for predominantly uphill avalanches to propagate? As mentioned above, this occurs when grains toppling downwards destabilise a sizeable number of grains above them. From a percolation point of view, this would require that bonds linking unstable sites be connected in an *upwards* direction (hence the use of the term 'directed percolation'). From Eq. (5.23) it can be shown that an instability at site (i, j) is transferred uphill when the height decrease $\delta z(i, j)$ at that site satisfies one of the two following conditions:

$$\delta z(i, j) > 2(s_c - s(i + 1, j)),$$
$$\delta z(i, j) > 2(s_c - s(i, j + 1)). \quad (5.24)$$

This can be used [126] to write mean field equations for the compound probability $T^\uparrow$ for the uphill transfer of slope instability:

$$T^\uparrow = p(\delta z = 2)p(s_c - s < 1)$$
$$+ p(\delta z = 1 + \alpha)p\left(s_c - s < \frac{1}{2}(1 + \alpha)\right)$$
$$+ p(\delta z = 2\alpha)p(s_c - s < \alpha), \qquad (5.25)$$

where the *p*s represent the probabilities of the events in parentheses. These probabilities can be computed directly from simulations [165]; it is found that $T^\uparrow$ varies near-monotonically between 0.45 (for tilt angles where triangular avalanches are predominant) and 0.55 (for tilt angles where uphill avalanches are the norm).

Our first comment on these results is that they are not a priori suggestive of a sharp transition; this is, however, hardly surprising given that the equations are strictly in mean field, with no fluctuations included. However, it is significant that there appears at least to be a suggestion that for upwardly directed bond connectivity probabilities of 0.55, uphill avalanching will occur. It is well known that the threshold for *directed bond percolation* on a square lattice is 0.64 [170], so it would appear that the value of 0.55 obtained in the above is lower than expected. A tentative explanation for this phenomenon is that since downhill ('triangular') avalanches propagate well before the onset of uphill avalanches [165], the percolation cluster in the downwards direction is already infinite. In such cases, it is known that [171] the value of the directed percolation threshold is reduced to that of the corresponding *undirected* problem, which happens to be 0.5 for square lattices.

With this clarification in place, we suggest that the transition to predominantly uphill avalanching occurs when $T^\uparrow = 0.5$, i.e. *when there is an infinite cluster of bonds connecting unstable sites uphill from the point of initiation*. This is in rough agreement with simulation results [126, 165]; better agreement can only be obtained by going beyond mean field and including the effects of fluctuations.

6

Compaction of disordered grains in the jamming limit: sand on random graphs

Granular compaction is characterised by a *competition between fast and slow degrees of freedom* [69]; far from the jamming limit, individual grains can quickly move into suitable voids in their neighbourhood. As the jamming limit is approached, however, voids which can accommodate whole grains become more and more rare; a cooperative rearrangement of grain clusters is required to fill the partial voids which remain. Such collective processes are necessarily slow, and eventually lead to *dynamical arrest* [149, 150].

The modelling of granular compaction has been the subject of considerable effort. Early simulations of shaken hard sphere packings [61, 62, 130], carried out in close symbiosis with experiment [172, 173], were followed by lattice-based theoretical models [75, 174, 175]; the latter could not, of course, incorporate the reality of a disordered substrate. Mean-field models [176] which could incorporate such disorder could not, on the other hand, impose the finite connectivity of grains included in Refs. [61, 62, 75, 130, 174, 175]. It was to answer the need of an analytically tractable model which incorporated *finitely connected grains on fully disordered substrates* that random graph models of granular compaction were first introduced by Berg and Mehta [152, 153].

A random graph [177] consists of a set of nodes and bonds, with the bonds connecting each node at random to a finite number of others, thus, from the point of view of connectivity, appearing like a finite-dimensional structure. Each bond may link two sites (a graph) or more (a so-called hypergraph). Why are random graphs useful for modelling granular physics? First, random graphs [177] are the simplest structures containing nodes with a *finite* number of neighbours. Clearly, real grains are always connected to a finite number of neighbours as evinced by their finite coordination numbers [61]; that this is a key property of grains with important physical consequences, ranging from kinetic constraints [178] to the

Sand on random graphs

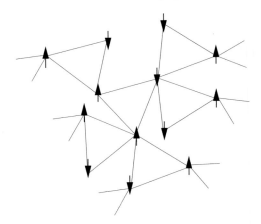

Fig. 6.1 A part of a random graph with triplets of sites forming plaquettes illustratng its local treelike structure (no planarity or geometric sense of distance are implied).

cascade dynamics of granular compaction [152, 153, 172, 173] is less immediately obvious. These issues will be further discussed in this chapter. Second, random graphs are among the simplest fully *disordered* constructs where, despite the existence of defined neighbourhoods of a site, no global symmetries exist. Disorder is an equally key feature of granular matter, even at the highest densities; its consequences include the presence of a range of coordination numbers [61, 62, 130] for any sandpile, corresponding to *locally varying neighbourhoods* of individual grains, a feature which can be incorporated via *locally fluctuating connectivities* [152, 153] in random graphs.

With these rationales in place, we now define some key concepts. Formally, a random graph [177] of N nodes and average connectivity c is constructed by considering all $N(N-1)/2$ possible bonds between the nodes and placing a bond on each of them with probability c/N. In other words, the connectivity matrix C_{ij} is sparse and has entries 1 (bond present) and 0 (bond absent), which are independent and identically distributed variables with probability c/N and $1 - c/N$ respectively. The resulting distribution of local connectivities is Poissonian with mean and variance c. The resulting structure is *locally* tree-like but has loops of length of order $\ln(N)$. Although there is no geometric concept of distance (in a finite-dimensional space), a chemical distance may be defined by determining the minimum number of steps it takes to go from one given point to another.

In order to define the models of granular compaction discussed in this chapter, hypergraphs with plaquettes connecting three or more nodes need to be constructed first. Choosing $C_{ijk} = 1(0)$ randomly with probability $2c/N^2 \, (1 - 2c/N^2)$ results in a random three-hypergraph, where the number of plaquettes connected to a site is distributed with a Poisson distribution of average c. An illustration of part of such a graph is shown in Fig. 6.1. Next, a specific spin model will be defined on this graph. Spin models on random graphs have been investigated for many years

[179], being halfway between infinite-connectivity models and finite-dimensional models; this leads to their having the analytic accessibility of the former within the framework of mean-field theory, as well as the finite connectivity of the latter. Interest in these models has intensified lately since they occur in the context of random combinatorial optimization problems [180] and inroads have been made towards their analytic treatment beyond replica-symmetry.

Having motivated our choice of random graphs as a basis, we proceed below to describe the first [152, 153] of many spin models of granular compaction.

6.1 The three-spin model: frustration, metastability and slow dynamics

The guiding factor in this choice of spin model is that it be the simplest model with frustration, metastability and slow dynamics; we will discuss the last two later, but remark at the outset that *geometrical frustration* is crucial to any study of granular matter. This concerns the fruitless *competition* between grains which try – and fail – to fill voids in the jamming limit, due either to geometric constraints on their mobility, or because of incompatibilities in shape or size. Our way of modelling this is via multi-spin interactions on plaquettes on a random graph [152, 153]. We choose in particular a three-spin Hamiltonian on a random graph (see Fig. 6.1) where N binary spins $S_i = \pm 1$ interact in triplets:

$$H = -\rho N = - \sum_{i<j<k} C_{ijk} S_i S_j S_k. \tag{6.1}$$

Here, the variable $C_{ijk} = 1$ with $i < j < k$ denotes the presence of a plaquette connecting sites i, j, k, while $C_{ijk} = 0$ denotes its absence. As mentioned above, choosing $C_{ijk} = 1(0)$ randomly with probability $2c/N^2 (1 - 2c/N^2)$ results in a random graph, where the number of plaquettes connected to a site is distributed with a Poisson distribution of average c – this models the *locally varying connectivities* between grains on a disordered substrate. The connection with granular compaction is made in accordance with Edwards' hypothesis [15], which assigns a thermodynamic 'energy' to the volume of a granular system: we, in our turn, interpret the *local contribution to the energy in different configurations of the spins as the volume occupied by grains in different local orientations*, in Eq. 6.1.

This Hamiltonian has been studied on a random graph in various contexts [181, 182]. It has a trivial ground state where all spins point up and all plaquettes are in the configuration $+ + +$ giving a contribution of -1 to the 'energy'. Yet, *locally*, plaquettes of the type $- - +, - + -, + - -$ (satisfied plaquettes) also give the same contribution; however, covering the graph with these mixed states will typically result in *frustration* of some of the interfacial spins. The *competition* between satisfying plaquettes locally and globally, given this degeneracy of the four configurations of plaquettes with $s_i s_j s_k = 1$, thus results in *frustration*. Also, since there are many possible ways of using the mixed states $--+, -+-, +--$ to cover the graph, which correspond to the various *local minima* on the 'energy' landscape of the system, there is a large entropy associated with these low-lying, nearly degenerate 'energy' states. On the other hand, there is only one way of covering the graph with the $+++$ state, which of course corresponds to the *global minimum* of the 'energy'. It is therefore more probable that a typical minimisation of the 'energy' in Eq. 6.1 will lead to a mixed state, one of the many optimal arrangements of the $--+, -+-, +--$ states; however, any such optimal state will not be the global minimum, and will always be in a state of frustration, due to mutual dissatisfaction of some plaquettes. The system will always try to 'do better', and its dynamics as it creeps around between its available local minima will be extremely slow; hence the observation that *frustration leads to slow dynamics*.

This mechanism has a suggestive analogy in the concept of geometrical frustration of granular matter, if we think of plaquettes as granular clusters. When grains are shaken, they rearrange locally, but locally dense configurations can be mutually incompatible. Voids could appear between densely packed clusters due to mutually incompatible grain orientations between neighbouring clusters resulting from their frustration. We suggest that the process of compaction in granular media thus also consists of a competition between the compaction of local clusters and the minimisation of voids globally, and that this feature of Eq. 6.1 is thus a very physical ingredient of the random graphs model.

Another key feature of this model [152, 153] is the existence of *metastable states*. We note from Fig. 6.2, which illustrates the phase space of a plaquette of three spins, that *two* spin flips are required to take a given plaquette from one satisfied configuration to another; an energy barrier thus has to be crossed in any intermediate step between two satisfied configurations. This has a mirror image in the context of granular dynamics, where compaction follows a temporary dilation; for example, a grain could form an unstable ('loose') bridge with other grains before it collapses into an available void beneath the latter [33, 61, 62, 130]. The extension to the potential energy landscape of a granular system is obvious; maxima, or barriers, separate the local minima of the system. The existence of this mechanism, by which an energy barrier has to be crossed in going from one metastable state to another, in the random graph model [152, 153], is thus also very important for modelling the reality of granular compaction.

6.2 How to tap the spins? – dilation and quench phases

The tapping algorithm used here is a simplified version of the tapping dynamics used in cooperative Monte Carlo simulations of sphere shaking [61, 62, 130]. We

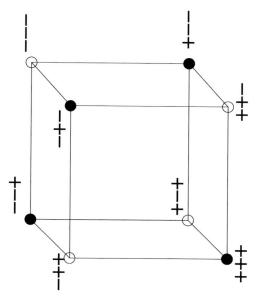

Fig. 6.2 The phase space of three spins connected by a single plaquette. Configurations of energy −1 (the plaquette is satisfied) are indicated by a black dot, those
of energy +1 (the plaquette is unsatisfied) are indicated by a white dot.

treat each tap as consisting of two phases. First, during the *dilation* phase, grains
are provided with free volume to move into; next, in the *quench* phase, they are
allowed to relax until a mechanically stable configuration is reached.

More technically, the dilation phase is modelled by a single sequential Monte
Carlo sweep of the system at a dimensionless temperature Γ. A site i is chosen at
random and flipped with probability 1 if its spin s_i is antiparallel to its local field
h_i, with probability $\exp(-h_i/\Gamma)$ if it is not, and with probability 0.5 if $h_i = 0$. This
procedure is repeated N times. Sites with a large absolute value of the local field h_i
thus have a low probability of flipping into the direction against the field; such spins
may be thought of as being highly *constrained* by their neighbours. The dynamics
of this 'thermal' dilation phase differs from the 'zero-temperature' dynamics used
in [183] where a certain fraction of spins is flipped regardless of the value of their
local field. The choice used here [152, 153] reflects the following physics: *if grains
are densely packed ('strongly bonded' to their neighbours), they are unlikely to be
displaced during the dilation phase of vibration.*

The grains are then allowed to relax via a $\Gamma = 0$ quench, which lasts until the
system has reached a *blocked* configuration[1] where each site i has $s_i = \text{sgn}(h_i)$ or

---

[1] Even in the presence of frustration, a blocked state can be suitably defined: it merely implies that the grain is
aligned with its *net* local field, i.e., it is connected to more unfrustrated than frustrated clusters.

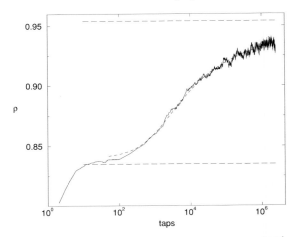

Fig. 6.3 Compaction curve at connectivity $c = 3$ for a system of 10^4 spins (one spin is flipped at random per tap). The data stem from a single run with random initial conditions and the fit (dashed line) follow (6.2) with parameters $\rho_\infty = 0.971$, $\rho_0 = 0.840$, $D = 2.76$ and $\tau = 1510$. The long-dashed line (top) indicates the approximate density 0.954 at which the dynamical transition occurs, the long-dashed line (bottom) indicates the approximate density 0.835 at which the fast dynamics stops, the *single-particle relaxation threshold*.

$h_i = 0$: thus, each grain is either aligned with its local field, or it is a 'rattler' [124]. Thus, at the end of each tap (dilation + quench), the system will be in a physically stable configuration [61, 62, 130].

6.3 Results I: the compaction curve

Among the most important of the results obtained with this model is the compaction curve obtained by tapping the model granular medium for long times. This is shown in Fig. 6.3, where three regimes of the dynamics can be identified. In the first regime, *fast individual dynamics* predominates, while in the second, one sees a logarithmic growth of the density via *slow collective* dynamics. The last regime consists of *system-spanning density fluctuations* in the jamming limit, where quantitative agreement with experiment [184] allows one to propose a *cascade theory of compaction during jamming*.

6.3.1 Fast dynamics till SPRT: every grain for itself!

At the end of the first tap, each grain is connected to more (or as many) unfrustrated than frustrated clusters. This is a direct result of the first tap being a zero-temperature quench: any site where this was not the case would simply flip its spin. More generally, a *fast* dynamics occurs in this regime whereby *single* grains *locally* adopt the orientation that, finally, optimises their density; this density ρ_0 has been termed [152, 153] the *single-particle relaxation threshold* (SPRT). The issue of this threshold value of the density, reached after a quench from random starting conditions, is highly nontrivial; its resolution involves the basins of attraction of the zero-temperature dynamics.

The problem may be illustrated by considering a single site i connected to $2k_i$ other sites and subject to the local field $h_i = 1/2 \sum_{jk} C_{ijk} s_j s_k$. For random initial conditions, the values of $l_i = h_i s_i$ are binomially distributed with a probability of $C_{(k_i-l_i)/2}^{k_i}(1/2)^{k_i}$ if $k_i - l_i$ is even and zero if it is odd. If $l_i < 0$, zero-temperature dynamics will flip this spin, turn l_i to $-l_i$ and turn $(k_i \pm l_i)/2$ satisfied (dissatisfied) plaquettes connected to it into dissatisfied (satisfied) ones. This will cause the l_j of $k_i \pm l_i$ neighbouring sites to decrease (increase) by 2. This dynamics stops when all sites have $l \geq 0$, giving $\rho_0 = 1/(3N) \sum_i l_i$.

Of course this issue is complicated by correlations between the local fields of neighbouring sites; if we neglect these correlations, however, a simple population model of N units, each with a Poisson distributed value of k_i, and a value of l_i distributed according to the initial binomial distribution, is obtained. At each step a randomly chosen element with negative l_i has its l_i inverted, and $k_i \pm l_i$ randomly chosen elements have their values of l decreased (increased) by 2 until $l_i \geq 0 \; \forall i$. This simplistic model works surprisingly well at low values of the connectivity c (with an error of about 10% up to $c = 6$) [152, 153], but obviously fails completely at large values of c or in fully connected models, where the role of correlations is overwhelmingly important.

In principle the differential equations describing the population dynamics could be solved analytically. Here we simply report the results for running the population dynamics numerically with $N = 10^4$ at $c = 3$: this yields a value of the SPRT, $\rho_0 = 0.835$ (shown as a dotted line in Fig. 6.3) which is *much higher* than the value of the density ρ (0.49) of a *typical* blocked configuration.

This is quite an extraordinary result; it implies that despite the exponential dominance of blocked configurations, random initial conditions *preferentially select* a higher density corresponding to the SPRT ρ_0. This prediction of an overshoot in the density achieved by fast dynamics has also, strikingly, been confirmed in independent lattice-based models [75, 174] of granular compaction, and points to a strong *non-ergodicity* in the fast dynamics of individual grains. We will discuss this further later on.

Another significant feature of this regime is that a fraction of spins is left with local fields exactly equal to zero, which thus keep changing orientation [185]. These are manifestations in this model of 'rattlers' [124], i.e. grains which keep changing their orientation within well-defined clusters [61, 62, 130]. They will later be used as a tool to probe the statistics of blocked configurations [152, 153].

To summarize: each grain reaches its *locally* optimal configuration via fast individual dynamics, resulting in the attainment of the SPRT density. All dynamics after this point is perforce collective.

6.3.2 *Slow dynamics of granular clusters: logarithmic compaction*

The second phase of dynamics in the compaction curve is *fully collective*: it removes some of the remaining frustrated plaquettes as clusters slowly rearrange themselves. A logarithmically slow compaction results [75, 173, 174], leading from the SPRT density ρ_0 to the asymptotic density ρ_∞. The resulting compaction curve may be fitted, with D and τ being characteristic constants, to the well-known logarithmic law [173]:

$$\rho(t) = \rho_\infty - (\rho_\infty - \rho_0)/(1 + 1/D \, \ln(1 + t/\tau)). \tag{6.2}$$

This can be written more transparently as

$$1 + t(\rho)/\tau = \exp\left\{ D \frac{\rho - \rho_0}{\rho_\infty - \rho} \right\}, \tag{6.3}$$

a form which makes clear that the dynamics becomes slow (logarithmic) as soon as the density reaches ρ_0. Although most grains are firmly held in place by their neighbours in this regime, *cascade-like* changes of orientation can occur. For example, if some grains changed orientation during the dilation phase, this would change the constraints on their neighbours; importantly, the freer dynamics of rattlers could also alter local fields in their neighbourhood, and cause previously blocked grains to reorient. Reorientation in cascades [152, 153] would then ensue, leading to collective granular compaction up to the asymptotic density ρ_∞. This has been identified [152, 153] with the density of random close packing [10] and associated with a *dynamical phase transition* [186, 187] as follows:

Traditionally, the dynamical transition is marked by the appearance of an exponential number of valleys in the free-energy landscape and thus a breaking of ergodicity [186, 188]. In the event that the dynamics is thermal, equilibration times diverge at the temperature corresponding to the dynamic transition. Cooling the system down gradually from high temperatures will also result in the system falling out of equilibrium at the dynamical transition temperature. Furthermore, the energy will get stuck at the energy at which the transition occurs.

Since this phenomenon is the result of the drastic change in the geometry of phase space, it is not surprising that one finds it also in the athermal dynamics dealt with here. Either gradually decreasing the tapping amplitude Γ or tapping at a low amplitude for a long time will get the system to approach the density ('energy') at which the dynamical transition occurs. The corresponding density calculated

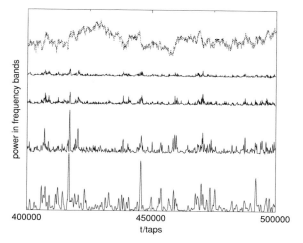

Fig. 6.4 The density fluctuations as a function of time resulting from 1024 taps are plotted as the topmost trace. The successive plots are of the power spectrum against time, in different frequency octaves. The power in the first octave (frequency 1/(1024 taps) -2/(1024 taps)) is the bottom-most trace, second octave (frequency 2/(1024 taps) - 4/(1024 taps)), above it, and so on to the top. Note that the fluctuations of the power in the different frequency bands are strongly correlated; they correspond to sudden changes in the density (top-most trace).

analytically [152, 153] is marked with a horizontal line in Fig. 6.3 and agrees well with the numerical value of the asymptotic density reached by the tapping dynamics. Note that, by definition, this density is the highest density reachable by local dynamics, before the system orders – it is thus natural to identify it with the random close packing density of a granular system.

6.3.3 Cascades at the dynamical transition

As mentioned above, free-energy barriers rise up causing the dynamics to slow down according to (6.2) as the density increases. The point where the height of these barriers scales with the system size marks a *breaking of the ergodicity of the dynamics*; an exponential number of valleys appears in the free-energy landscape at the dynamical transition, shown as a horizontal line in Fig. 6.3. Also shown in Fig. 6.3 are marked fluctuations around the logarithmic compaction law, especially as the jamming limit is approached; their correlations over several octaves have been the subject of detailed experimental investigations [184]. To compare the results (Fig. 6.3) of the model of [152, 153] with experiment, we follow the experimental analysis [184], taking Fourier transforms of the timeseries $\rho(t)$ to plot their power spectrum against time in Fig. 6.4; note that this is plotted in different frequency bands, exactly as in [184].

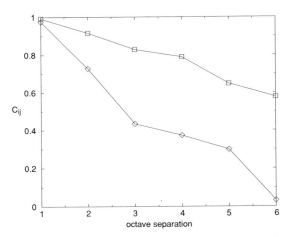

Fig. 6.5 The rescaled covariances of the power fluctuations as are plotted as a function of the octave separation for both the ferromagnetic 3-spin model (squares) and the parking-lot model (diamonds). The definition of these quantities is provided in the text; a high value of the rescaled covariance indicates strong correlations of the power-fluctuations of two given frequency bands.

The results [152, 153] indicate that, as in the experimental data [184], there are 'bursts' in the power spectrum fluctuations: the decomposition of these bursts over several octaves shows that they are caused by strong correlations of noise power over a wide range of frequencies. Importantly, the correlation matrices obtained are in quantitative agreement with experiment [184], as will be discussed further below.

The bursts in the power fluctuations, both in experiment [184] and theory [152, 153], are typically present in all the frequency bands when they occur (Fig. 6.4), indicating that the fluctuations in the the power spectrum are correlated over a wide range of frequencies. Qualitatively, this indicates that the density fluctuations near the dynamical transition of Fig. 6.3 are correlated over a wide length of time (and hence length) scales, both in theory and experiment. To make the agreement more quantitative, we plot in Fig. 6.5 (as in the corresponding experimental data [184]), the average of the correlation matrix C_{ij} as a function of octave separation $|i - j|$,

where $C_{ij} := M_{ij} \sqrt{\dfrac{M_{ii} M_{jj}}{(M_{ii} - 1)(M_{jj} - 1)}}$ probes the non-Gaussian components of

the correlation of the noise power in the ith and jth octaves. M_{ij} is defined as the covariances of noise power fluctuations $\langle \delta O_i \delta O_j \rangle / \sqrt{\langle (\delta O_i)^2 \rangle \langle (\delta O_j)^2 \rangle}$ for $i \neq j$ and $M_{ii} := \langle (\delta O_j)^2 \rangle / \sum_{k \in i} \langle P_k \rangle^2$, where the average is over 5000 time steps in the asymptotic regime, δO_i are the power fluctuation around the average in the ith octave and P_k, $k \in i$ is the power in the kth frequency bin in octave i. Note the slow decay of the results; interestingly, this decay mirrors that of the experimental data analysed in [184], where the average C_{ij} decays slowly with octave separation (particularly in the bottom and top of the sample) with C_{ij} decaying to about 0.6 over six decades. We plot for comparison in Fig. 6.5 the corresponding results for another widely-used model in granular media, the so-called 'parking-lot' model [189]; here particles of a certain size are desorbed from and absorbed by a surface at random, subject to a no-overlap constraint. Here, by contrast with the experimental data, there is a very quick decay of the correlation matrix to zero.

Clearly, the 3-spin model discussed here reproduces experimental data in a way that the parking-lot model does not. What is the reason for this difference, i.e. which precise ingredient of the 3-spin model produces this agreement? To answer this question, we first notice from Fig. 6.5 that 'bursts' found in the time-series of the density are responsible for the corresponding correlated bursts of noise power over a wide range of frequencies. Also, it is known [152, 153] that in spin models with *finite* connectivity, such bursts arise quite naturally due to *cascades* of spin-flips, where the flipping of a single spin alters the local fields acting on its neighbouring sites. The configuration of the spins on these sites may then no longer be locally stable, causing them to flip in turn. The first spin thus acts as a 'plug' releasing the neighbouring spins and setting off a *cascade* of successive spin-flips. A plug may also be composed of two or more sites, which need to have their spins flipped before neighbouring spins are released.

Putting all of this together, we see that in *the 3-spin model, bursts in the density fluctuations are due to cascades of spin-flips which arise from the change in local fields caused by the flipping of a single spin (or several spins); this instability propagates through ever larger neighbourhoods, causing correlated bursts in noise power fluctuations.* The crucial ingredients in the 3-spin model would therefore seem to be finite connectivity, as well as correlations that permit the unleashing of a cascade process. Fully connected models [176] (where each spin interacts with all spins in the system, but with an interaction energy scaling as $1/\sqrt{N}$) will clearly not manifest this since they have infinite connectivity. More interestingly, the parking-lot model, despite its finite connectivity, does not have correlations which are suitable for the modelling of at least this aspect of granular media – the creation and filling of gaps by particles does not cause the appearance of further gaps in this model, in the way spins flips can trigger a cascade.

We use the above insights to argue [152, 153] that *in real granular media, the observed correlations [184] of density fluctuations are due to a cascade process in granular compaction near the jamming limit.* Here, orientational/positional changes in strongly constrained grains give rise to propagating instabilities, leading to a near-global rearrangement of the granular medium. Pictorially, the movement of a single grain in this regime is only possible as the consequence of a system-wide cooperative motion of grains: this leads to sharp changes in overall density, and to the observed 'bursts' in the power spectrum of density fluctuations [184].

To put these results in perspective, we recall the results of a previous chapter, which suggested that occasionally, a first-order jump to a 'crystalline' state over significant lengthscales can occur in a granular system [131–133]. We suggest that such jumps could occur when the correlations over different length- and timescales become such that this purely local dynamics can result in a global minimisation of voids; put another way, the search for a 'better' local minimum of potential energy of the system can, at least in part of the system, result in the global minimum of packing, the 'crystalline' state, being reached.

6.4 Results II: realistic amplitude cycling – how granular media jam at densities lower than close-packed

This random graphs model has also been used to simulate *amplitude cycling*, an experimental [173] protocol on tapped granular media. Here, the granular medium is tapped at a given amplitude Γ for a time τ, after which its amplitude is changed by an infinitesimal $\delta\Gamma$; this process is repeated for cycles of increasing and decreasing Γ. The control parameter turns out to be the so called 'ramp rate', which is the ratio $\delta\Gamma/\tau$; this is a measure [75, 174] of the 'equilibration' allowed to the granular medium. Clearly, low values of this will correspond to quasistatic processes, while large values will correspond to near-adiabaticity.

The most simple-minded application of this protocol in the model results in the scenario of Fig. 6.6. At both high and low cycling rates, the density ρ first reaches the SPRT ρ_0, increases with increasing amplitude, and decreases again at large values of Γ. Thereafter, ρ always decreases with increasing Γ. The part of the curve where Γ is increased for the first time has been termed [172, 173] the *irreversible branch*; the *reversible branch* refers to the trajectory traced out by all successive increases and decreases of tapping amplitude. The results of Fig. 6.6, in agreement with many other models [75, 132, 174, 190], suggest that as the ramp rate is decreased, the system will eventually attain the RCP density ρ_∞. In particular, these models predict that in the limit of near-zero ramp rates, the irreversible branch disappears, with ρ becoming a single-valued function of Γ.

This prediction is in direct contradiction to the experimental results of [172, 173]; these suggest that, at least for experimentally realisable times, low-amplitude shaking does *not* result in RCP being reached. Instead, they suggest that some grains in the jamming limit of a granular assembly are so strongly constrained that they will *never* be displaced by low-amplitude taps.

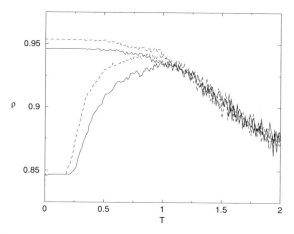

Fig. 6.6 Amplitude cycling: Γ is varied in both directions between 0 and 2 at two rates $\delta\Gamma = 10^{-4}, 10^{-5}$ per tap, with $\tau = 1$. The lower ramp rate (shown by the dotted line) results in a higher final density.

In order to model the above experimental scenario, let us modify the simplest picture of amplitude cycling presented above. First, we realise that in the dynamics of the model described thus far, sites with a high local field (corresponding to *strongly constrained* grains) may be flipped at any finite value of Γ with a correspondingly *small but finite* probability. This is what eventually leads the system to the RCP density ρ_∞.

To prevent this drift to RCP, one could naively think of introducing a threshold in local fields such that spins with fields above this threshold are not flipped. It turns out [152, 153], however, that this is insufficient; the orientational dynamics of neighbouring spins will always loosen the constraints on previously blocked spins in the end, and lower their local fields below any given threshold. The above implies that the constraints on grains are not related to *orientational* frustration alone; it was suggested [152, 153] that they might also be *mechanical* in nature, related to force networks [158, 159] between grains. Further, it seemed reasonable to suppose that such effectively immobile [96] 'blockages' could only be removed kinetically by the imposition of large intensity (Γ) vibrations.

Accordingly, the concept of *low-amplitude pinning* of grains was introduced [152, 153]: assign to each site i a real number r_i between zero and one, such that only grains with $r_i < \Gamma$ (mechanically constraining forces less than external vibration intensity) would be free to move. This modification could in principle lead to a lower value of ρ_∞ after amplitude cycling: for example, spin plaquettes (granular clusters) generated during the high-amplitude part of the cycle, would be effectively immobile at lower amplitudes, leading to wasted space.

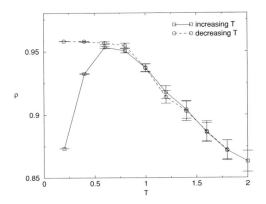

Fig. 6.7 The asymptotic density for tapping amplitudes ranging from $\Gamma = 0.2$ to $\Gamma = 2$ in steps of 0.2. The density measured after 10^7 taps at each amplitude and convergence to a steady-state each time was checked.

However, this too is insufficient to stop the evolution of the system to ρ_∞. It turns out [152, 153] that jamming at lower densities can only be achieved if low-amplitude pinning is *combined* with the choice of an extremely low ramp rate, via a large 'equilibration time' τ. If this is done, grains are allowed to 'equilibrate' at each amplitude, thus making sure that a steady-state of the density is always reached. This leads, finally, to the low-amplitude immobilisation of granular clusters created at high amplitudes, to the consequent blocking of voids, and hence to 'jamming' [172, 173, 191] at densities lower than RCP (see Fig. 6.7).

The results [152, 153] of amplitude cycling on the modified model thus indicate that it is important to have *mechanical pinning as well as long equilibration times for jamming to occur*.[2] In fact, the random configurations of immobile spins at each value of Γ can be viewed [152, 153] as an additional quenched disorder, and their effect on neighbouring mobile spins as an additional random local field.

These results demonstrate also a rather fundamental difference between excitations in glassy systems and granular media. In glasses, one would expect the configurations of spins reached at high values of temperature and subsequently frozen, *immediately* to alter the behaviour of the system at lower values of temperature. In athermal media like granular systems, however, it is important to allow *equilibration* at each value of the shaking intensity Γ, in order even to begin to observe the hysteretic effects which result in jamming at lower-than-RCP densities.

---

[2] Note that it was the lack of mechanical pinning in Refs. [75, 174, 132, 190] which led, despite the use of very low ramp rates and large 'equilibration times' τ at each tap, to RCP being approached asymptotically, to jamming at densities lower than ρ_∞ *not* being observed.

6.5 Discussion

In the above, we have presented a *finitely connected* spin model on random graphs [152, 153], with the aim of examining the compaction of tapped granular media on a fully disordered substrate. The SPRT density separates regions of cooperation and competition, each with its own distinctive features. Fast non-ergodic relaxation of individual grains terminated at the SPRT density, in the compaction curve; collective relaxation followed, manifest first by logarithmic compaction, and next by system-wide density fluctuations around RCP, both of which match – quantitatively – experimental results [172, 173, 184]. The model explains the latter in terms of a *cascade process of reorienting grains* that occurs near the jamming limit of granular matter. Also, in contradistinction to other models [75, 132, 174, 190], the results indicate that jamming at densities lower than RCP occurs as a result of *competition between mechanical and orientational frustration*, during amplitude cycling. We leave the discussion of the configurational entropies [152, 153, 192] of this model to a later chapter.

7

Shaking a box of sand I – a simple lattice model

7.1 Introduction

Vibrating sand results in very varied dynamics, ranging from glassy [132, 172, 173] to fluidised [193–197]. In much of this book, we will focus on the former, while a subsequent chapter will contain a review of the latter regime. Of course, it is important to have a theoretical understanding of how one regime gives way to the other; it is for this reason that the model discussed in this chapter is a simple model of a vibrated sandbox, which interpolates between the glassy and fluidised regimes through an extremely interesting intermediate regime, whose properties are not fully understood to date, and are the subject of current investigations.

The model [174] is based on the generalisation of an earlier cellular automaton (CA) model [22, 75, 83, 165] of an avalanching sandpile. This version of the model contains only near-neighbour interactions, with grains being in one of two orientational states. It shows *both* fast and slow dynamics in the appropriate regimes and, in its simplest form, reduces to an exactly solvable model in the frozen or jammed regime. Of course, in order to replicate the truly glassy behaviour of the jammed regime, one needs to introduce true long-range interactions. This will be done in the next chapter, where in particular the effect of grain shapes will be probed using an extension of this model [34, 35].

7.2 Definition of the model

As mentioned above, a principal motivation for this model is the addition of orientational rearrangement to the normal flow mechanism of CA models. In this spirit, consider a rectangular lattice of height H and width W with $N \leq HW$ grains located at its lattice points. Each 'grain' is a rectangle with sides 1 and $a \leq 1$, respectively, so that a grain which lies on its long(short) side is said to be

Granular Physics, ed. Anita Mehta. Published by Cambridge University Press. © A. Mehta 2007.

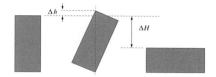

Fig. 7.1 A vertical grain needs to be tilted through the height Δh to reach the unstable equilibrium position and flop to the horizontal, while a horizontal grain needs to be tilted through an *additional* height ΔH to reach the vertical.

horizontal(vertical). Let this box be shaken with vibration intensity Γ. The most intuitively reasonable rules for its dynamics should involve (a) grains falling under gravity to a void below (without thresholds) or diagonally adjacent (provided appropriate height thresholds are met, as in usual CA models [65]); (b) grains flying directly or diagonally above themselves, provided they are given sufficient energy by the vibration intensity Γ; (c) grains flipping easily to their horizontal orientation, and flipping back less easily to their vertical one (Fig. 7.1); and (d) vertical grains being more unstable, they should add to the height of the column, and thus increase the grain's propensity to fall to sites which are diagonally below them, i.e. 'down the pile'.

We write below the mathematical formulation of these rules; consider a grain (i, j) in row i, column j whose height at any given time is given by $h_{ij} = n_{ij-} + a n_{ij+}$, with n_{ij-} the number of vertical grains and n_{ij+} the number of horizontal grains below (i, j). We give in the following a prescription [174] for the dynamics of this grain under shaking:

- If lattice sites $(i + 1, j - 1)$, $(i + 1, j)$, or $(i + 1, j + 1)$ are empty, grain (i, j) moves there with a probability $\exp(-1/\Gamma)$, in units such that the acceleration due to gravity, the mass of a grain, and the height of a lattice cell all equal unity.
- If the lattice site $(i - 1, j)$ below the grain is empty, it will fall down.
- If lattice sites $(i - 1, j \pm 1)$ are empty, the grain at height h_{ij} will fall to either lower neighbour, provided the height difference $h_{ij} - h_{i-1,j\pm1} \geq 2$.
- The grain flips from horizontal to vertical with probability $\exp(-m_{ij}(\Delta H + \Delta h)/\Gamma)$, where m_{ij} is the mass of the pile (consisting of grains of unit mass) above grain (i, j). For a rectangular grain, $\Delta H = 1 - a$ is the height difference between the initial horizontal and the final vertical state of the grain. Similarly, the *activation energy* for a flip reads $\Delta h = b - 1$, where $b = \sqrt{1 + a^2}$ is the diagonal length of a grain.
- The grain flips from vertical to horizontal with probability $\exp(-m_{ij} \Delta h/\Gamma)$.

In this form, although the interactions are nearest, or at most next-nearest neighbour, the model is not exactly solvable, because of the presence of voids in the system. The solutions in much of this chapter will therefore be largely numerical, with an analytical solution obtained only when voids are strictly absent.

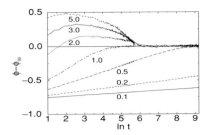

Fig. 7.2 Plot of $\phi - \phi_\infty$ versus $\ln t$, for different values of Γ, indicated on the curves. Note that ϕ_∞ decreases with increasing Γ, and is thus *distinct* for each curve.

7.3 Results I: on the packing fraction

In this section, we examine the behaviour of the packing fraction of the model, as a function of the vibration intensity Γ. Let N^- and N^+ be the numbers of vertical and horizontal grains in the box. The packing fraction ϕ is:

$$\phi = \frac{N^+ - aN^-}{N^+ + aN^-}, \tag{7.1}$$

which we use as an order parameter. The vertical orientation of a grain thus wastes space proportional to $1 - a$, relative to the horizontal one.

We examine the response of the packing fraction for typical parameter values ($\Delta H = 0.3$, $\Delta h = 0.05$) to shaking at varying intensities in Fig. 7.2. Clearly, the asymptotic values of the packing fraction ϕ_∞ will differ for each intensity; in fact, as was shown in earlier chapters, it decreases with increasing intensity [61, 62, 130]. The asymptotic packing fraction was determined for each intensity, and the difference $\phi - \phi_\infty$ plotted as a (logarithmic) function of time T in Fig. 7.2. Note that the initial packing fraction was the *same* in each case.

The dynamical response of the shaken sandbox [174] includes three distinct regions, each illustrated by representative curves in the figure. We note first a fluidised region (for $\Gamma \gg 1$), where we observe an initial increase (caused by a *nonequilibrium* and transient 'ordering' of grains in the boundary layer) of the packing fraction that quickly relaxes to the equilibrium values ϕ_∞ in each case. This overshooting effect in Fig. 7.2 increases with Γ, since grains ever deeper in the sandbox can now overcome their activation energy to relax to the horizontal. This *inhomogeneous relaxation* has been mentioned in earlier chapters, and observed in computer simulations of shaken spheres [131]. Very strikingly, the overshoot has also been observed in the context of a totally different model, the random graphs model of granular compaction [152, 153]. To date there is no satisfactory quantitative explanation for this overshoot, which seems firmly established in view of its model-independence.

Next, note an intermediate region (for $\Gamma \approx 1$), where the packing fraction remains approximately constant in the bulk, while the surface equilibrates via the fast dynamics of *single-particle relaxation*. The specific ϕ_∞ at which this occurs (0.917 here), is the *single-particle relaxation threshold density* (SPRT) observed in Ref. [152, 153]; nonequilibrium, non-ergodic, fast dynamics allows single particles locally to find their equilibrium configurations at this density. Analogous effects have been observed in recent experiments on colloids [124], where the correlated dynamics of *fast* particles was seen to be responsible for most relaxational behaviour before the onset of the glass transition. Once again, the non-ergodic dynamics in the vicinity of the SPRT is very poorly understood, at the time of writing this book. Additionally, this intermediate regime, with properties in between the jammed and fluidised states, with its balance of individual and collective dynamics, remains one of the most fascinating mysteries of granular matter.

Lastly, we see a frozen region (for $\Gamma \ll 1$), where the slow dynamics of the system results in a *logarithmic growth* of packing fraction with time:

$$\phi - \phi_\infty = b(\Gamma) \ln t + a, \tag{7.2}$$

where $b(\Gamma)$ increases with Γ, in good agreement with experiment [172, 173]. The slow dynamics has been identified in a previous chapter with a cascade process, where the free volume released by the relaxation of one or more grains allows for the ongoing relaxation of other grains in an extended neighbourhood. As Γ decreases, the corresponding ϕ_∞ increases asymptotically towards the jamming limit ϕ_{jam}, identified with a *dynamical phase transition* in a previous chapter [152, 153].

7.4 Results II: on annealed cooling, and the onset of jamming

We next investigate the analogue of 'annealed cooling', where Γ is increased and decreased cyclically, and the response of the packing fraction observed [172, 173]. The results obtained here are similar to those [131] seen using more realistic models of shaken spheres, but the simplicity of the present model allows for greater transparency.

Starting with the sand in a fluidised state, as in experiment [172, 173], the sandbox is submitted to taps at a given intensity Γ for a time t_{tap} and the intensity is increased in steps of $\delta\Gamma$; at a certain point, the cycle is reversed, to go from higher to lower intensities. The entire process is then iterated twice. Figure 7.3 shows the resulting behaviour of the volume fraction ϕ as a function of Γ, where an 'irreversible' branch and a 'reversible' branch of the compaction curve are seen, which meet at

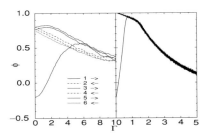

Fig. 7.3 Hysteresis curves. Left: $\Delta\Gamma = 0.1$, $t_{\text{tap}} = 2000$ time units. Right: $\Delta\Gamma = 0.001$, $t_{\text{tap}} = 10^5$ time units. Note the approach of the irreversibility point Γ^* to the 'shoulder' Γ_{jam}, as the ramp rate $\delta\Gamma/t_{\text{tap}}$ is lowered.

the 'irreversibility point' Γ^* [172, 173]. The left- and right-hand sides of Fig. 7.3 correspond respectively to high and low values of the 'ramp rate' $\delta\Gamma/t_{\text{tap}}$ [172, 173].

As the ramp rate is lowered, we note that the width of the hysteresis loop in the so-called reversible branch decreases. The 'reversible' branch is thus not reversible at all; hard-sphere simulations of shaken spheres [61, 62, 130] confirm the first-order, irreversible nature of the transition. As mentioned in an earlier chapter, the density may attain values that are substantially higher than random close packing, and quite close to the crystalline limit [131, 132]. An analogous transition has also been observed experimentally in the compaction of rods [133]. Note also that the 'irreversibility point' Γ^* (the shaking intensity at which the irreversible branch and the reversible branch meet) approaches Γ_{jam} (the shaking intensity at which the jamming limit ϕ_{jam} is approached), in agreement with results on other discrete models [190].

Of course, as mentioned in an earlier chapter, this feature is an idealisation; in reality the transition to the crystalline limit is not approached quite so smoothly as predicted here, since force networks act as mechanical impediments. However, it is interesting to see how the simple use of phenomenology in this model can get some of the results that are more cumbersomely obtained from hard-sphere simulations.

We next use the simplicity of the model to explore the onset of jamming, probing in between the regimes where fast and slow dynamics respectively predominate. This is explored via a configurational overlap function

$$\chi(t_{\text{ref}}, \Delta t) = \frac{1}{N} \sum_{i,j} \Theta[B_{i,j}(t_{\text{ref}}), B_{i,j}(t_{\text{ref}} + \Delta t)]. \tag{7.3}$$

Here $B_{i,j}(t)$ can take three distinct values depending on whether the lattice site (i, j) at time t is (a) empty, (b) occupied by a $+$ grain, or (c) occupied by a $-$ grain; $\Theta[X, Y] = 1 - \delta_{X,Y}$; i.e., $\Theta[X, Y] = 0$ if $X = Y$; and Δt is the time lag. The function $\chi(t_{\text{ref}}, \Delta t)$ is therefore 0 when configurations are identical at t_{ref} and $t_{\text{ref}} + \Delta t$

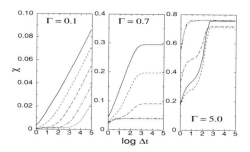

Fig. 7.4 Overlap functions $\chi(t_{\text{ref}}, \Delta t)$, Eq. (7.3), for $\Gamma = 0.1, 0.7, 5.0$. Line styles distinguish five reference times from $t_{\text{ref}} = 1$ (full line) to $t_{\text{ref}} = 10^4$ (dotted line). The time unit is defined as HW attempted Monte Carlo moves.

and takes larger values depending on the differences of the configurations at those times. Figure 7.4 shows results for different values of Γ, for $\Gamma = 0.1, 0.7, 5.0$.

The left-hand panel ($\Gamma = 0.1$) shows the logarithmic behaviour characteristic of ageing; here the leftmost (full) line shows the behaviour at $t_{\text{ref}} = 1$, while the dotted lines to the right of it represent increasing values of t_{ref}. We see clearly that 'older' systems change more slowly (as in life!); in other words, with increasing waiting times, $\chi(t_{\text{ref}}, \Delta t)$ increases much more slowly from zero. Clearly also, with an appropriate rescaling of time, an older system can be made to look like a younger one. This, and the logarithmic increase of $\chi(t_{\text{ref}}, \Delta t)$ with Δt, are classic symptoms of glassy dynamics, a point to which we will return later.

The right-hand panel ($\Gamma = 5$) shows the quick equilibration virtually independent of waiting times, which characterises the fluidised regime. Notice that for all waiting times t_{ref}, configurations begin to change rapidly; there is no ageing, no slow dynamics and every system finds its equilibrium quickly.

The middle panel ($\Gamma = 0.7$) exemplifies the behaviour characteristic of the transition between the two regimes, the intermediate phase referred to in the earlier section: this has features of both 'glassy' and fluidised regimes. On the one hand, younger systems find their equilibrium quickly, as in the fluidised case (χ rises quickly from zero); and on the other hand, there is an age-dependence, whereby older systems seemingly equilibrate to metastable configurations that are not very different from their starting positions, depending on the waiting time t_{ref}.

It should be borne in mind that these extremely interesting features come out of a very simple and physical model, and have at least a qualitative relationship with the phenomena that they are seeking to describe. However, care should be taken not to over-interpret these results. The ageing we see cannot be the result of real glassy dynamics, as there are no long-range interactions in the system, which is the reason that we have referred to its glassiness within quotes. Why it manifests ageing, and many other glassy phenomena, is still somewhat mysterious, although ongoing work at the time of writing this book is shedding some light on this issue.

7.5 Results III: when the sandbox is frozen

When there are no free voids within the sandbox, we refer to it as frozen or 'glassy' (taking care to use quotes, since, as mentioned above, there are no long-range inter-actions, and hence no real glassy behaviour in this model). Here, the model reduces [174] to an exactly solvable model of W independent columns of H noninteracting 'grains' $\sigma_n(t) = \pm 1$, with $\sigma = +1$ denoting a horizontal grain, and $\sigma = -1$ denot-ing a vertical grain. The orientation of the grain at depth n, measured from the top of the system, evolves according to a Markov dynamics with depth-dependent rates

$$\begin{cases} w(-1 \to +1) = \exp(-n\Delta h/\Gamma), \\ w(+1 \to -1) = \exp(-n(\Delta H + \Delta h)/\Gamma), \end{cases} \tag{7.4}$$

as $m_{ij} = n = H + 1 - i$.

This indicates clearly that the deeper a grain is in the sandbox, the less free it is to move, and also that the horizontal to vertical transition is more hindered than its reverse, both of which rules make sense. Other than their depth-dependence, where each grain carries the mass of the grains above it, there are no interactions between the grains. We emphasise this issue here, in order to contrast it with the situation of the next chapter, where long-range interactions involving grain orientations will be introduced into the model.

The order parameter describing the mean orientation, which we hereafter refer to as 'orientedness', $\overline{M}(t) = (1/H) \sum_{n=1}^{H} M_n(t)$, with $M_n(t) = \langle \sigma_n(t) \rangle$, is related to the packing fraction of Eq. (7.1) as

$$\overline{M} = \frac{(1+a)\phi - (1-a)}{1 + a - (1-a)\phi}. \tag{7.5}$$

At equilibrium, the orientedness profile is given by

$$M_{n,eq} = \tanh\left(n/(2\xi_{eq})\right), \tag{7.6}$$

while the local equilibration time diverges exponentially with depth n as

$$\tau_{n,eq} \approx \exp(n/\xi_{dyn}). \tag{7.7}$$

In other words, the equilibration rate of an n-dependent order parameter will be $\sim \xi_{dyn} \ln t$ (cf. the overlap function χ defined in the earlier subsection or Eqs. 7.12, 7.13 to follow). These expressions involve two characteristic lengths of the model, the equilibrium length ξ_{eq} and the dynamical length ξ_{dyn}, which read

$$\xi_{eq} = \frac{\Gamma}{\Delta H}, \quad \xi_{dyn} = \frac{\Gamma}{\Delta h}. \tag{7.8}$$

In the scaling regime where the height H and both lengths ξ_{eq} and ξ_{dyn} are large, the mean orientedness is

$$\overline{M}_{\text{eq}} \approx (2\xi_{\text{eq}}/H) \ln \cosh(H/(2\xi_{\text{eq}})). \tag{7.9}$$

For $H \ll \xi_{\text{eq}}$, we have

$$\overline{M}_{\text{eq}} \approx H/(4\xi_{\text{eq}}) \ll 1. \tag{7.10}$$

In this case, the system is very weakly ordered, even at equilibrium. For $H \gg \xi_{\text{eq}}$, we have

$$\overline{M}_{\text{eq}} \approx 1 - (2 \ln 2)\xi_{\text{eq}}/H. \tag{7.11}$$

Now, the system is strongly ordered at equilibrium, except for its top skin layer, whose depth is of order ξ_{eq}. The length ξ_{eq} is therefore the length up to which disorder persists in the granular material *when it has attained equilibrium*.

As the equilibration time diverges exponentially with the depth, orientational order propagates down the system logarithmically slowly. More specifically, for a large but finite time t, only a top layer up to an 'ordering length' $\Lambda(t)$ has equilibrated, with

$$\Lambda(t) \approx \xi_{\text{dyn}} \ln t. \tag{7.12}$$

We have $M_n(t) \approx M_{n,\text{eq}}$ for $n \ll \Lambda(t)$, whereas $M_n(t) \approx 0$ for $n \gg \Lambda(t)$. The most ordered grains are situated at a depth comparable to $\Lambda(t)$; the length ξ_{dyn} therefore determines the length *to which order has propagated* in the granular material in the 'glassy' regime.

Grains at depth $\Lambda(t)$ have a maximum orientedness

$$M_{\text{max}}(t) \approx \tanh\left((\omega/2) \ln t\right), \tag{7.13}$$

where [174]

$$\omega = \frac{\xi_{\text{dyn}}}{\xi_{\text{eq}}} = \frac{\Delta H}{\Delta h} = \frac{1-a}{b-1} \tag{7.14}$$

is the ratio of both characteristic lengths. For grains where there are nearly equivalent orientations ($a \sim 1$), even at equilibrium, there will be a large number of 'disordered' configurations in the top layer, since these will be almost equivalent to the strictly ordered one.

Both lengths ξ_{eq} and ξ_{dyn} have, in experimental terms, the interpretation of *the depth of the boundary layer* in a vibrated granular system; in the first case, this description applies when equilibrium has been reached, while in the second case, this applies to the nonequilibrium evolution of a vibrated granular bed.

Table 7.1 *Two different nonequilibrium regimes*

	Regime I $\xi_{eq} \ll \Lambda(t) \ll H$ $(\omega \ln t \gg 1)$	Regime II $\Lambda(t) \ll \xi_{eq},\ H$ $(\omega \ln t \ll 1)$
$\overline{M}(t)$	$\Lambda(t)/H$	$[\Lambda(2t)]^2 /(4H\xi_{eq})$
$\overline{S}(t, s)$	$1 - [\Lambda(t) - \Lambda(s)]/H$	$1 - [\Lambda(2(t - s))] /H$
$\overline{C}(t, s)$	$1 - [2\Lambda(t) - \Lambda(t + s)]/H$	$1 - [\Lambda(2(t - s))] /H$

7.6 Results IV: two nonequilibrium regimes

Let us recall that $\Lambda(t)$ is the equilibration length of this model; clearly then, in order for it to exhibit interesting nonequilibrium or ageing effects, the system size must be much less than this length, i.e. one must have $\Lambda(t) \ll H$.

The two-time quantities we investigate to explore ageing are the full two-time correlation function,

$$S_n(t, s) = \langle \sigma_n(t)\sigma_n(s)\rangle, \tag{7.15}$$

and the connected one,

$$C_n(t, s) = S_n(t, s) - M_n(t)M_n(s), \tag{7.16}$$

with $0 \leq s$ (waiting time) $\leq t$ (observation time). In terms of the overlap function of Eq. (7.3), these can be written as

$$\overline{S}(t_{\text{ref}} + \Delta t, t_{\text{ref}}) = 1 - 2\chi(t_{\text{ref}}, \Delta t). \tag{7.17}$$

We are led to consider two different non-equilibrium regimes, which result from different ratios of the two characteristic lengths of the system. In each case, the mean observables can be expressed in terms of these lengths alone (see Table 7.1). In Regime I, the maximal ordering is very close to perfect, as $1 - M_{\text{max}}(t) \sim t^{-\omega} \ll 1$. This is the conventional frozen regime (to which the data in Fig. 7.4 correspond). The top layer of the system is strongly ordered, most of the grains are flat, and likely to stay that way: the ageing phenomenon corresponds to the slow ordering attempts of grains deeper in the bulk, quantified by the logarithmic growth of the ordering length $\Lambda(t)$. Table 7.1 shows that the mean orientedness is nothing but the fraction $\Lambda(t)/H$ of the system that has equilibrated. The two-time correlations are non-stationary, and they involve $\Lambda(s)$, $\Lambda(t)$ and $\Lambda(t + s)$.

In Regime II, the maximal ordering is very weak, as $M_{\text{max}}(t) \approx (\omega/2) \ln t \ll 1$. This regime exists only for $\omega \ll 1$, i.e., $a \to 1$ in the geometrical model. It corresponds to an even slower dynamics, since now any attempts at ordering are hindered additionally by a strong probability that a horizontal grain will flip to the vertical orientation. Table 7.1 shows that the mean orientedness involves the square of the ordering length, while the two-time correlations do not exhibit *any* non-stationary features characteristic of ageing, at least to leading order, in this scaling regime.

The physical difference between the two scenarios is comprehensible in terms of orientational modes of shaped grains. Recall that in Regime I, $\xi_{eq} \ll \xi_{dyn}$ so that $\omega \gg 1$; from Eq. (7.14), this implies that one side of the rectangular grain is much larger than the other, so there is a strong tendency to prefer one of the two orientations on stability grounds. In a spirit of generalisation, we say therefore that when grains are very asymmetrically shaped, and there is a strong preferred orientation, the nonequilibrium regime of granular dynamics will carry all the usual characteristics of ageing.

Now recall that that in Regime II, $\xi_{eq} \gg \xi_{dyn}$ so that $\omega \ll 1$; from Eq. (7.14), this implies that the grain is nearly square ($a \sim 1$), and any grain flip is easily reversed by a corresponding flop! Generalising once again, we suggest that where grains are symmetrically shaped and the restoring 'force' to get to a particular orientation is weak, the signatures of ageing will be hard to detect even in a highly nonequilibrium dynamical regime.

Even bearing in mind that the present model is an extremely simplified representation of a granular medium, it would be interesting to test these speculations experimentally: would ageing experiments carried out on differently shaped grains of identical materials give different results?

7.7 Discussion

In conclusion, the simplicity of the sandbox model makes it a useful conceptual tool for probing the dynamical responses of vibrated sand, from the fluidised to the frozen regimes. In the latter case, the model is exactly solvable, which allows one to describe the by now well-established picture of logarithmic compaction, in terms of two characteristic lengths, as well as providing an interesting insight into shape-dependent ageing. The improvement of this necessarily qualitative picture by the addition of more realistic and complex interactions, while still retaining the overall conceptual simplicity of the model, constitutes the subject matter of the next chapter.

8

Shaking a box of sand II – at the jamming limit, when shape matters!

In this chapter we extend the model of the previous chapter in two different directions; the first and most important aim is to introduce long-range interactions with a view to obtaining properly glassy behaviour, and the second is to explore the role of *grain shapes* in granular compaction.

The model [34, 35] is based on the following picture. Consider a box of sand in the presence of gravity in the jamming limit. Adopting, as in the previous chapter, a lattice-based viewpoint, we visualise this box as being constituted of rows and columns of grains. When this box is shaken along the direction of gravity, the predominant dynamical response of the sandbox is known to be in the vertical direction – recall the on- and off-lattice computer simulation results presented in earlier chapters [61, 62, 75, 130, 174] which show that correlations in the transverse plane (i.e. along rows of grains) are negligible compared to these. Another important aspect of the jamming limit is that grain-sized voids are typically absent. The dominant dynamical mechanism in this regime is therefore grain reorientation *within* each column to minimise the size of the partial voids that persist. We thus focus on a column model of grains in the jamming limit [34, 35].

We now extend the concept of disorder to include the effect of grain shapes. Each ordered grain occupies one unit of space, while each disordered grain occupies $1 + \epsilon$ units of space, with ϵ a measure of the *partial void trapped by misorientation*. In changing the notation of wasted space from a as in the previous chapter to ϵ here, we have in mind that, unlike the way in which we envisaged a as the aspect ratio of a rectangular grain there, ϵ can take any value, rational or irrational, in this chapter. It can, in this chapter, equally represent the integer-value mismatch of rectangular grains, as the possibly irrational value of the void space generated when an irregular grain is packed in its most misoriented way. Of course, in restricting ourselves to a two-state model, we are greatly simplifying the picture, since in real life irregular

Granular Physics, ed. Anita Mehta. Published by Cambridge University Press. © A. Mehta 2007.

grains can have a multiplicity of orientational states. We proceed (with this caveat in mind), however, with the view to understanding as much as possible of this minimal model.

The basis of the dynamics in this model is strictly local, as in real life. Each grain feels a local field from all the grains above it (we have in mind that the topmost grains are the freest to move, while those below are increasingly 'weighed under'), and in the jamming limit, it responds to this local field by minimising the void space available to it. There is an interesting parallel here with the economic history of families – those families with a history of financial prudence gift their descendants with resources that they can choose to waste, while in the obverse case, the descendants have to mop up the debts of their progenitors. Likewise in this column model – we will see that if the space wasted by all the grains above a particular one is minimal, the local field on it will be small, and it will be relatively free to choose either an ordered or a disordered state with respect to this field. Conversely, if most of the space resources of the column have already been used up by upper grains, the grain will be constrained to adopt only that orientation which will mop up the extra void space, and transmit a lower value of local field to grains below itself.

8.1 Definition of the model

In the column model of [34, 35] grains are indexed by their depth n measured from the free surface. Each grain can be in one of two orientational states – ordered $(+)$ or disordered $(-)$ – the 'spin' variables $\{\sigma_n = \pm 1\}$ thus uniquely defining a configuration. As in the random graphs model [152, 153] presented in an earlier chapter, a local field h_n constrains the temporal evolution of spin σ_n, such that excess void space is minimised.

In the presence of a vibration intensity Γ, grains reorient with an ease that depends on their depth n within the column (grains at the free surface must clearly be the freest to move!), as well as on the local void space h_n available to them. The main extension of the model described in the earlier chapter [174] is in the transition probabilities, which are now written as

$$w\left(\sigma_n = \pm \rightarrow \sigma_n = \mp\right) = \exp\left(-n/\xi_{\text{dyn}} \mp h_n/\Gamma\right). \tag{8.1}$$

The dynamical length ξ_{dyn} [34, 35, 174] is, as before, the boundary layer of the column; within this dynamics are *fast*, while well beyond it they are *slow*. The local field h_n is a measure of *excess void space* [21]:

$$h_n = \epsilon\, m_n^- - m_n^+, \tag{8.2}$$

where m_n^+ and m_n^- are respectively the numbers of $+$ and $-$ grains above grain n. The definition, Eq. (8.2), is such that a transition from an ordered to a disordered state for grain n is *hindered* by the number of voids that are already above it, as might be expected for an ordering field in the jamming limit.

In the $\Gamma \to 0$ limit of zero-temperature dynamics [152, 153], the probabilistic rules (8.1) become deterministic: the expression $\sigma_n = \text{sign } h_n$ (provided $h_n \neq 0$) determines the *ground states* of the system. *Frustration* [148] manifests itself for $\epsilon > 0$, which leads to a rich ground-state structure, whose precise nature depends on whether ϵ is rational or irrational. The connection with the model presented in the earlier chapter [174] is obtained by setting $\epsilon < 0$; this corresponds to a *complete absence of frustration and a single ground state of ordered grains*, as obtained there.

For irrational ϵ, no local field h_n can ever be zero (see (8.2)). Noting that irrational values of ϵ denote shape irregularity, we conclude that the *excess void space is nonzero even in the ground state of jagged grains*. Their ground state, far from being perfectly packed, turns out [34, 35] to be quasiperiodic.

Regularly shaped grains correspond to rational $\epsilon = p/q$, with p and q mutual primes. We see from (8.2) that now, some of the h_n can vanish; these correspond, as noted in a previous chapter, to 'rattlers'. A rattler at depth n thus has a perfectly packed column above it, so that it is free to choose its orientation [34, 35, 124, 152, 153]. For regular grains in their ground state, rattlers occur periodically (as in crystalline packings!) at points such that n is a multiple of the period $p + q$.[1] Every ground state is thus a random sequence of two patterns of length $p + q$, each containing p ordered and q disordered grains; this degeneracy leads to a *zero-temperature configurational entropy* or *ground-state entropy* $\Sigma = \ln 2/(p + q)$ per grain.

8.2 Zero-temperature dynamics: (ir)retrievability of ground states, density fluctuations and anticorrelations

Regular and irregular grains behave rather differently when submitted to zero-temperature dynamics. The (imperfect) but unique ground state for irregular grains is rapidly retrieved; the perfect (and degenerate) ground states for regular grains never are, resulting in *density fluctuations*.

We recall the rule for zero-temperature dynamics:

$$\sigma_n \to \text{sign } h_n. \tag{8.3}$$

---

[1] For example, when $\epsilon = 1/2$, each disordered grain 'carries' a void half its size; units of perfect packing must be permutations of the triad $+ - -$, where two 'half' voids from each of the disordered grains are perfectly filled by an ordered grain. The *stepwise compacting* dynamics [34, 35] selects only two of these patterns, $+ - -$ and $- + -$.

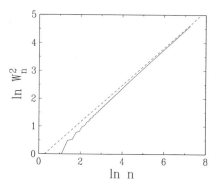

Fig. 8.1 Log–log plot of $W_n^2 = \langle h_n^2 \rangle$ against depth n, for zero-temperature dynamics with $\epsilon = 1$. Full line: numerical data. Dashed line: fit to asymptotic behaviour leading to (8.4) (after [34, 35]).

Starting with irregular grains (with a given irrational value of ϵ) in an initially disordered state, one quickly recovers the ground state with zero-temperature dynamics. The ground state in fact propagates ballistically from the free surface to a depth $L(t) \approx V(\epsilon) t$ [34, 35] at time t, while the rest of the system remains in its disordered initial state. When $L(t)$ becomes comparable with ξ_{dyn}, the effects of the free surface begin to be damped. In particular, for $t \gg \xi_{dyn}/V(\epsilon)$ we recover the logarithmic coarsening law $L(t) \approx \xi_{dyn} \ln t$, also seen in other theoretical models [152, 153, 174] of the slow relaxation of tapped granular media [172, 173]. To recapitulate, *the ground state for irregular grains is quickly (ballistically) recovered with zero-temperature dynamics, until the boundary layer ξ_{dyn} is reached; below this, the column is essentially frozen, and coarsens only logarithmically.*

For regular grains with rational ϵ, the local field h_n in (8.3) vanishes for rattlers. Their dynamics is stochastic even at zero temperature, since they have a choice of orientations: a simple way to update them is according to the rule $\sigma_n \to \pm 1$ with probability $1/2$. This stochasticity results in an intriguing dynamics even well within the boundary layer ξ_{dyn}, while the dynamics for $n \gg \xi_{dyn}$ is, as before, logarithmically slow [34, 35].

In what follows, we will focus on the fast dynamics within the boundary layer. The main result is that zero-temperature dynamics does not drive the system to any of its degenerate ground states, but instead engenders a *fast relaxation to a nontrivial steady state*, independent of initial conditions, which consists of *unbounded density fluctuations*. This recalls density fluctuations close to the jamming limit [152, 153, 172, 173], in other studies of granular compaction.

Figure 8.1 shows the variation of these density fluctuations as a function of depth n:

$$W_n^2 = \langle h_n^2 \rangle \approx A\, n^{2/3}, \quad A \approx 0.83. \tag{8.4}$$

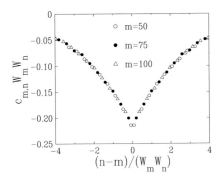

Fig. 8.2 Scaling plot of the orientation correlation function $c_{m,n}$ for $n \neq m$ in the zero-temperature steady state with $\epsilon = 1$, demonstrating the validity of (8.5) and showing a plot of (minus) the scaling function F (after [34, 35])

The fluctuations are approximately Gaussian, with a definite excess at *small* values: $|h_n| \sim 1 \ll W_n$. We recall that non-Gaussianness was also observed in experiments on density fluctuations in tapped granular media [184]; in the theory here, we interpret it in terms of grain (anti)correlations. If grain orientations were fully uncorrelated, one would have the simple result $\langle h_n^2 \rangle = n\epsilon$, while (8.4) implies that $\langle h_n^2 \rangle$ grows much more slowly than n.

It turns out that, at least within a dynamical cluster of radius $n^{2/3}$ [34, 35], the orientational displacements of each grain are *fully anticorrelated*. Figure 8.2 shows that the orientation correlations $c_{m,n} = \langle \sigma_m \sigma_n \rangle$ scale as [34, 35]

$$c_{m,n} \approx \delta_{m,n} - \frac{1}{W_m W_n} F\left(\frac{n-m}{W_m W_n} \right), \qquad (8.5)$$

where the function F is such that $\int_{-\infty}^{+\infty} F(x)\,dx = 1$. We find also that, within such a dynamical cluster, the fluctuations of the orientational displacements are *totally screened*: $\sum_{n \neq m} c_{m,n} \approx -c_{m,m} = -1$. These results recall the *anticorrelations in grain displacements* observed in independent simulations of shaken hard spheres close to jamming [61, 62, 130] that were presented in Fig. 3.6; there they corresponded to compaction via bridge collapse, as upper and lower grains in bridges [21] collapsed onto each other, releasing void space. Again, this model-independent observation confirms the robustness of the phenomenon: grain displacements are typically anticorrelated near jamming.

8.3 Rugged entropic landscapes: Edwards' or not?

The most remarkable feature of the column model is, arguably, the rugged landscape of microscopic configurations visited during the steady state of zero-temperature

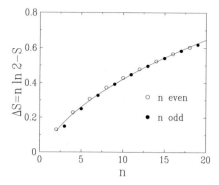

Fig. 8.3 Plot of the measured entropy reduction ΔS in the zero-temperature steady state with $\epsilon = 1$, against $n \leq 19$. Symbols: numerical data, for $t \sim 10^9$ and $n \approx 20$. Full line: fit $\Delta S = (62 \ln n + 53)10^{-3} n^{1/3}$.

dynamics (for regular grains); this is all the more striking because the macroscopic entropy is *flat*, in agreement with Edwards' hypothesis [15].

The entropy of the steady state of zero-temperature dynamics is defined by the usual Boltzmann formula,

$$S = - \sum_{\mathcal{C}} p(\mathcal{C}) \ln p(\mathcal{C}), \qquad (8.6)$$

where $p(\mathcal{C})$ is the probability that the system is in the orientation configuration $\mathcal{C}$ in the steady state, and the sum runs over all the 2^n configurations of a system of n grains. This can be estimated theoretically by using (8.4). Consider n as a fictitious discrete time, with the local field h_n as the position of a random walker at time n. For a free lattice random walk of n steps, one has $\langle h_n^2 \rangle = n$, as all configurations are equiprobable, so that the entropy reads $S_{\text{flat}} = n \ln 2$. For a column of regularly shaped grains, this model [34, 35] predicts instead $\langle h_n^2 \rangle = W_n^2 \ll n$; the entropy S of the random walker is therefore reduced with respect to S_{flat}. The entropy reduction [198] $\Delta S = S_{\text{flat}} - S = n \ln 2 - S$ can be estimated [34, 35] to be

$$\Delta S \sim \sum_{m=1}^{n} \frac{1}{W_m^2} \sim n^{1/3}. \qquad (8.7)$$

Evaluating the steady-state entropy S numerically, using (8.6) and measuring all configurational probabilities $p(\mathcal{C})$, we find (see Fig. 8.3) that ΔS is small; for example, for $n = 12$, we have $\Delta S \approx 0.479$, in good agreement with the results of Eq. (8.7). This is a convincing demonstration that anticorrelations (see previous section) lead to relatively small corrections to the overall dynamical entropy of the steady state, which is *flat*, in agreement with Edwards' hypothesis [15].

To investigate the effect of the constraints, we plot the normalised configurational probabilities $2^{12} p(\mathcal{C})$ for a column of 12 grains against the $2^{12} = 4096$

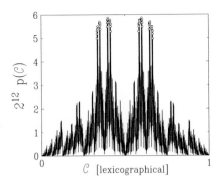

Fig. 8.4 Plot of the normalised probabilities $2^{12} p(\mathcal{C})$ of the configurations of a column of 12 grains in the zero-temperature steady state with $\epsilon = 1$, against the configurations $\mathcal{C}$ in lexicographical order. The empty circles mark the $2^6 = 64$ ground-state configurations, which turn out to be the most probable (after [34, 35]).

configurations $\mathcal{C}$ in Fig. 8.4, which are labelled in 'lexicographical' order (i.e. as $+ + + + + + + + + + + +, - + + + + + + + + + + +, + - + + + + + + + + + +$, etc.) Note that the actual values of the configurational probabilities $p(\mathcal{C})$ are microscopically small! At this microscopic scale, however, the entropic landscape is startlingly rugged; some configurations are clearly visited far more often than others. It turns out that the most visited configurations are the ground states of the system (empty circles). We suggest that this behaviour is generic, i.e., *the dynamics of compaction in the jammed state leads to a microscopic sampling of configuration space which is highly non-uniform, so that its ground states are visited most frequently*. The model [34, 35] thus provides a natural reconciliation between, on the one hand, the intuitive perception that not all *microscopic* configurations can be equally visited during compaction in the jamming limit, that the most compact configurations should be the most visited; and, on the other, the flatness hypothesis of Edwards, which states that for large enough systems, the *macroscopic* entropic landscape of visited configurations is flat [15].

The dynamical entropy generated by the random graphs model [152, 153] of a previous chapter is also reconcilable with Edwards' flatness [15], at least in the jamming limit discussed above. This was explored via rattlers (sites i such that the local field $h_i = 0$) in the blocked configurations generated after each tap. We have seen above that they have a rather crucial role to play in the density fluctuations of this column model [34, 35]; it turns out that they are also a good probe of Edwards' flatness under the tapping dynamics of the random graphs model [152, 153].

If blocked states at a given density are equiprobable, theoretical arguments given in [152, 153] show that a plot of the fraction of connected rattlers versus the density should reproduce this. Figure 8.5 shows the results for four single runs of plotting

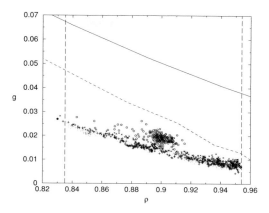

Fig. 8.5 The fraction g of connected rattlers during four runs of a tapped random graph [152, 153] with $N = 1000$, $c = 3$ at $T = 0.4$ (dots), $T = 0.56$ (+), $T = 0.7$ (×) and $T = 1.5$ (circles). The solid and dashed lines correspond respectively to annealed and quenched theoretical values corresponding to Edwards' flatness. The vertical lines indicate the approximate values for ρ_0 (left line) and ρ_∞ (right line).

the fraction of rattlers g against density ρ, at increasing amplitudes of vibration Γ. The dashed line and full lines correspond respectively to quenched and annealed replica symmetric averages for g, assuming Edwards' flatness. We notice that there is a reasonable congruence of all the numerical results and the (theoretically more accurate) quenched average at the asymptotic density ρ_∞. Thereafter, there are systematic divergences with lower density and higher Γ.

We can draw the following conclusions from this. First, at the jamming limit near RCP (i.e. the asymptotic density ρ_∞ – see Chapter 7), the dynamically generated entropies are flat, in accord with Edwards' hypothesis [15], as well as with the results of the column model [34, 35]. Second, as we move to the regimes of higher vibration and lower density, the entropic landscape gets rougher – one can imagine a process whereby the roughening visible on microscopic scales near jamming (see Fig. 8.4) begins to increase to macroscopic scales as one moves away from jamming. In this regime, we observe that configurations which are dynamically accessed by tapping (see the symbols in Fig. 8.5) correspond to *higher than typical densities* (dashed and full lines in Fig. 8.5) – we recall from Chapter 7 that this occurs when non-ergodic fast dynamics dominate granular relaxation. Putting all of this together, we conclude that also, according to the random graphs model of [152, 153], entropies of configurations reached by slow (ergodic) dynamics near the jamming limit manifest Edwards' flatness [15]; when, however, fast (non-ergodic) dynamics predominate (e.g. for higher tapping amplitudes and lower densities), there are systematic deviations from flatness.

Configurational entropies of strongly nonequilibrium models with slow dynamics, are, however, not generically flat. To demonstrate this, we present results for a

Fig. 8.6 A typical pattern of surviving clusters on the square lattice for the cluster aggregation model of Ref. [200]. Black (resp. white) squares represent $\sigma_n = 1$ (resp. $\sigma_n = 0$), i.e., surviving (resp. dead) sites. The left panel shows a 150^2 sample, while the right panel is enlarged (40^2) for clarity.

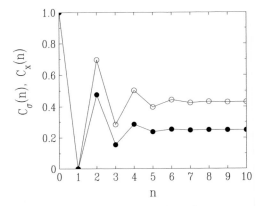

Fig. 8.7 Plot of correlation functions for the cluster aggregation model of Refs. [200] against the distance n along the chain. Empty symbols: correlation $C_\sigma(n)$ of the survival index. Full symbols: correlation $C_x(n)$ of the reduced mass.

model of nonequilibrium aggregation, which despite its origins in cosmology [199] turns out to have applications in the gelation of stirred colloidal solutions [200]. This 'winner-takes-all' model of cluster growth, whereby the largest cluster always wins, manifests both fast and slow dynamics. In mean field, the slow dynamical phase results in at most one surviving cluster at asymptotic times; however, on finite lattices, there can be many *metastable* clusters which survive forever, provided they are each isolated from the others (Fig. 8.6).

We remark that this 'isolation' of surviving sites implies a very strong anticorrelation between neighbouring sites in this model; that is, each survivor must have voids around it, or run the risk of dying out. These anticorrelations are manifest in Fig. 8.7, both for cluster survival and cluster mass on a one-dimensional version of the model. The presence of such anticorrelations and of competition between slow and fast dynamics in a nonequilibrium context suggests strong analogies between this model [200] and the random graphs [152, 153] and column [34, 35] models. We might therefore naively expect some version of Edwards' flatness to hold; however, the results [200] suggest that it does *not*.

In conclusion, we emphasise that Edwards' flatness in the landscape of configurational entropies is *not* the generic fate of strongly nonequilibrium models with slow dynamics, even when they have many features in common. The similarity between the column model [34, 35] of this chapter, and the random graphs model of [152, 153] discussed in Chapter 6 is thus all the more remarkable; *both models manifest Edwards' flatness in the jamming limit, deviating from it whenever free volume constraints are relaxed.*

8.4 Low-temperature dynamics along the column: intermittency

Finally, in this chapter, we investigate the low-temperature dynamics of the column model. For rational ϵ, the presence of a finite but low shaking intensity merely increases the magnitude of density fluctuations [172, 173], given that the zero-temperature dynamics is in any case stochastic. However, for irrational ϵ, low-temperature dynamics introduces an *intermittency in the position of a surface layer*; this has recently been observed in experiments on vibrated granular beds [201].

This happens as follows [34, 35]: when the shaking amplitude Γ is such that it does not distinguish between a very small void h_n and the strict absence of one, the site n 'looks like' a point of perfect packing. The grain at depth n then has the freedom to point the 'wrong' way; we call such sites *excitations*, using the thermal analogy. The probability of observing an excitation at site n scales as $\Pi(n) \approx \exp(-2|h_n|/\Gamma)$. The uppermost site n such that $|h_n| \sim \Gamma \ll 1$ will be the 'preferred' excitation; it is propagated ballistically (cf. the zero-temperature irrational ϵ dynamics of Section 8.2) until another excitation is nucleated above it. Its instantaneous position $\mathcal{N}(t)$ denotes the layer at which shape effects are lost in thermal noise, i.e., *it separates an upper region of quasiperiodic ordering from a lower region of density fluctuations* (cf. Eq. (8.4)).

Figure 8.8 shows a typical sawtooth plot of the instantaneous depth of this layer, $\mathcal{N}(t)$, for a temperature $\Gamma = 0.003$. The *ordering length*, defined as $\langle \mathcal{N} \rangle$, is expected to diverge at low temperature, as excitations become more and more rare; we find in fact [34, 35] a divergence of the ordering length at low temperature of the form $\langle \mathcal{N} \rangle \sim 1/(\Gamma |\ln \Gamma|)$. This length is a kind of finite-temperature equivalent of the 'zero-temperature' length ξ_{dyn}, as it divides an ordered boundary layer from a lower (bulk) disordered region. Within both these boundary layers (ξ_{dyn} and $\langle \mathcal{N} \rangle$), fast dynamics predominate, while for column depths beyond these, slow dynamics set in.

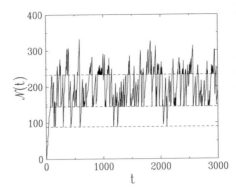

Fig. 8.8 Plot of the instantaneous depth $\mathcal{N}(t)$ of the ordered layer, for $\epsilon = \Phi$ (the golden mean) and $\Gamma = 0.003$. Dashed lines: leading nucleation sites given by Fibonacci numbers (bottom to top: $F_{11} = 89$, $F_{12} = 144$, $F_{13} = 233$) (after[34, 35]).

8.5 Discussion

We have discussed the effect of shape in granular compaction near the jamming limit, via a column model of grains [34, 35]. The main conclusions are that jagged (irregular) grains are characterised by optimal ground states, which are easily retrievable, while smooth (regular) grains cannot retrieve their ground states of perfect packing; in the latter case, even zero-temperature dynamics results in density fluctuations. Also, while slow dynamics predominate deep inside the column model of compacting grains, fast dynamics gives rise to strikingly rough configurational landscapes and surface intermittency.

We have in the above focused largely on the fast dynamical behaviour of the top of a column of grains in the jamming limit, alluding to an increasingly slower dynamics as we get to frozen grains at its bottom. In reality, of course, near the middle of the column, grains feel both the effects of the free surface and the frozen base. The physics of this [134] is being probed currently. Initial results indicate that, as in simulations [131], some of the most interesting dynamical effects such as nucleation and intermittency in compaction occur in this intermediate zone.

9

Avalanches with reorganising grains

When grains are deposited on a sandpile, avalanches result. These have much in common with many other varieties of avalanche; for example, snow or rocks, or even the stress releases that result in earthquakes. The unifying phenomenon in all these cases is that of a *threshold instability*: an overburden builds up, typically that due to surface roughening, to the point where this threshold is crossed, and grains are released in an avalanche. Avalanches can be classed in two main categories; those that do not have intrinsic time or length scales, and those that do. Avalanches relevant to granular media belong to the second category, and we shall discuss their characteristics in depth. We will, however summarise some of the known characteristics of the first category, referring readers who are interested in more details to ref. [65] on the subject.

9.1 Avalanches type I – SOC

Bak, Tang and Wiesenfeld, in their now famous theory of self-organised criticality (SOC), suggested [65] that extended systems were marginally stable, such that the slightest overburdening would cause avalanching; a sandpile at its so-called 'critical' angle of repose was held to be paradigmatic of this. Although this turned out to be, in retrospect, the wrong paradigm, the explorations that surrounded it in fact greatly enriched the physics of granular avalanches. We touch briefly on the important features of SOC here. Bak *et al.* claimed that such avalanches had no intrinsic time or length scale; thus, avalanches of all sizes are equally probable in the theory of SOC, with a power spectrum describable by a $1/f$ power law. While experiments [72, 74] have shown that typical avalanche traces are not so simply described, we describe in the following paragraph the kind of physical scenario that might lead to such statistics.

Granular Physics, ed. Anita Mehta. Published by Cambridge University Press. © A. Mehta 2007.

Consider a sandpile whose surface is constrained to be smooth and elastic, on which grains are deposited in the limit of low inertia; grains will land on the surface, and possibly dislodge other surface grains, leading to a chain reaction where an avalanche flows down the pile. Importantly, grains are not allowed to stick to the surface of the pile, and cannot create roughness; the surface remains as smooth after the avalanche as it was before. This mechanism is evidently stochastic; depending on the first landing point of the deposited grain, and the local nature of the surface, avalanches of variable sizes can be generated. Also, there is a minimal interaction of the deposited grains with the surface; since the surface is constrained to stay smooth, the implication is that as many grains leave the surface on average as are deposited on it. No 'bumps' are allowed to form, as they are unsustainable by the surface; equally, the deposited grains have low inertia, and do not 'fluidise' the grains on the surface on impact. The surface remains essentially unchanged, so that all pre-existing thresholds hold for avalanching; the angle of repose is close to being unique.

In practice, sandpile surfaces are rough, angles of repose are non-unique, and deposited grains interact strongly with the substrate. These give rise to the second category of avalanches – Type II – specific to granular media which will be discussed later in this chapter. To put this in context, we review below the standard results on Type I avalanches.

9.1.1 Review of sandpile cellular automata – Type I

In general, lattice grain models, in which particles are simply represented by regular objects in discrete geometries, are powerful computational tools for studying granular dynamics. Their discrete nature and geometrical parallelism are significant advantages; on the other hand, they require considerable interpretation and analysis.

The development of lattice grain models follows from lattice gas models of fluid flows [202, 203]. There, for a particular set of collision rules, the coarse-grained and long-time behaviour of the lattice gas has been shown to have a precise mapping onto the solutions of the Navier–Stokes equations for incompressible fluid flow. An equivalent correspondence has not been made for the lattice grain methods, so that a unique set of lattice based collision rules is not firmly established for granular flow models. However, some models, such as that of [117], incorporate space filling and inelastic interactions, acting as valuable indicators for good continuum models of granular flow [7].

The most celebrated lattice grain model concerns the flow of grains down the sloping surface of a sandpile [65]. The simplest nontrivial sandpile model consists of monodisperse, unit square grains stacked in columns on a one-dimensional base of length L. The instantaneous state of the sandpile is described by a set of column heights $z_i \geq 0, 1 \leq i \leq L$. In turn, column heights can be used to define local slopes s_i such that

$$s_i = z_i - z_{i-1}, i > 1, \tag{9.1}$$

with $s_1 = z_1$. At each timestep, one grain is added at column i for $1 \leq i \leq L$ such that:

$$z_i \rightarrow z_{i+1}, s_i \rightarrow s_{i+1}, \tag{9.2}$$

and, if $i < L$,

$$s_{i+1} \rightarrow s_{i+1} - 1. \tag{9.3}$$

If, after the addition of a grain, the local slope s_i exceeds some threshold slope s_c, then n_f grains fall from the surface of column i onto columns below itself. In *local* sandpile models, grains falling from column i land on column $i - 1$, but in nonlocal models, falling grains may be distributed over all columns $[1, i - 1]$, with grains able to exit the sandpile from column 1. The number of grains falling at each event, n_f, may either be constant (as in the 'limited' model of Kadanoff *et al.* [77]) or determined dynamically (as in the 'unlimited' model of [77]).

Falling grains cause changes in several column heights (i and $i - 1$ in local models), which could lead to the generation of supercritical slopes, $s > s_c$, elsewhere. More grains could now fall, leading to a chain reaction. When the sandpile returns to a state where all slopes are subcritical, the event chain for one timestep is said to be complete. The number of falling events, n_s, and the number of grains which exit the pile, n_x, are both measures of the size of an avalanche that ensues.

It is known that the order in which the columns are updated is unimportant, so that series or parallel updates are equally efficient. The evolution of the sandpile model may be computed solely in terms of the discrete local slope variables, s_i, using an integer or bit representation, leading to a cellular automaton model [203].

Kadanoff *et al.* [77] have shown that the avalanche distribution function of models of this type is typically multifractal. There are no special avalanche sizes and, therefore, these manifest SOC. Nonlocal and/or unlimited models, different values for n_f and higher dimensionalities do not lead to any substantive change from SOC behaviour, although they do lead to a change of universality class [77].

These simple models, however, fail to explain the dominance of large avalanches in real sandpiles [72, 74]. In the next section we focus on these, and describe in detail a cellular automaton model [75, 83, 168] which leads to their generation.

9.2 Avalanches type II – granular avalanches

Avalanches are the signatures of instabilities on an evolving sandpile: spatiotemporal roughness is alternately built up and smoothed away in the course of avalanche flow. We present below an intuitive picture of avalanching in sandpiles, pointing out that it could be relevant to other scenarios (e.g. granular flows along an inclined plane [166], or sediment consolidation [204]).

As deposition occurs on a sandpile surface, clusters of grains grow unevenly at different positions and roughness builds up until further deposition renders some of these unstable. They then start 'toppling', so that grains from unstable clusters flow down the sandpile, knocking off grains from other clusters. The net effect of this is to 'wipe off' protrusions (where there is a surfeit of grains at a cluster) and to 'fill in' dips, where the oncoming avalanche can disburse some of its grains. Typically, small avalanches build up surface roughness, while large avalanches have a smoothing effect; in the latter case, a rough precursor surface typically leads to avalanche onset, and subsequently, an overall smoothing of the surface. This result is rather robust, having been found independently using a variety of models, which will be presented in this and succeeding chapters. In the next chapter, a coupled map lattice model demonstrating stick–slip dynamics [22] will be discussed, while in the following one, continuum equations coupling surface to bulk relaxation [95, 96] will be presented.

In this chapter, we focus on a cellular-automaton model [75, 83, 168] of an evolving sandpile to look in more depth at the mechanisms by which a large avalanche smooths the surface. This sandpile model is a 'disordered' and non-abelian version of the basic Kadanoff cellular automaton [77]; in the present model grain 'flip' is added to the grain flow which is already present in the Kadanoff model.

The disordered model sandpile[1] is built from rectangular lattice grains that have aspect ratio $a \leq 1$ arranged in columns i with $1 \leq i \leq L$, where L is the system size. Each grain is labelled by its column index i and by an orientational index 0 or 1, corresponding respectively to whether the grain rests on its larger or smaller edge.

The dynamics of this model, which have been described at length elsewhere [83, 168], are:

- Grains are deposited on the sandpile with fixed probabilities of landing in the 0 or 1 position.
- The incoming grains, as well as all the grains in the same column, can then 'flip' to the other orientation stochastically (with probabilities which decrease exponentially with

---

[1] Note that while the representation of disorder in this model is identical to that of the model [174] presented in Chapter 7, its dynamics are entirely different.

depth from the surface). This is a way of introducing a time-dependent disorder into the problem.

- Column heights are then computed as follows: the height of column i at time t, $h(i, t)$, can be expressed in terms of the instantaneous numbers of 0 and 1 grains, $n_0(i, t)$ and $n_1(i, t)$ respectively:

$$h(i, t) = n_1(i, t) + an_0(i, t). \tag{9.4}$$

- Finally, grains fall to the next column down the sandpile (maintaining their orientation as they do so) if the height difference exceeds a specified threshold as in the Kadanoff model [77]. At this point, avalanching occurs.

The presence of the flipping mechanism – 'annealed disorder' – leads, for large enough system sizes, to a preferred size of large avalanches [75], while in the absence of disorder, scale-invariant avalanche statistics are observed. Below, the evolving state of the sandpile surface is correlated with the onset and propagation of avalanches.

9.2.1 Dynamical scaling for sandpile cellular automata

It is customary in the study of generalised surfaces to examine the widths generated by kinetic roughening [169], and then establish properties related to *dynamical scaling*. This procedure can be generalised to include the kinetic roughening of sandpile cellular automata. The hypothesis of dynamical scaling for sandpile surfaces [83] reads, in terms of the surface width W of the sandpile:

$$W(t) \sim t^{\beta}, \qquad t \ll t_{\text{crossover}} \equiv L^z; \tag{9.5}$$
$$W(L) \sim L^{\alpha}, \qquad L \to \infty. \tag{9.6}$$

Thus, to start with, roughening occurs at the CA sandpile surface in a time-dependent way; after an initial transient, the width scales asymptotically with time t as t^{β}, where β is the *temporal roughening* exponent. This regime is appropriate for all times less than the crossover time $t_{\text{crossover}} \equiv L^z$, where $z = \alpha/\beta$ is the dynamical exponent and L the system size. After the surface has *saturated*, i.e. its width no longer grows with time, the *spatial roughening* characteristics of the mature interface can be measured in terms of α, an exponent characterising the dependence of the width on L.

The surface width $W(t)$ for a sandpile automaton is defined in terms of the mean-squared deviations from a suitably defined mean surface; the instantaneous mean surface of a sandpile automaton is thus defined as the surface about which the sum of column *height* fluctuations vanishes. Clearly, in an evolving surface, this must be a function of time; hence all quantities in the following analysis will be presumed to be instantaneous.

The mean slope $\langle s(t) \rangle$ defines expected column heights, $h_{av}(i, t)$, according to

$$h_{av}(i, t) = i \langle s(t) \rangle, \tag{9.7}$$

where it is assumed that column 1 is at the bottom of the pile. Column height deviations are defined by

$$dh(i, t) = h(i, t) - h_{av}(i, t) = h(i, t) - i \langle s(t) \rangle. \tag{9.8}$$

The mean slope must therefore satisfy

$$\Sigma_i[h(i, t) - i \langle s(t) \rangle] = 0 \tag{9.9}$$

since instantaneous height deviations about it vanish; thus[2]

$$\langle s(t) \rangle = 2\Sigma_i[h(i, t)]/L(L + 1) \tag{9.10}$$

The instantaneous width of the surface of a sandpile automaton, $W(t)$, can be defined as:

$$W(t) = \sqrt{\Sigma_i[dh(i, t)^2]/L}, \tag{9.11}$$

which can in turn be averaged over several realisations to give $\langle W \rangle$, the average surface width in the steady state.

Another quantity of interest is the height–height correlation function, $C(j, t)$; this is defined by

$$C(j, t) = \langle dh(i, t)dh(i + j, t) \rangle/\langle dh(i, t)^2 \rangle, \tag{9.12}$$

where the mean values are evaluated over all pairs of surface sites separated by j lattice spacings:

$$\langle dh(i, t)dh(i + j, t) \rangle = \Sigma_i(dh(i, t)dh(i + j, t))/(L - j) \tag{9.13}$$

for $0 \leq j < L$. This function is symmetric and can be averaged over several realisations to give the time-averaged correlation function $\langle C(j) \rangle$.

9.2.2 Qualitative effects of avalanching on surfaces

Figure 9.1(a) shows a time series for the mass of a large ($L = 256$) evolving disordered sandpile automaton.[3] The series has a typical quasiperiodicity [74]. The vertical line in Fig. 9.1(a) denotes the position of a particular 'large' event, while

---

[2] Note that this slope is distinct from the quantity $\langle s'(t) \rangle = h(L, t)/L$ that is obtained from the average of all the local slopes $s(i, t) = h(i, t) - h(i - 1, t)$, about which *slope* fluctuations would vanish on average.
[3] Throughout this chapter we refer to disordered sandpiles described in reference [168] with parameters $z_0 = 2$, $z_1 = 20$ and $a = 0.7$, unless otherwise stated.

in Fig. 9.1(b), the marked 'second peak' [83] in the avalanche size distribution is composed of such large avalanches.

The large avalanche highlighted in Fig. 9.1(a) is pictured in Fig. 9.1(c). The initiation site is marked by an arrow and the outline of the surface before and after the avalanche (corresponding to about 5 per cent of the mass of the sandpile) shaded in black. We note that this is an example of an 'uphill' avalanche described in Chapter 5 [165, 166]; the avalanche propagates downwards, of course, but it destabilises particles *above* its point of initiation. The inset shows the relative motion of the surface during this event; the signatures of smoothing by large avalanches are already evident, as the precursor state in the inset is much rougher than the final state.

In Fig. 9.1(d), the grain-by-grain picture of the aftermath pile is superposed on the precursor pile, which is shown in shadow. An examination of the aftermath pile and the precursor pile [83] shows that the propagation of the avalanche across the upper half of the pile has left only a very few disordered sites in its wake (i.e. the majority of the remaining sites are of type 0) whereas the lower half (which was undisturbed by the avalanche) still contains many disordered, i.e. type 1 sites in the boundary layer. This suggests that larger avalanches rid surface layers of their disorder-induced roughness, a fact that is borne out by more quantitative investigations below.

Detailed studies have revealed [83] that the very largest avalanches, which are system-spanning, remove virtually all disordered sites from the surface layer; one is then left with a normal 'ordered' sandpile, where the avalanches have their usual scaling form for as long as it takes for a layer of disorder to build up. When the disordered layer reaches a critical size, another large event is unleashed; this is the underlying reason for the quasiperiodic form of the time series shown in Fig. 9.1(a). This picture is confirmed by a totally different model, the coupled-map lattice model of a reorganising sandpile [22], to be discussed in the next chapter.

The sequence of Figs. 9.1 a–d. is shown also for an ordered pile – Fig. 9.2 a–d and a small disordered pile – Fig. 9.3 a–d. The small disordered pile has a mass–time series (Fig. 9.3a) that is midway between the scale-invariance of the ordered pile (Fig. 9.2a) and the quasiperiodicity of the large disordered pile (Fig. 9.1a). The avalanche size distribution of the small disordered pile (Fig. 9.3b) is likewise intermediate between that of the ordered pile (which shows the scale invariance observed by Kadanoff *et al.* [77]) and the two-peaked distribution characteristic of the disordered pile [75, 83]. This suggests that a crossover length must exist, after which the fully non-invariant scale characteristics of real sandpiles would begin to manifest; interestingly, such a crossover length in the mass time series has indeed been observed in experiment [74].

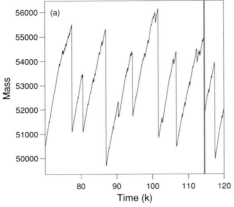

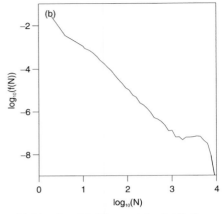

(a) A time series of the mass. The vertical line indicates the position in this series of the large avalanche illustrated in c, d.

(b) A log–log plot of the event size distribution.

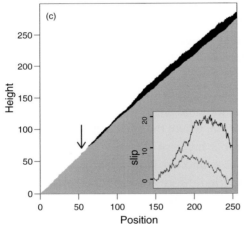

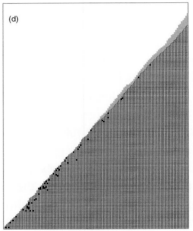

(c) An illustration of a large wedge shaped avalanche; a lighter aftermath pile has been superposed onto the dark precursor pile, and an arrow shows the point at which the event was initiated. The inset shows the relative positions of the two surfaces and their relationship to a pile that has a smooth slope.

(d) A detailed picture of the internal structure in the aftermath of a large avalanche event. The individual grains of the aftermath pile (for columns $1 - 128$ of the sandpile with $L = 256$) are superposed on the gray outline of the precursor pile.

Fig. 9.1 Statistics for a model sandpile ($L = 256$) with a structurally disordered surface layer

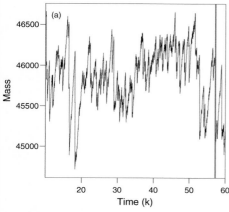

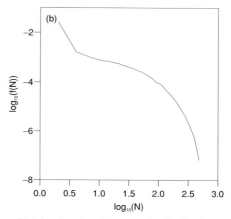

(a) A time series of the mass. The vertical line indicates the position in this series of the large avalanche illustrated in c, d.

(b) A log–log plot of the event size distribution.

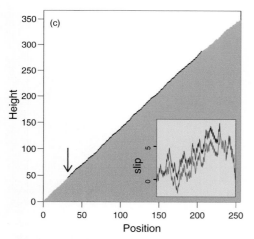

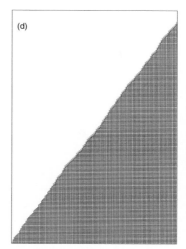

(c) An illustration of a large wedge shaped avalanche; a lighter aftermath pile has been superposed onto the dark precursor pile, and an arrow shows the point at which the event was initiated. The inset shows the relative positions of the two surfaces and their relationship to a pile that has a smooth slope.

(d) A detailed picture of the internal structure in the aftermath of a large avalanche event. The individual grains of the aftermath pile (for columns $1-128$ of the sandpile with $L = 256$) are superposed on the gray outline of the precursor pile.

Fig. 9.2 Statistics for a model sandpile ($L = 256$) without structural disorder.

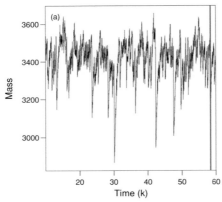

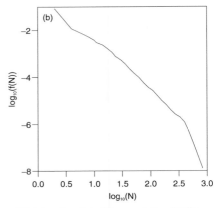

(a) A time series of the mass. The vertical line indicates the position in this series of the large avalanche illustrated in c, d.

(b) A log–log plot of the event size distribution.

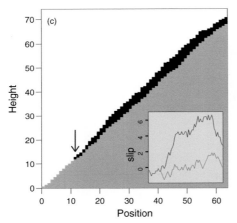

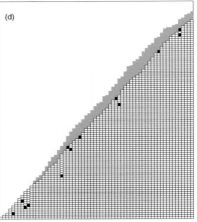

(c) An illustration of a large wedge shaped avalanche; a lighter aftermath pile has been superposed onto the dark precursor pile, and an arrow shows the point at which the event was initiated. The inset shows the relative positions of the two surfaces and their relationship to a pile that has a smooth slope.

(d) A detailed picture of the internal structure in the aftermath of a large avalanche event. The individual grains of the aftermath pile (for columns $1 - 128$ of the sandpile with $L = 256$) are superposed on the gray outline of the precursor pile.

Fig. 9.3 Statistics for a *small* model sandpile ($L = 64$) with a structurally disordered surface layer

In both small and large disordered piles, we see evidence of large 'uphill' avalanches which shave off a thick boundary layer containing large numbers of disordered sites, and leave behind a largely ordered pile (see Figs. 9.1 c–d and Figs. 9.3 c–d). By contrast, the ordered pile loses typically two commensurate

Table 9.1 *Instantaneous properties of disordered model sandpiles*

L	State of pile	Packing fraction ϕ	Slope	Width
256	Before	0.997	1.15	4.45
256	After	1.000	1.07	2.06
64	Before	0.991	1.17	1.41
64	After	0.998	1.04	1.15

layers even in the largest avalanche, with a correspondingly unexciting aftermath state left behind in its wake (Figs. 9.2c –d).

Thus, even at a qualitative level, there is a post-avalanche smoothing of the sandpile surface, beyond a crossover length; interestingly, this is also the prediction of continuum equations [96] modelling the dynamics of sandpile surfaces, which will be discussed in Chapter 11. The discrete model of the current chapter [83, 168] reveals that this smoothing is achieved by the removal of disorder, the implications of which will be discussed in the concluding section.

9.2.3 The effect of avalanching on sandpile surfaces – some observations of material properties

We present here the results of investigations [83] of some material properties of the sandpile [205] in pre- and post-avalanche configurations. From these we can draw the following conclusions.

- The *mean slope* of the disordered sandpile peaks (see Table 9.1) before a large avalanche and drops immediately after; this statement is true for events of any size and thus remains trivially true for the ordered sandpile.
- The *packing fraction* ϕ of the disordered sandpile increases after a large event, i.e. effective consolidation occurs during avalanching (see Table 9.1). This consolidation via avalanching mirrors that which occurs when a sandpile is shaken with low-intensity vibrations [54, 61, 172].
- However, a far deeper statement can be made about the comparison of the surface width for pre- and post-large event sandpiles; Table 9.1 shows that the surface width goes down considerably during an event, once again suggesting that a rough precursor pile is smoothed by the propagation of a large avalanche.

Table 9.2 shows the results of investigations of the dependence of various material properties of a disordered sandpile on the aspect ratio of the grains [206]; while Fig. 9.4 illustrates the variation of the corresponding avalanche size distribution.

Table 9.2 *Properties of model sandpiles*

Aspect ratio	Packing fraction ϕ	Slope	Width
0.6	0.997	1.44	2.33
0.65	0.997	1.42	2.34
0.7	0.997	1.12	3.76
0.75	0.997	1.19	3.74
0.8	0.998	1.17	2.35
0.9	0.999	1.25	2.29
0.95	1.000	1.32	2.37
1.0	1.000	1.40	2.41

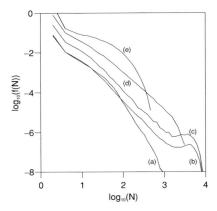

Fig. 9.4 Exit mass event size distributions for a disordered sandpile model with $L = 256$, and $a = (a)0.6, (b)0.7, (c)0.75, (d)0.85, (e)1.0$. The curves have each been shifted by 0.5 to make them distinct.

There is a transition as aspect ratios of 0.7 are approached from above or below; sandpiles with these 'critical' aspect ratios manifest strong disorder [168] in the sense of:

• a 'second peak' in the avalanche size distribution denoting a preferred size of large avalanches;
• large surface widths denoting an increased surface roughness;
• a strong correlation between interfacial roughness and avalanche flow, since the mean surface width varies dramatically in the pre- and post- large event piles.

Clearly, sandpiles containing grains with aspect ratios close to unity act essentially as totally ordered piles [77]; there is, however, a significant symmetry in the shape of the avalanche size distribution curves above and below the transition region (see Fig. 9.4 a and d). This somewhat surprising non-monotonicity could shed some light on the still unexplained results of the Oslo 'rice pile' experiments [206] which had indicated that rice grains in a pile manifested SOC, while spheres did not. The effects of granular shape on dynamics are extremely nontrivial, as the results presented in Chapters 7 and 8 have indicated. The non-monotonic dependence on aspect ratio of the exit mass distribution in Fig. 9.4 is yet another indication of this; even in this toy model of a sandpile, there appears to be a retrieval of certain aspects of scale-invariance as the aspect ratio of its rectangular grains is increased or decreased away from 0.7. The next question is, what is so special about the number 0.7? A possible answer is that it is a rather 'asymmetric' fraction of 1, so that grains with this as aspect ratio will not pack well in general.[4] This will lead typically to surface roughening (bumps will form on the surface, as grains 'stack' rather than 'pack'), so that conditions will be the least propitious for SOC to be observed, as discussed in the beginning of this section. Extending this to grains of arbitrary shape, we speculate that *the more asymmetric/'rough' the grain, the more likely one would be to see surface roughening, and hence the absence of scale invariance*.

The nature of the distribution of disorder is also an important factor in determining the nature of the avalanche distribution. Disorder may be absent, localised in a boundary layer, or uniformly distributed throughout the sample. It turns out [83] that the avalanche statistics that are obtained as a result of the distribution of disorder [75] show an interesting parallel with the results of Fig. 9.4. The case of total order in Fig. 9.4 (marked 'e') corresponds evidently to the absence of disorder also in the results of [75]. The cases where there is a marked absence of scale-invariance in Fig. 9.4 (marked 'b' and 'c') correspond in Ref. [75] to the case where the disorder is concentrated in a boundary layer; in both these cases there is a second peak in the avalanche size distribution, corresponding to large avalanches. Finally, the intermediate cases of weak disorder (characterised by a flatter size distribution of avalanche sizes which is, nevertheless, *not* scale-invariant) in Fig. 9.4 (marked 'a' and 'c') correspond to the case where disorder is uniformly distributed through the sandpile, in the results of Ref. [75].

Additionally we present, in Fig. 9.5, the time-dependent mass–mass correlation function of a particular disordered sandpile,[5] averaged over many realisations, corresponding to different values of the time t_0. A strong feature emerges; after the start from unity, the correlation function of mass decays for a while, and then rises again. This characteristic time (corresponding to about 5000 units in the plot of Fig. 9.5,

---

[4] This is, of course, even more true of irrational aspect ratios dealt with in Chapter 8 – here we are only concerned with a few, typically rational, values of aspect ratio.
[5] The parameters of this sandpile have been chosen to correspond to strongly non-scale-invariant avalanche statistics, with a second peak in the avalanche size distribution.

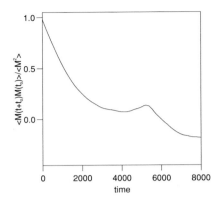

Fig. 9.5 The time-dependent mass–mass correlation function for a disordered
model sandpile with $L = 256$.

for mass correlations is, we suggest, approximately the *time between large, system-
spanning avalanches*. Our reasoning goes as follows: small avalanches, which correspond in length and time to non-system-spanning events, will be averaged out
between different realisations. The only trace that could possibly survive would
correspond to the system size; hence the time between the origin and the observed
peak corresponds to the time between large, system-spanning avalanches. The mass
of the pile decays overall during the release of grains in such an avalanche, only to
rise again as enough grains are deposited on the surface to unleash the next large
avalanche. The dynamics of this 'build-up and release' mechanism will be further
explored by the model of the next chapter [22].

We next present in Fig. 9.6 the (normalised) *equal-time* height–height corre-
lation function $\langle dh(r + r_0)dh(r_0)\rangle/\langle dh(r_0)^2\rangle$ for a large disordered sandpile, also
averaged over many realisations, i.e. values of r_0. As we might expect from its
definition, this should give us an idea of the roughness of the sandpile surface – it,
and related functions, will be extensively discussed in Chapter 11 from a theoretical
point of view. We see from the figure that there are positive height correlations over
about 80 columns in this case, after which there are negative correlations. Naively,
this distance should correspond to the distance between a 'bump' and a 'dip' in the
sandpile surface.

This issue is probed more quantitatively via the inset of Fig. 9.6,
where the (related) function $1 - \langle dh(r + r_0)dh(r_0)\rangle/\langle dh(r_0)^2\rangle \equiv \langle (dh(r + r_0) -
dh(r_0))^2\rangle/2\langle dh(r_0)^2\rangle$ is plotted. From the definition Eq. 9.6 of the spatial roughen-
ing exponent α, we have [207]:

$$\langle (dh(r + r_0) - dh(r_0))^2\rangle \sim |r|^{2\alpha}. \tag{9.14}$$

The function plotted in the inset of Fig. 9.6 should clearly manifest a similar r-
dependence. A linear fit to points with $r < 30$ (shown by the line in the inset of

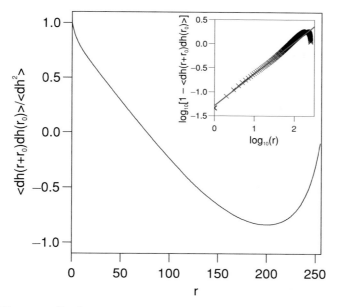

Fig. 9.6 The normalised equal-time correlation function of column height deviations for a disordered model sandpile with $L = 256$. The inset illustrates a fit to a
related function, going as $r^{0.67}$ (r is measured in lattice units).

Fig. 9.6) indicates a power-law dependence of the form $r^{0.67}$ for $r \ll L$, imply-
ing that $\alpha \sim 0.34$. More light will be shed on this particular value of the spatial
roughening exponent in the next subsection.

9.2.4 Spatial and temporal roughening of sandpile surfaces

As mentioned earlier in this section, the hypothesis of dynamical scaling for sandpiles assumes that the roughening process occurs in two stages. First, the surface
roughening is time-dependent; then once the roughness becomes temporally constant, the surface is said to saturate, and attain its asymptotic roughness over the
whole of the system. Measurements of the corresponding temporal roughening
exponent β (defined in Eq. (9.5)) and spatial roughening exponent α (defined in
Eq. (9.6)) have been carried out [83] for both ordered and disordered piles, whose
results are as follows.

- For disordered sandpiles, $\beta = 0.42 \pm 0.05$, while for ordered sandpiles, $\beta = 0.17$
  ± 0.05.
- For ordered sandpiles, $\alpha = 0.356 \pm 0.05$, irrespective of the length scale over which this
  is measured (as expected!). For disordered sandpiles below a crossover size of $L_c = 90$,
  $\alpha = 0.356 \pm 0.05$; for large disordered sandpiles with $L \gg L_c$, $\alpha = 0.723 \pm 0.04$.

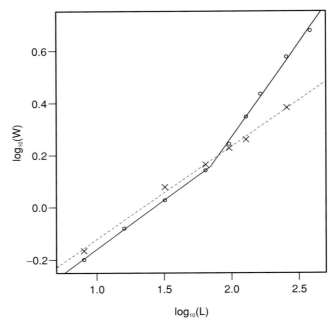

Fig. 9.7 A log–log plot of the surface widths, W, against the system size, L, for model sandpiles; (x) ordered piles, (o) disordered piles. Widths are measured in lattice units and all points have error bars that are ~ 0.02.

- Based on the above values, the dynamical exponent $z \equiv \alpha/\beta$ is 1.72 ± 0.29 for the disordered, and 2.09 ± 0.84 for the ordered sandpiles. Thus, within experimental error, the dynamical exponent is unchanged by disorder.

These results [83, 168] are exhibited in terms of the variation of the surface width W, as a function of system size L, in Fig. 9.7. The figure shows clearly the crossover in α as a function of system size, for disordered sandpiles. The spatial roughening exponent for disordered and ordered sandpiles (shown in the figure for comparison) start off the same, and then, after sizes of the crossover length L_c have been exceeded, the disordered pile is rougher, i.e. it manifests a larger value of α as indicated above. The fact that, for lengths less than L_c, disordered and ordered piles manifest the *same* roughness exponent α immediately explains the observations of the previous subsection. There, a spatial roughening exponent of $\alpha \sim 0.37$ (near the value for the ordered sandpile!) was obtained over a lengthscale of $r \sim 30$ in a fully disordered pile, via measurements of the height–height correlation function (Fig. 9.6). This is only to be expected, since, in the inset of Fig. 9.6, α was evaluated over a length $r \sim 30$, which is less than the crossover length L_c; thus *even in the presence of full disorder, the measured α was closer to that corresponding to the ordered sandpile*.

The existence of this crossover length has been variously interpreted as a length related to reorganisation in the boundary layer of a sandpile [69] or to variations in the angle of repose in a (disordered) sandpile [70]. Disorder appears to be crucial for the existence of such experimentally observed [74] crossovers, since, for example, ordered sandpile models [208, 209] show no crossover in their measurements of α. We note also that while, within error bars, the dynamical exponent z does not change when disorder is introduced into a sandpile, the disordered pile is clearly rougher with respect to both temporal and spatial fluctuations (α and β are both higher). This may be explained by the fact that, for an ordered sandpile, no abnormally large events occur, which is reflected by the lower values of fluctuations and exponents relative to the disordered case.

9.3 Discussion and conclusions

We have presented a thorough investigation of the effects of avalanching on a sand-pile surface, focusing on the interrelationship between the nature of the avalanches and the surfaces they leave behind. We have also postulated a principle of dynamical scaling for sandpile surfaces, and measured the roughening exponents for a sample disordered sandpile. Finally, we have related the characteristics of avalanching in the model system to those obtained experimentally.

There are several questions, unanswered in the above, which are worthy of investigation. These include the dependence of the crossover length L_c on the microscopics – e.g. grain shape, texture, restitution, friction – in the pile, as well as the more macroscopic distribution of disorder (i.e. whether boundary or uniform) [75]. Other questions, such as the dependence of the avalanche statistics on dynamics, or a further investigation of the stick–slip mechanism [210] known to be at the basis of the quasiperiodic avalanche time series, will be addressed in the next chapter.

10

From earthquakes to sandpiles – stick–slip motion

In this chapter, we seek to explain the nature of experimentally observed [28] avalanche statistics from a more event-based point of view than in the earlier chapter. In some ways, the difference between the two approaches is akin to that between Monte Carlo and molecular dynamics approaches. In the last chapter (as in Monte Carlo simulations), the dynamics is 'simulated' – real grains do not, after all, topple as a result of height thresholds – with a view to matching only the end results of, in this case, avalanche statistics. In this chapter, we try to model an (albeit simplified) version of the real dynamics that occurs when grains avalanche. Interestingly, though both approaches are totally different, the results are robustly similar – we find via both approaches the prediction of a special scale for large avalanches, and, in this chapter, propose a dynamical mechanism which leads to their being unleashed.

10.1 Avalanches in a rotating cylinder

Here we describe a model [22] of an experimental situation which forms the basis of many traditional as well as modern experiments; a sandpile in a rotating cylinder. Consider the dynamics of sand in a half-cylinder that is rotating slowly around its axis. Supposing that the sand is uniformly distributed in the direction of the axis, we are dealing with an essentially one-dimensional situation. The driving force arising from rotation continually affects the stability of the sand at all positions in the pile and is therefore distinct from random deposition. Both surface flow and internal restructuring are included as mechanisms of sandpile relaxation; we focus on a situation where reorganisation within the pile dominates the flow. Finally, we look at the effect of random driving forces in the model and compare the results with those from other models.

Granular Physics, ed. Anita Mehta. Published by Cambridge University Press. © A. Mehta 2007.

10.2 The model

Since the effect of grain reorganisation driven by slow tilt is most naturally visualised from a continuum viewpoint, the model [22] incorporates grains which form part of a continuum. Column heights, h_i, are real variables, while column numbers, $1 < i < L$, are discrete as usual. We consider granular driving forces, f_i, that include, in addition to a term that drives the normal surface flow, a contribution that is proportional to the deviation of the column height from an 'ideal' height; this ideal height is a simple representation of a natural random packing of the grains in a column, so that columns which are taller (shorter) than ideal would be relatively loosely (closely) packed, and driven to consolidate (dilate) when the sandpile is perturbed externally. Thus:

$$f_i = k_1(h_i - iaS_0) + k_2(h_i - h_{i-1} - aS_0), \quad i \neq 1, \tag{10.1}$$

where h_i are the column heights, k_1 and k_2 are constants, a is the lattice spacing, and iaS_0 is the ideal height of column i. We note that

- the first term, which depends on the absolute height of the sandpile, corresponds to a force that drives column compression or expansion towards the ideal height. Since we normally deal with columns which are more dilated than their normal height, we will henceforth talk principally about column compression;
- the second term is the usual term driving surface flow, which depends on local slope, or height differences; the offset of S_0 is the ideal slope from which differences are measured.

Equation (10.1) suggests the redefinition of heights $az_i \equiv h_i - iaS_0$ which leads to the dimensionless representation:

$$f_i = (k_1/k_2)z_i + (z_i - z_{i-1}), \quad i \neq 1. \tag{10.2}$$

When column i is subject to a force greater than or equal to the threshold force f_{th}, the height changes are as follows:

$$z_i \rightarrow z_i - \delta z,$$
$$z_{i-1} \rightarrow z_{i-1} + \delta z', i \neq 1. \tag{10.3}$$

The column-height changes that correspond to a typical relaxation event described by Eq. (10.2) are illustrated in Fig. 10.1. A height δz removed from column i (due to a local driving force that exceeds the threshold force) leads to a flow of grains with total height increment $\delta z'$ from column i onto column $i - 1$, and a *consolidation* of the grains in column i, which decreases its height by $(\delta z - \delta z')$. This clearly expresses the action of two relaxation mechanisms – reorganisation

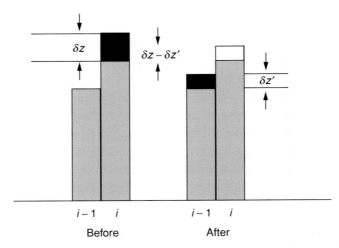

Fig. 10.1 A schematic diagram showing the column height changes that describe a single relaxation event in the CML sandpile model.

and flow. The decomposition of the relaxation, that is, a particular choice for δz and $\delta z'$, is discussed below.

Such a coupling between the column heights may lead to the propagation of instabilities along the sandpile and hence to an avalanche. Since avalanches have also been discussed widely in the context of earthquakes [210], a discrete model of earthquakes, put forward by Nakanishi [211], is chosen to highlight those features which are common to sandpiles and earthquakes. In this spirit, the force relaxation function is chosen [22] to be [211]:

$$f_i - f_i' = f_i - f_{th}(((2 - \delta f)^2/\alpha)/((f_i - f_{th})/f_{th} + (2 - \delta f)/\alpha) - 1), \quad (10.4)$$

where f_i and f_i' are the granular driving forces on column i before and after a relaxation event. This function has a minimum value ($= \delta f f_{th}$) when $f_i = f_{th}$, and increases monotonically with increasing f_i; this form models the stick–slip friction associated with sandpiles and earthquakes. For driving forces f_i below the threshold force f_{th}, nothing happens; but for forces that exceed this threshold, the size of relaxation events increases in proportion to the excess force. Accordingly, the minimum value of the function (10.1) is known as the minimum event size, and its initial rate of increase, $\alpha = d(f_i - f_i')/d(f_i - f_{th})$ at $f_i = f_{th}$, is called the *amplification* [211]. In the sandpile model, amplification refers to the phenomenon whereby grains collide with each other during an avalanche so that their inertial motion contributes to its buildup; thus α is an expression of grain inertia. Using

Eq. 10.2, the map can be rewritten in terms of the driving forces as [22]:

$$f_i - f_i' = 2\delta z / \Delta,$$
$$\delta z' = (\delta z)/(1 + k_1/k_2), \qquad (10.5)$$

$$f_{i-1}' - f_i = f_{i+1}' - f_{i+1} = -\Delta(f_i' - f_i)/2, \quad i \neq 1 \text{ or } L. \qquad (10.6)$$

In both sandpile and earthquake models, the amount of redistributed force at a relaxation event is governed by the parameter $\Delta = 2(1 + k_1/k_2)/[1 + (1 + k_1/k_2)^2]$; since the undistributed force is 'dissipated', $(1 - \Delta)$ becomes the dissipation coefficient [22]. Note, however, that in the sandpile model, this dissipation is linked to nonconservation of the sandpile volume arising from the compression of columns towards their ideal heights; here, $(1 - \Delta)$ is therefore linked to the phenomenon of granular *consolidation*.

Boundary conditions appropriate to a sandpile in a rotating cylinder – open at $i = 0$ and closed at $i = L$ – are used. Equations (10.4) and (10.6) give a prescription for the evolution of forces f_i, $i = 1, L$, so that any forces in excess of the threshold force are relaxed according to (10.6) and redistributed according to (10.4). Alternatively, this sequence of events can be followed in terms of the redistribution of column heights according to (10.3) and (10.5).

We will see below that for all $\Delta \neq 1$, the largest part of the volume change during relaxation occurs as a result of consolidation; the quantity of interest is thus the difference between the old and new configurations, rather than the mass exiting the sandpile [75]. A measure of this change is the quantity

$$\ln M = \ln \Sigma_i (f_i - f_i') = \ln[\Sigma_i ((k_1/k_2)(z_i - z_i') + z_L - z_L']. \qquad (10.7)$$

Here, z_i' is the height of column i immediately after a relaxation event; this quantity is the analogue of the event magnitude in earthquake models [210, 211]. We will discuss the variation of this quantity as a function of model parameters in the next subsection; in a later subsection, we will compare the response of the rotated sandpile model with that of the same model subjected to random deposition.

10.3 Results

10.3.1 Rotated sandpile

For a sandpile in a rotating cylinder, tilt results in continuous changes of slope over the surface (in contrast to the case of random deposition, where slope changes are local and discontinuous [75]). To model this, the coupled map lattice (CML) model of [22] is driven continuously.

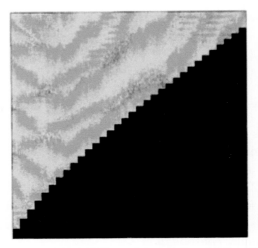

Fig. 10.2 The shape of a critical CML model sandpile with $L = 32$ and $\Delta = 0.95$. The line indicates the 'ideal' column heights.

From a configuration in which all forces f_i are less than the threshold force, elements of height z_i^+ are added onto each column with

$$z_i^+ = i(f_{\text{th}} - f_j)/(1 + jk_1/k_2), \quad i = 1, L, \tag{10.8}$$

where $f_j = \max(f_i)$. This transformation describes the effect of rotating the base of the sandpile with a constant angular speed until a threshold force arises at column j.[1] The response to the tilting is, as described above, a flow of particles down the slope as well as reorganisation of particles within the sandpile.

The predominant effect of the model is to cause volume changes by consolidation, rather than to generate surface flow. Using the relation between force and column height (Eq. (10.2)), and integrating from the left, we can construct the shape of a critical sandpile which has driving forces equal to the threshold force on all of its columns; in terms of the variable $\zeta \equiv (1 + \sqrt{1 - \Delta^2})/\Delta$, the critical sandpile has column heights z_i^c given by

$$z_i^c = f_{\text{th}}[1 - \zeta^{-1} \exp(1 - \zeta)(i - 1)a]/(\zeta - 1), \quad \Delta < 1. \tag{10.9}$$

This shows that for all $\Delta < 1$, the critical sandpile starts at $i = 1$ with a slope greater than S_0, and subsequently the slope decreases until it becomes steady at S_0 for $i \gg 1$, where the constant deviation of the column heights from their ideal values is given by $f_{\text{th}}\Delta/(1 - \Delta + \sqrt{1 - \Delta^2})$ (Fig. 10.2).

---

[1] Note that this is distinct from the external driving force in the earthquake model of [211], which would correspond to the *uniform* addition of height elements across the sandpile surface.

It has been verified [22] by simulation that the corresponding state is an attractor. Note that this sandpile shape is quite distinct from that generated by standard lattice sandpiles, and is close to the S-shaped sandpile observed in rotating cylinder experiments [71].

From Eq. (10.1), it is clear that any value of steady slope which differs from S_0 would lead to a linear growth in the first term – this is therefore unstable. Thus, stability enforces solutions where the average slope, for $i \gg 1$, is S_0. For a truly critical pile, the second term in Eq. (10.1) is identically zero for $i \gg 1$ so that, except in the small i region, the threshold force that drives relaxation arises solely from the compressive component. This predominance of the compressive term then leads to column height changes that are typically $\sim f_{th}\Delta\delta f$ and, in the parameter range under consideration, are small compared to the 'column grain size' $S_0 a$ (the average step size in a lattice slope with gradient S_0 and column width a). In other words, *typical events are likely to be due to internal rearrangements generating volume changes that are small fractions of 'grain sizes'*. They can be visualised as the slow intracluster rearrangement of grains, rather than the surface flow events in standard lattice models [65]. In a slowly rotating cylinder, it is to be expected [69, 71] that such reorganisational events within the sandpile will outweigh the surface flow of avalanches.

The steady state response of the driven sandpile may be represented as a sequence of events, each of which corresponds to a set of column height changes. Each avalanche is considered to be instantaneous, so that the temporal separation of consecutive events is defined by the driving force (10.8). We choose a timescale in which the first column has unit growth rate, and begin each simulation at $t = 0$ with a sandpile containing columns which have small and random deviations from their natural heights; also, we set $a = S_0 = 1$ to fix the arbitrary horizontal and vertical length scales, and we fix $f_{th} = 1$ to define units of 'force'. The dynamics of events do not depend explicitly on these choices.

In Fig. 10.3, we plot the distribution function per unit time and length R $\log(M)$ against $\log(M)$, for sandpiles with size $L = 512$ and parameter values $\delta f = 0.01$, $\alpha = 2, 3, 4$, and $\Delta = 0.6, 0.85, 0.95$. We note in particular the small value of δf, and mention that the results are qualitatively unaffected by choosing δf in the range $0.001 < \delta f < 0.1$; given its interpretation in terms of the smallest event size, this reflects the choice of the quasistatic regime, where small cooperative internal rearrangements predominate over large single-particle motions. The distribution functions in Fig. 10.3 indicate a scaling behaviour in the region of small magnitude events and, for larger magnitudes, frequencies that are larger than would be expected from an extension of the same power law.

Also, the phase diagram in the $\Delta - \alpha$ plane indicates qualitatively distinct behaviour for low-inertia, strongly consolidating (low α and Δ) systems

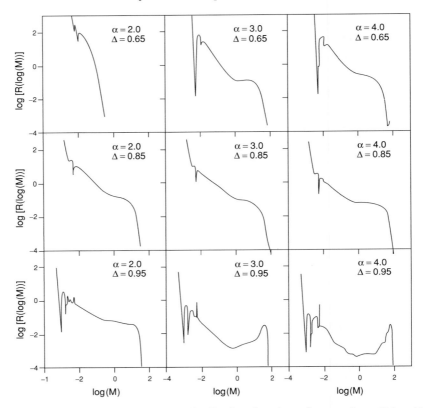

Fig. 10.3 A logarithmic plot of the distribution function of event sizes, $R \log M$, for 10^7 consecutive events in a CML model sandpile with $L = 512$ and parameter values $\delta f = 0.01$, $\alpha = 2, 3, 4$, and $\Delta = 0.6, 0.85, 0.95$.

where the magnitude distribution function has a single peak, and high-inertia, weakly consolidating (high α and Δ) systems for which the magnitude distribution has a clearly distinct second peak. We explain these qualitatively as follows:

The rotation of the sandpile causes a uniform increase of the local slopes and a preferential increase of absolute column heights in the upper region of the sandpile. The sandpile is thus driven towards its critical shape where relaxation events are triggered locally. These events will be localised ('small') or cooperative ('large'), depending on α and Δ. For strongly consolidating systems with small amplification α, a great deal of excess volume is lost via consolidation, and the effect of surface granular flow is small; in these circumstances, the propagation and buildup of an instability is unlikely, so that events are in general localised, uncorrelated and hence small. This leads to the appearance of the single peak in the distributions in the upper left corner of Fig. 10.3. Alternatively, for weakly consolidating systems with large amplification, surface flow is large, dilatancy predominates, and there are

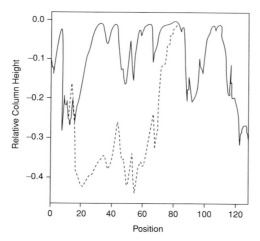

Fig. 10.4 A plot of column heights, relative to their critical heights, for a CML model sandpile with $L = 128$ and parameter values $\delta f = 0.01$, $\alpha = 3$, and $\Delta = 0.85$. The full (dotted) line shows the configuration before (after) a large event.

many space-wasting configurations; this situation favours (see Fig. 10.11) the pre-dominance of large avalanches, which are manifested by the appearance of a second peak in the distributions at the lower right corner of Fig. 10.3. In principle, these large avalanches would be halted by strong configurational inhomogeneities such as a 'dip' on the surface, where the local driving force (10.1) is far below thresh-old; however, simulations show [22] that such configurations are rare in sandpiles that are close to criticality, leading to the second peak in this parameter range.

The results of Fig. 10.3 are reminiscent of those presented [83] in the previous chapter: there the transition from scale invariance (power-law behaviour) to its absence (a second peak corresponding to large avalanches) was explained by the increasing presence of 'evolving disorder' [75]. In this chapter, we specify that the form of this evolving disorder involves physical parameters like inertia and dilatancy [1] – since strongly consolidating materials are by definition weakly dilatant. We will discuss these issues in depth, later in this chapter.

Next, in Fig. 10.4, the relative column heights z_i are plotted against the distance of the column from the axis of rotation. The solid line denotes the configuration before, and the dotted line that after, a large avalanche; we note that a section of the sandpile has 'slipped' quite considerably during the event. Note the similarity with the inset of Fig. 9.1(c), corresponding to the results of an entirely different model.

Figure 10.5 shows a time series of avalanche locations that occur for a model sandpile in the two-peak region. The large events are almost periodic and each one is preceded by many small precursor events; this is reminiscent of Fig. 9.1a in the previous chapter, and fleshes out in space the time-dependent content of that diagram [83]. In addition, it is apparent that large avalanches tend to occur repeatedly at or around the same regions of the sandpile, whose location changes only very slowly compared to the interval between the large avalanches; these correlations in both the positions and the times of large events are often referred to as 'memory' [69, 71, 106]. We provide below an insight, via this model [22], into the nature of configurational memory.

Note that the relaxation function (10.4) is a smooth function of the excess force $f_i - f_{th}$ (which depends on the surface configuration of the pile). If the surface is rough, large values of excess force will be generated; this will in its turn lead to large events.[2] After such large events, configurations will be smoothed out by virtue of the relaxation function Eq. (10.4). These smoother configurations will not generate large forces, reducing the initiation probability of a new event in the same region of the sandpile, until, once again, the whole region is again driven towards its critical configuration, thus generating quasiperiodic events in the time series. In particular, in strongly dilatant material (large Δ) large dips or bumps on the surface (which are able to halt the progress of large avalanches) persist, remaining as significant features (albeit weakened and/or displaced) even after a large event. Quasiperiodic large events can then repeatedly disrupt those regions of the system corresponding to such surface inhomogeneities, especially if grain inertia is high (large α), thus manifesting *configurational memory*. In the opposite parameter regime, when inertia is weak and consolidation effective (small α and Δ), small uncorrelated events occur, which do not mark the surface. For moderate values of α and Δ, of course, both small and large events will be seen (Fig. 10.5).

The shape of the critical sandpile, which we discussed earlier, leads to another interesting feature, namely, an intrinsic size dependence. As mentioned before, the shape is characterised by the length of the decreasing slope region and the constant deviation of column heights from their ideal values in the steady slope region which follows (Fig. 10.2). The length Λ of the decreasing slope region at small i has a finite extent given by

$$\Lambda \sim (1 + \sqrt{1 - \Delta^2})/(1 - \Delta + \sqrt{1 - \Delta^2}).$$

(10.10)

Interestingly, this can be made an arbitrary fraction of the sandpile by an appropriate choice of system size. Such an intrinsic size dependence resulting from competition between surface and bulk relaxation is of particular interest, since it has been observed experimentally [72, 74].

As mentioned in the introductory section, we would expect few events to result in mass exiting the pile, as this model is one in which internal volume reorganisations dominate surface flow. Thus, mass will exit a pile either via the propagation of large

---

[2] Recall that large events here mean extensive reorganisations, rather than surface flow events, as pointed out in the beginning of this chapter.

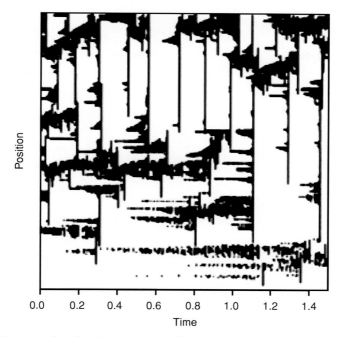

Fig. 10.5 A plot showing the locations of relaxation events (changes in column heights), that occur during an interval of length 1.5 which begins at $t = 10^4$, for a CML model sandpile with $L = 256$ and parameter values $\delta f = 0.01$, $\alpha = 3$ and $\Delta = 0.85$.

events (which occur for α and Δ large) or if surface flow is significant (typically events initiated in the decreasing slope region Λ).

Figure 10.6 shows the logarithm of the exit mass size distribution function, $f \log(m_x)$, for a sandpile of size $L = 128$ with $\delta f = 0.01$, $\alpha = 4$ and $\Delta = 0.95$. While the absolute magnitudes of the event sizes are suppressed in comparison to Fig. 10.3, we see the two-peak behaviour consistent with the corresponding event size distribution function. The large second peak indicates that a significant proportion of the exit mass is due to large events referred to above; also, we have checked that this is the only part of the distribution that survives for larger system sizes, in agreement with the length dependence above. Finally, we mention that the two-peak behaviour obtained for the exit mass distribution of the rotated sandpile model discussed here is in satisfying agreement with that presented in the earlier chapter for the cellular automaton model of a deposited sandpile [75, 83].

10.3.2 Sandpile driven by random deposition

The perturbation more usually encountered in sandpiles is random deposition. We may replace the organized addition (10.8) with the random sequential addition of

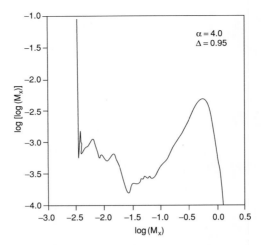

Fig. 10.6 A logarithmic plot of the distribution function of exit mass sizes, f log m_x, for a sandpile of size $L = 128$ with $\delta f = 0.01$, $\alpha = 4$ and $\Delta = 0.95$ for 10^7 consecutive events. The exit mass m_x is the sum of height increments $\delta z'$ that topple from the first column during an event.

height elements, $z_i^+ = z_g$, onto columns $i = 1, \ldots, L$. When the added elements are small compared to the minimum event size, so that $z_g \ll f_{th}\delta f$, random addition is statistically equivalent to uniform addition, which was the case considered in the context of earthquakes [211].

The distribution of event sizes, shown by the full lines (corresponding to $z_g = 0.01$) in Fig. 10.7, is then not markedly distinct from that shown in Fig. 10.3 for the rotational driving force, and both are similar to the distributions presented in [211]. In most of the parameter ranges we consider, the event size distribution functions are size-independent, indicating that intrinsic properties of the sandpile are responsible for their dominant features, which include a second peak representing a preferred scale for large avalanches.

Figure 10.8 shows the corresponding time series of event locations – note that random driving leads to events which are relatively evenly spread over the sandpile and to repeated large events, each preceded by their precursor small events. Note also that after the passage of a large event and/or catastrophe over a region, there is an interval before events are generated in response to the deposition; this underlines the picture referred to earlier, whereby large events leave their signatures on the landscape in the form of dips, for instance. These configurational sinks are associated in the model with forces well below threshold, so that grains deposited on them will, for a while, not cause any relaxation events until the appropriate thresholds are reached. If we now increase the size of the incoming height elements so that they are larger than the minimum event size but are still small compared to

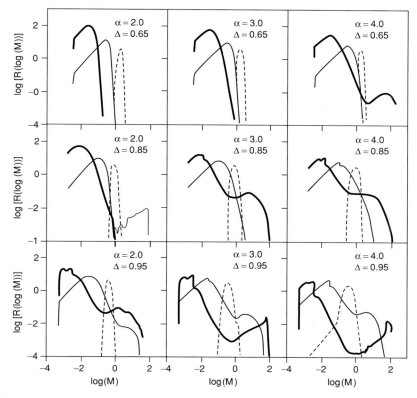

Fig. 10.7 A logarithmic plot of the distribution function of event sizes, $R \log M$ (full lines) for 10^7 consecutive events in a randomly driven CML model sandpile with $L = 512$ and parameter values $\delta f = 0.01$, $\alpha = 2, 3, 4$, $\Delta = 0.6, 0.85, 0.95$ and $z_g = 0.01$. Faint lines show the corresponding distribution functions for $z_g = 0.1$ and 1.0.

the column grain size (i.e. $f_{th}\delta f < z_g < aS_0$), there are two direct consequences. First, the driving force leads to local column height fluctuations $\sim z_g$, so that the surface is no longer smooth; these height fluctuations play the role of additional random barriers which impede the growth of avalanches, thus reducing the probability for large, extended events. Second, given that the added height elements are much larger than the minimum event size, their ability to generate small events is also reduced; the number of small events therefore also decreases. The size distributions for this case consequently have a domed shape with apparently two scaling regions. This case is illustrated by one of the fainter lines in Fig. 10.7 (corresponding to $z_g = 0.1$). Note that for large α and Δ, the large events, being more persistent, are able to overcome the configurational barriers ($\sim z_g$ fluctuations in column heights) referred to above, and we still see a second peak indicating the continued presence of a preferred avalanche size.

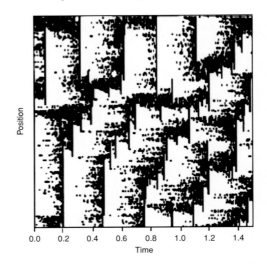

Fig. 10.8 A plot showing the locations of relaxation events (changes in column heights) that occur during an interval of length 1.5 which begins at $t - 10^3$, for a randomly driven CML model sandpile with $L = 256$ and parameter values $\delta f = 0.01$, $\alpha = 3$, $\Delta = 0.85$ and $z_g = 0.01$.

As the perturbation strength becomes even stronger, $(z_g > a S_0)$, so that column height fluctuations are comparable with the column grain size, there are frequent dips on the landscape, which can act as configurational traps for large events. All correlations between events begin to be destroyed and relaxation takes place locally, giving a narrow range of event sizes determined only by the size of the deposited grains. This situation is illustrated by the second faint line in Fig. 10.7, which corresponds to $z_g = 1.0$.

Finally, and for completeness, we link up with the familiar scaling behaviour of lattice sandpiles [65]; as mentioned before, scale invariance is recovered when the driving force is proportional to slope differences alone and no longer contains the second mechanism of compression and/or reorganisation, so that $k_1 = 0$ and $\Delta = 1$. In order, more specifically, to match up with the local and limited $n_f = 2$ model of Kadanoff *et al.* [77], we start with the randomly driven model and

- set $k_1 = 0$ and $\Delta = 1$ in Eq. (10.1);
- set $z_g = 1$;
- choose a relaxation function $f_i - f_i' = f_{th} \delta f$; this is a constant independent of f_i for $f_i > f_{th}$, and in particular contains no amplification;
- set $\delta f = 4.0$, so that each threshold force causes a minimum of two 'grains' to fall onto the next column at every event, so that $n_f = 2$.

This special case of the CML model is then identical with the scale-invariant model of Kadanoff *et al.* [77].

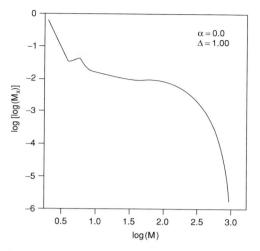

Fig. 10.9 A logarithmic plot of the distribution function of exit mass sizes, f $\log(m_x)$, for 10^7 consecutive events in a randomly driven CML model sandpile with $L = 512$ and parameter values $\delta f = 4$, $\Delta = 1$ and $z_g = 1$.

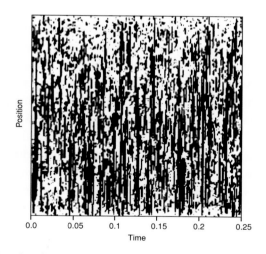

Fig. 10.10 A plot showing the locations of relaxation events (changes in column heights) that occur during an interval of length 0.25 which begins at $t = 10^3$, for a randomly driven CML model sandpile with $L = 256$ and parameter values $\delta f = 4$, $\Delta = 1$ and $z_g = 1$.

Figure 10.9 shows the smooth (scaling) exit mass size distribution in the limit of no dissipation, and the corresponding spatial distribution of scaling events is shown in Fig. 10.10. Uncorrelated events are observed over many sizes, indicating a return to scale invariance.

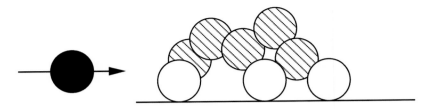

Fig. 10.11 A schematic diagram illustrating the mechanism for large-avalanche formation. When Δ is large, there is a great deal of undissipated volume in the cluster, resulting in the upper (shaded) grains being unstable to small perturbations. When α is large, the black grain hitting the cluster has large inertia so that a large avalanche results when it dislodges the shaded grains.

10.4 Discussion

Here we have provided a decorated lattice model to represent grain and cluster couplings in a sandpile [69]. This coupled map lattice model [22] has two important parameters; α (amplification), which is a measure of inertia, and Δ, which is a measure of dilatancy.

The main result is that for large α and Δ there is a preferred size for large avalanches, which is manifested as a second peak in the distribution of event sizes (Fig. 10.3). In terms of a simple picture this is because, for large α and Δ, grains have enough inertia to speed past available traps, with a large amount of dilatancy – i.e. unrelaxed excess volume – on the surface. Locally, such excess volume can be visualised as being trapped inside a cluster like that in Fig. 10.11. When an oncoming high-inertia grain (the dark grain in the figure) hits such a dilatant cluster, its metastable grains will be knocked off, and a large avalanche unleashed. For small α and Δ we see, by contrast, mainly small events leading to a single peak in Fig. 10.3; we visualise this by imagining slowly moving grains (low inertia) drifting down the surface, locking into voids and dissipating excess volume efficiently. This qualitative picture also indicates that initiated avalanches will be terminated relatively rapidly, leading to many small events.

Also, regions of the sandpile which look like Fig. 10.11 are wiped clean by the effect of large avalanches, so that further deposition or rotation has no effect for a while. However, for large α and Δ, excess void space will be created around the same region after a number of small events have occurred (Fig. 10.5). These spatiotemporal correlations result in a quasiperiodic repetition of large avalanches around the same regions of the sandpile, resulting in configurational memory [67, 68].

We have examined the response of this CML model [22] to random deposition, with particular reference to the size z_g of the deposited grains. Three distinct regimes are observed.

- When z_g is of the order of a 'minimum event', i.e., it is comparable to the smallest fractional change in volume caused by a reorganising grain, the response is similar to that of rotation (Fig. 10.7).
- When z_g is intermediate between the minimum event size and the typical column grain size of the sandpile, reorganisations of grains corresponding to the smallest volume changes are ruled out; on the other hand, there are moderately sized barriers ($\sim z_g$) across the landscape impeding the progress of large events. The appropriate size distribution in Fig. 10.7 has, consequently, a shape which lacks the extreme small and large events of the previous case.
- When z_g is larger than the column grain size of the sandpile, large configurational barriers are generated by deposition, and these act as traps for large events. Correlations between events are destroyed, leading to a narrow distribution of event sizes corresponding to local responses to deposition.

The central analogy between this work and work on spring-block models of earthquakes [210, 211] is that small avalanches build up configurational stress (here, via surfaces which look like Fig. 10.11) which then leads, quasiperiodically, to large, stress-releasing avalanches. We are also able, via this analogy, to model friction and inertia for sandpiles, emulate the stick–slip dynamics that results when grains contact each other during avalanches, and obtain satisfying agreement with both experiments [74] and independent theoretical models [83].

We note that the limit $\Delta = 1$ is a special case; this corresponds to the situation with no reorganisation ($k_1 = 0$) and describes a sandpile which is constantly at an ideal density. The granular driving force no longer has a compressive component and, as for standard sandpile models [65], depends only on height differences. The approach to this limit is also of interest, involving a discontinuous transition to a regime in which the critical sandpile has a constant slope ($S_0 + f_{th}$). The neighbourhood of the limit $\Delta \sim 1$ is a region of very weak dissipation, and, as has been seen in other deterministic nonlinear dynamical systems [51, 52], could well be characterised by complex periodic motion at large times; this has been argued to be especially relevant to models with periodic boundaries [212]. It turns out that for the regions of parameter space explored here, such periodic motion features in sequences containing up to 5×10^7 events [22].

Interesting extensions of this work would include its application to the stratified segregation known [55] to exist when rotating cylinders are shaken, as well as a deeper investigation of the intrinsic layer Λ, with a view to explaining interesting boundary-layer phenomena in sandpiles via lattice models.

11

Coupled continuum equations: the dynamics of sandpile surfaces

11.1 Introduction

11.1.1 Some general remarks

The two previous chapters have dealt, in different guises, with the post-avalanche smoothing of a sandpile which is expected to happen in nature [213]. It is clear what happens physically: an avalanche provides a means of shaving off roughness from the surface of a sandpile by transferring grains from bumps to available voids [22, 69, 83], and thus leaves in its wake a smoother surface. However, surprisingly little research has been done on this phenomenon so far, despite its ubiquity in nature, ranging from snow to rock avalanches.

In particular, what has not attracted enough attention in the literature is the qualitative difference between the situations which obtain when sandpiles exhibit intermittent and continuous avalanches [151]. In this chapter we examine both the latter situations, via coupled continuum equations [95, 96] of sandpile surfaces. These were originally envisaged [69] as the local version of coupled equations that had been written down using global variables in [42]; subsequently, many versions were introduced in the literature [214, 215] to model different situations. The use of these equations has also since been diversified into many areas, including ripple formation [216] and the propagation of sand dunes [217], about which we will have something to say at the end of this chapter.

In order to discuss this, we introduce first the notion that granular dynamics is well described by the competition between the dynamics of grains moving independently of each other and that of their collective motion within clusters [69]. A convenient way of representing this is via coupled continuum equations with a specific coupling between mobile grains ρ and clusters h on the surface of a sandpile [95]. This represents a formal outline of the most general situation of the coupling between

Granular Physics, ed. Anita Mehta. Published by Cambridge University Press. © A. Mehta 2007.

surface and bulk in a sandpile; specific terms can now be modified to model specific scenarios. In general, the complexity of sandpile dynamics leads us to equations which are coupled, nonlinear and noisy: these equations present challenges to the theoretical physicist in more ways than the obvious ones to do with their detailed analysis and/or their numerical solutions.

11.1.2 Sand in rotating cylinders; a paradigm

A particular experimental paradigm that we choose to put the discussions in context is that of sand in rotating cylinders [218]. In the case when sand is rotated slowly in a cylinder, intermittent avalanching is observed; thus sand accumulates in part of the cylinder to beyond its angle of repose [70] and is then released via an avalanche process across the slope. This happens intermittently, since the rotation speed is less than the characteristic time between avalanches. By contrast, when the rotation speed exceeds the time between avalanches, we see continuous avalanching on the sandpile surface. Though this phenomenon has been observed [70] and analysed physically [151] in terms of avalanche statistics, we are not aware of measurements which measure the characteristics of the resulting surface in terms of its smoothness or otherwise. What we focus on here is precisely this aspect, and make predictions for future experiments.

In the regime of intermittent avalanching, we expect that the interface will be the one defined by the 'bare' surface, i.e. the one defined by the relatively immobile clusters across which grains flow intermittently. This then implies that the roughening characteristics of the h profile should be examined. The simplest of the three models we discuss in this chapter (an exactly solvable model referred to hereafter as Case A) as well as the most complex one (referred to hereafter as Case C) treat this situation, where we obtain in both cases an asymptotic smoothing behaviour in h. When on the other hand, there is continuous avalanching, the flowing grains provide an effective film across the bare surface and it is therefore the species ρ which should be analysed for spatial and temporal roughening. In the model hereafter referred to as Case B we look at this situation, and obtain the surprising result of a gradual crossover between purely diffusive behaviour and hypersmooth behaviour. In particular, the analysis of Case C reveals the presence of hidden length scales whose existence was suspected analytically, but not demonstrated numerically in earlier work [95, 219].

The normal procedure for probing temporal and spatial roughening in interface problems is to determine the asymptotic behaviour of the interfacial width with respect to time and space, via the single Fourier transform. Here only one of the variables (x, t) is integrated over in Fourier space, and appropriate scaling relations are invoked to determine the critical exponents which govern this behaviour.

However, it turns out that this leads to ambiguities for those classes of problems where there is an absence of simple scaling, or to be more specific, where multiple length scales exist [220]. In such cases we demonstrate that the double Fourier transform (where *both* time and space are integrated over) yields the correct answers. This point is illustrated by Case A, an exactly solvable model that we introduce; we then use it to understand Case C, a nonlinear model where the analytical results are clearly only approximations to the truth.

11.2 Review of scaling relations for interfacial roughening

In order to make some of these ideas more concrete, we now review some general facts about rough interfaces [221]. Three critical exponents, α, β and z, characterise the spatial and temporal scaling behaviour of a rough interface. They are conveniently defined by considering the (connected) two-point correlation function of the heights,

$$S(x - x', t - t') = \langle h(x, t)h(x', t')\rangle - \langle h(x, t)\rangle\langle h(x', t')\rangle. \qquad (11.1)$$

We have

$$S(x, 0) \sim |x|^{2\alpha} \quad (|x| \to \infty) \quad \text{and} \quad S(0, t) \sim |t|^{2\beta} \quad (|t| \to \infty),$$

and more generally

$$S(x, t) \approx |x|^{2\alpha} F\left(|t|/|x|^z\right)$$

in the whole long-distance scaling regime (x and t large). The scaling function F is universal in the usual sense; α and $z = \alpha/\beta$ are respectively referred to as the roughness exponent and the dynamical exponent of the problem. In addition, we have for the full structure factor which is the double Fourier transform $S(k, \omega)$,

$$S(k, \omega) \sim \omega^{-1} k^{-1-2\alpha} \Phi(\omega/k^z),$$

which gives in the limit of small k and ω,

$$S(k, \omega = 0) \sim k^{-1-2\alpha-z} \quad (k \to 0) \quad \text{and} \quad S(k = 0, \omega) \sim \omega^{-1-2\beta-1/z} \quad (\omega \to 0). \qquad (11.2)$$

The scaling relations for the corresponding single Fourier transforms are

$$S(k, t = 0) \sim k^{-1-2\alpha} \quad (k \to 0) \quad \text{and} \quad S(x = 0, \omega) \sim \omega^{-1-2\beta} \quad (\omega \to 0). \qquad (11.3)$$

In particular, we note that the scaling relations for $S(k, \omega)$ (Eq. (11.2)) always involve the simultaneous presence of α and β, whereas those corresponding to $S(x, \omega)$ and $S(k, t)$ involve these exponents *individually*. Thus, in order to evaluate the double Fourier transforms, we need in each case information from the growing as well as the saturated interface (the former being necessary for β and the latter for α), whereas for the single Fourier transforms, we need only information from the saturated interface for $S(k, t = 0)$ and information from the growing interface for $S(x = 0, \omega)$. On the other hand, the information that we will get out of the double Fourier transform will provide a more unambiguous picture in the case where multiple length scales are present, something which cannot easily be obtained in every case with the single Fourier transform.

In the sections to follow, we present, analyse and discuss the results of Cases A, B and C respectively. We then reflect on the unifying features of these models, and make some educated guesses on the dynamical behaviour of real sandpile surfaces. Finally, we present as an example of the use of these equations, a study of the dynamics of aeolian sand ripples [216].

11.3 Case A: the Edwards–Wilkinson equation with flow

The first model involves a pair of linear coupled equations, where the equation governing the evolution of clusters ('stuck' grains) h is closely related to the very well-known Edwards–Wilkinson (EW) model [85]. The equations are:

$$\frac{\partial h(x, t)}{\partial t} = D_h \nabla^2 h(x, t) + c \nabla h(x, t) + \eta(x, t), \tag{11.4}$$

$$\frac{\partial \rho(x, t)}{\partial t} = D_\rho \nabla^2 \rho(x, t) - c \nabla h(x, t), \tag{11.5}$$

where the first of the equations describes the height $h(x, t)$ of the sandpile surface at (x, t) measured from some mean $\langle h \rangle$, and is precisely the EW equation in the presence of the flow term $c \nabla h$. The second equation describes the evolution of flowing grains, where $\rho(x, t)$ is the local density of such grains at any point (x, t). As usual, the noise $\eta(x, t)$ is taken to be Gaussian, so that:

$$\langle \eta(x, t) \eta(x', t') \rangle = \Delta^2 \delta(x - x') \delta(t - t'),$$

with Δ the strength of the noise. Here, $\langle \cdots \rangle$ refers to an average over space as well as over noise.

11.3.1 Analysis of the decoupled equation in h

For the purposes of analysis, we focus on the first of the two coupled equations (Eq. (11.4)) presented above,

$$\frac{\partial h}{\partial t} = D_h \nabla^2 h + c \nabla h + \eta(x, t),$$

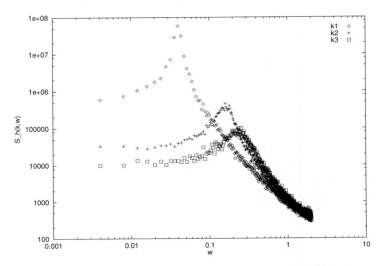

Fig. 11.1 The correlation function $S_h(k_i, \omega)$ against ω for three different wavevectors $k_1 = 0.02(\diamondsuit), k_2 = 0.08(+)$ and $k_3 = 0.12(\square)$ with parameters $c = 2.0, D_h = 1.0$ and $\Delta^2 = 1.0$. The positions of the peaks are given by $\omega_1 = 0.04, \omega_2 = 0.16$ and $\omega_3 = 0.24$, as expected from Eq. (11.6).

noting that this equation is essentially decoupled from the second. (This statement is, however, not true in reverse, which has implications to be discussed later.) We note that this is entirely equivalent to the Edwards–Wilkinson equation [85] in a frame moving with velocity c,

$$x' = x + ct, \quad t' = t,$$

and would on these grounds expect to find only the well-known EW exponents $\alpha = 0.5$ and $\beta = 0.25$ [85]. This would be verified by naive single Fourier transform analysis of Eq. (11.4), which yields these exponents via Eq. (11.3).

Equation (11.4) can be solved exactly as follows. The propagator $G(k, \omega)$ is

$$G_h(k, \omega) = (-i\omega + D_h k^2 + ikc)^{-1}.$$

This can be used to evaluate the structure factor

$$S_h(k, \omega) = \frac{\langle h(k, \omega)h(k', \omega')\rangle}{\delta(k + k')\delta(\omega + \omega')}.$$

which is the Fourier transform of the full correlation function $S_h(x - x', t - t')$ defined by Eq. (11.1). The solution for $S_h(k, \omega)$ so obtained is:

$$S_h(k, \omega) = \frac{\Delta^2}{(\omega - ck)^2 + D_h^2 k^4}. \tag{11.6}$$

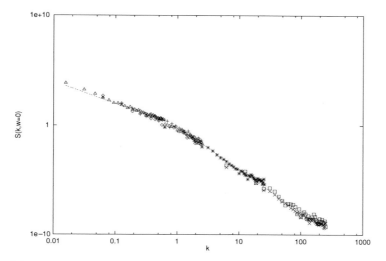

Fig. 11.2 The double Fourier transform, $S(k, \omega = 0)$, obtained from Eq. (11.4) (Case A) for the h–h correlation function, showing the crossover from high to low k. The different markers in the figure correspond to different grid sizes Δx to sample distinct regions of k space; thus the markers $\triangle$, $\times$ and $\square$ correspond to decreasing grid sizes and increasing wavevector ranges. The parameters used in the calculation are $c = D_h = \Delta^2 = 1.0$ and the characteristic wavevector is $k_0 = c/D_h = 1.0$. The dashed line is a plot of $S_h(k, \omega = 0)$ vs k for Case A with appropriate parameters, to serve as a guide to the eye.

This is illustrated in Fig. 11.1, while representative graphs for $S_h(k, \omega = 0)$ and $S_h(k = 0, \omega)$ are presented in Figs. 11.2 and 11.3 respectively.

It is obvious from Eq. (11.6) that $S_h(k, \omega)$ does not show simple scaling. More explicitly, if we write

$$S_h^{-1}(k, \omega = 0) = \frac{\omega_0^2}{\Delta^2} \left(\frac{k}{k_0}\right)^2 \left[1 + \left(\frac{k}{k_0}\right)^2\right]$$

with $k_0 = c/D_h$, and $\omega_0 = c^2/D_h$, we see that there are two limiting cases:

- for $k \gg k_0$, $S_h^{-1}(k, \omega = 0) \sim k^4$; using again $S_h^{-1}(k = 0, \omega) \sim \omega^2$, we obtain $\alpha_h = 1/2$ and $\beta_h = 1/4$, $z_h = 2$ via Eqs. (11.2).
- for $k \ll k_0$, $S_h^{-1}(k, \omega = 0) \sim k^2$; using the fact that the limit $S_h^{-1}(k = 0, \omega)$ is always ω^2, this is consistent with the set of exponents $\alpha_h = 0$, $\beta_h = 0$ and $z_h = 1$ via Eqs. (11.2).

The first of these contains no surprises, being the normal EW fixed point [85], while the second represents a new 'smoothing' fixed point.

We now explain this smoothing fixed point via a simple physical picture. The competition between the two terms in Eq. (11.4) determines the nature of the fixed point observed: when the diffusive term dominates the flow term, the canonical EW

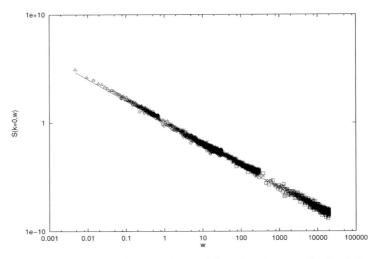

Fig. 11.3   The double Fourier transform, $S(k = 0, \omega)$, vs ω obtained from Eq. (11.4) (Case A) for the h–h correlation function. The different markers in the figure correspond to different grid sizes Δt to sample distinct regions of ω space; thus the markers $\triangle$, $\times$ and $\square$ correspond to decreasing grid sizes and increasing frequency ranges. The solid line is a plot of $S_h(k = 0, \omega)$ vs ω for Case A with appropriate parameters, to serve as a guide to the eye. The parameters are $c = D_h = \Delta^2 = 1.0$.

fixed point is obtained, in the limit of large wavevectors k. On the contrary, when the flow term predominates, the effect of diffusion is suppressed by that of a travelling wave whose net result is to penalise large slopes; this leads to the smoothing fixed point obtained in the case of small wavevectors k. We emphasise, however, that this is a toy model of smoothing, which will be used to illuminate the discussion of models B and C below.

11.3.2 Some caveats

We realise from the above that the interface h is smoothed because of the action of the flow term which penalises the sustenance of finite gradients ∇h in Eq. (11.4). However, Eq. (11.4) is effectively decoupled from Eq. (11.5), while Eq. (11.5) is manifestly coupled to Eq. (11.4). In order for the coupled Eqs. (11.4) to qualify as a valid model of sandpile dynamics, we would need to ensure that no instabilities are generated in either of these by the coupling term $c\nabla h$.

In this spirit, we look first at the value of ρ averaged over the sandpile, as a function of time (Fig. 11.4a). We observe that the incursions of $\langle \rho \rangle$ into negative values are limited to relatively small values, suggesting that the addition of a constant background of ρ exceeding this negative value would render the coupled system meaningful, at least to a first approximation. In order to ensure that

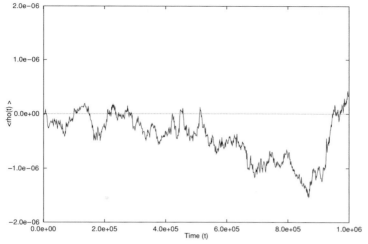

(a) Variation of $\langle\rho(t)\rangle$ with time t. Here $\langle\rho(t)\rangle$ is the average over the sandpile surface of 100 sample configurations.

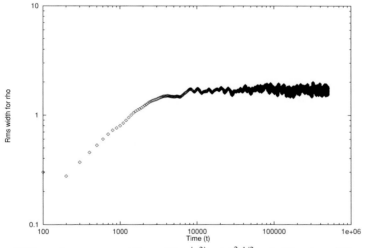

(b) The root mean square width $\rho_{\text{rms}}(t) = \left(\langle\rho^2\rangle - \langle\rho\rangle^2\right)^{1/2}$ against time t over 100 sample configurations.

Fig. 11.4 Statistical behaviour of density as a function of time. The grid size $\Delta t = 0.005$ and $c = \Delta^2 = D_h = 1.0$.

this average does not involve wild fluctuations, we examine the fluctuations in ρ, viz. $\sqrt{\langle\rho^2\rangle - \langle\rho\rangle^2}$ (Fig. 11.4b). The trends in that figure indicate that this quantity appears to saturate, at least up to computationally accessible times. Finally we look at the *minimum* and *maximum* value of ρ at any point in the pile over a large range of times (Fig. 11.4c); this appears to be bounded by a modest (negative) value of

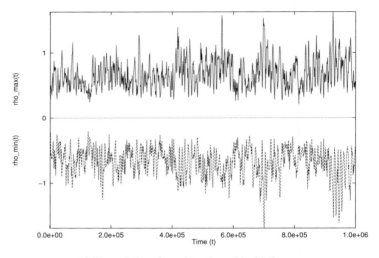

(c) The variation of $\rho_{\max}(t)$ and $\rho_{\min}(t)$ with time t.

Fig. 11.4 (*cont.*)

'bare' ρ. Our conclusions are thus that the fluctuations in ρ saturate at computation-ally accessible times and that the negativity of the fluctuations in ρ can always be handled by starting with a constant ρ_0, a constant 'background' of flowing grains, which is more positive than the largest negative fluctuation.

Physically, then, the above implies that, at least in the presence of a constant large density ρ_0 of flowing grains, it is possible to induce the level of smoothing corresponding to the fixed point $\alpha = \beta = 0$. This model is thus one of the simplest possible ways in which one can obtain a representation of the smoothing of the 'bare surface' that is frequently observed in experiments on real sandpiles after intermittent avalanche propagation [213].

11.4 Case B: when moving grains abound

These model equations, first presented in [95], involve a simple coupling between the species h and ρ, where the transfer between the species occurs only in the presence of the flowing grains and is therefore relevant to the regime of continuous avalanching when the duration of the avalanches is *large* compared to the time between them. The equations are:

$$\frac{\partial h(x, t)}{\partial t} = D_h \nabla^2 h(x, t) - T(h, \rho) + \eta_h(x, t), \tag{11.7}$$

$$\frac{\partial \rho(x, t)}{\partial t} = D_\rho \nabla^2 \rho(x, t) + T(h, \rho) + \eta_\rho(x, t), \tag{11.8}$$

$$T(h, \rho) = -\mu \rho (\nabla h), \tag{11.9}$$

where the terms $\eta_h(x, t)$ and $\eta_\rho(x, t)$ represent Gaussian white noise as usual:

$$\langle \eta_h(x, t)\eta_h(x', t')\rangle = \Delta_h^2 \delta(x - x')\delta(t - t'),$$
$$\langle \eta_\rho(x, t)\eta_\rho(x', t')\rangle = \Delta_\rho^2 \delta(x - x')\delta(t - t'),$$

and the $\langle \cdots \rangle$ stands for average over space as well as noise.

A simple physical picture of the coupling or 'transfer' term $T(h, \rho)$ between h and ρ is the following: flowing grains are added in proportion to their local density to regions of the interface which are at less than the critical slope, and vice versa, *provided that the local density of flowing grains is always nonzero.* This form of interaction becomes zero in the absence of a finite density of flowing grains ρ (when the equations become decoupled) and is thus the simplest form appropriate to the situation of continuous avalanching in sandpiles. We analyse in the following the profiles of h and ρ consequent on this form.

It turns out that a singularity discovered by Edwards [222] three decades ago in the context of fluid turbulence is present in models with a particular form of the transfer term T; the above is one example, while another example is the model due to Bouchaud *et al.* (BCRE) [214], where

$$T = -\nu\nabla h - \mu\rho(\nabla h)$$

and the noise is present only in the equation of motion for h. This singularity, the so-called infrared divergence, largely controls the dynamics and produces unexpected exponents.

11.4.1 Numerical analysis

We focus now on the numerical results for Case B. The coupled equations in this section and the following one were numerically integrated using the method of finite differences. Grids in time and space were kept [96] as fine-grained as computational constraints allowed so that the grid size in space Δx was chosen to be in the range $(0.1, 0.5)$, whereas that in time was in the range Δt $(0.001, 0.005)$. Thus the instabilities associated with the discretisation of nonlinear continuum equations were avoided and convergence was checked by keeping Δt small enough such that the quantities under investigation were independent of further discretisation. These results were also checked for finite size effects. In the calculations of this section $D_h = D_\rho = 1.0$ and $\mu = 1$, with the results being averaged over several independent configurations. The exponents α and β and the corresponding error bars were calculated from the slopes of the fitted straight lines, $-(1 + 2\beta)$ and $-(1 + 2\alpha)$ respectively.

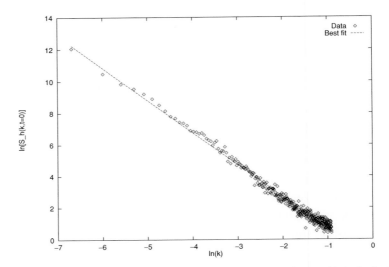

Fig. 11.5 Log–log plot of the single Fourier transform $S_h(k, t = 0)$ vs k obtained from Eqs. (11.7)–(11.9) (Case B). The best fit has a slope of $-1 - 2\alpha_h = -2.03 \pm 0.014$.

On discretising Eqs. (11.7)–(11.9), the divergences that were previously observed in [95] were found once again. These have since become a field of study in their own right [223]. These divergences are a direct representation of the infrared divergence mentioned above, and we follow here a parallel course to [95] in regulating these via an explicit regulator, replacing the function $\mu\rho\nabla h$ by the following:

$$
\begin{aligned}
T &= +1 & &\text{for} & \mu\rho(\nabla h) &> 1, \\
&= \mu\rho(\nabla h) & &\text{for} & -1 \le \mu\rho(\nabla h) &\le 1, \\
&= -1 & &\text{for} & \mu\rho(\nabla h) &< -1.
\end{aligned}
$$

The Fourier transform $S_h(k, t = 0)$ (Fig. 11.5) is consistent with a spatial roughening exponent $\alpha_h \sim 0.501 \pm 0.007$ via the observation of

$$S_h(k, t = 0) \sim k^{-2.03\pm0.014},$$

and the Fourier transform $S_h(x = 0, \omega)$ (Fig. 11.6) is consistent with a temporal roughening exponent $\beta_h \sim 0.465 \pm 0.008$ via the observation of

$$S_h(x = 0, \omega) \sim \omega^{-1.93\pm0.017}.$$

Hence the value $z_h \sim 1.07$ is obtained.

The full structure factor $S_h(k, \omega)$ has been calculated at two different k points and Fig. 11.7 displays the results. The solid and dashed lines in Fig. 11.7 are plots for $k = 0.1$ and $k = 0.2$ with $\Gamma_0 = 0.4$ and 0.5 respectively. The spatial structure

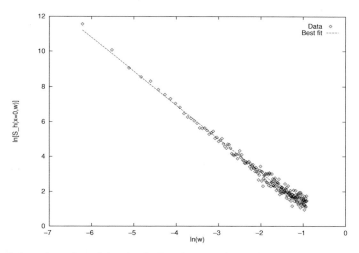

Fig. 11.6 Log–log plot of the single Fourier transform $S_h(x = 0, \omega)$ vs ω for Case B obtained from Eqs. (11.7)–(11.9). The best fit shown in the figure has a slope of $-1 - 2\alpha_h = 1.93 \pm 0.017$.

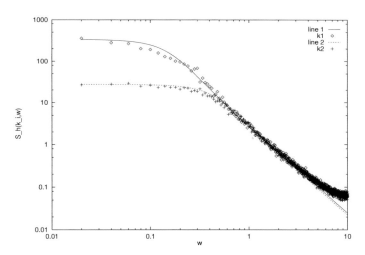

Fig. 11.7 The double Fourier transform $S_h(k_i, \omega)$ vs ω (Case B) calculated at two different wavevectors $k_i = 0.1(\diamondsuit), 0.2(+)$. The solid curves are theoretical estimates (computed in Ref. [96]) of the solutions, meant as a guide to the eye.

factor $S_h(k, \omega = 0)$ shows a power-law behaviour (Fig. 11.8) given by

$$S_h(k, \omega = 0) \sim k^{-3.40 \pm .029},$$

and the temporal structure factor $S_h(k = 0, \omega)$ shows a power-law behaviour (Fig. 11.9) given by

$$S_h(k = 0, \omega) \sim \omega^{-1.91 \pm .017}.$$

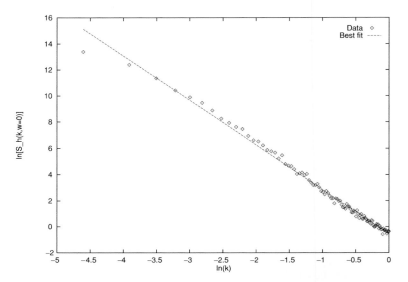

Fig. 11.8 Log–log plot of the double Fourier transform $S_h(k, \omega = 0)$ vs k (Case B) obtained from Eqs. (11.7)–(11.9). The best fit has a slope of $-(1 + 2\alpha_h + z_h) = -3.40 \pm 0.029$.

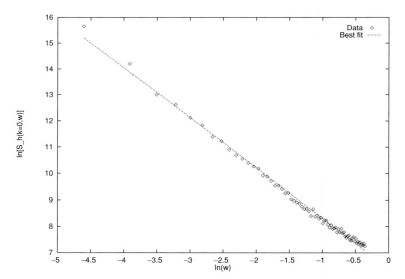

Fig. 11.9 Log–log plot of the double Fourier transform $S_h(k = 0, \omega)$ vs ω obtained from Eqs. (11.7)–(11.9) (Case B). The best fit displayed in the figure has a slope of $-(1 + 2\beta_h + 1/z_h) = -1.91 \pm 0.017$.

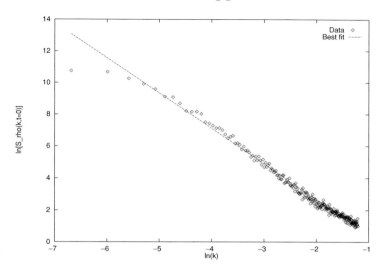

Fig. 11.10 Log–log plot of the single Fourier transform $S_\rho(k, t = 0)$ vs k (Case B) showing a crossover from a slope of $-1 - 2\alpha_\rho = 0$ at small k to -2.12 ± 0.017 at large k.

The single Fourier transform $S_\rho(k, t = 0)$ (Fig. 11.10) shows a crossover behaviour from

$$S_\rho(k, t = 0) \sim k^{-2.12 \pm 0.017}$$

for large wavevectors to

$$S_\rho(k, t = 0) \sim \text{constant}$$

as $k \to 0$. Note, however, that the simulations manifest, in addition to the above, the normal diffusive behaviour represented by $\alpha_\rho = 0.56$ at large wavevectors. The single Fourier transform in time $S_\rho(x = 0, \omega)$ (Fig. 11.11) shows a power-law behaviour

$$S_\rho(x = 0, \omega) \sim \omega^{-1.81 \pm 0.017}.$$

While the range of wavevectors in Fig. 11.10 over which crossover in $S_\rho(k, t = 0)$ is observed was restricted by the computational constraints [96], the form of the crossover appears conclusive. Checks (with fewer averages) over larger system sizes revealed the same trend.

11.4.2 Homing in on the physics: a discussion of smoothing in Case B

We focus in this section on the physics of the equations and the results. In the regime of continuous avalanching in sandpiles, the major dynamical mechanism is that of mobile grains ρ flowing into voids in the h landscape as well as the converse process of unstable clusters (a surfeit of ∇h above some critical value) becoming

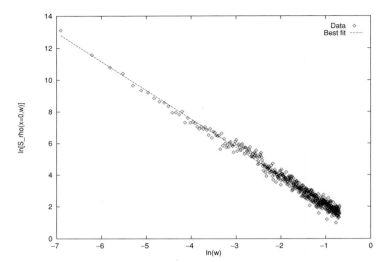

Fig. 11.11 Log–log plot of the single Fourier transform $S_\rho(x = 0, \omega)$ vs ω obtained from Eqs. (11.7)–(11.9) (Case B). The best fit has a slope of $-1 - 2\beta_\rho = -1.81 \pm 0.017$.

destabilised and adding to the avalanches. Results [96] for the critical exponents in h indicate no further spatial smoothing beyond the diffusive; however, those in the species ρ indicate a crossover from purely diffusive to an asymptotic hypersmooth behaviour. Thus, the claim for continuous avalanching is as follows.

Flowing grains play the major dynamical role, as all exchange between h and ρ takes place only in the presence of ρ. These flowing grains distribute themselves over the surface, filling in voids in proportion both to their local density and to the depth of the local voids. It is this distribution process that leads in the end to a strongly smoothed profile in ρ. Additionally, since in the regime of continuous avalanching, the effective interface is defined by the profile of the *flowing* grains, it is this profile that will be measured experimentally for, say, a rotating cylinder with high velocity of rotation.

11.5 Case C: tilt combined with flowing grains

The last case we discuss in this part of the chapter involves a more complex coupling [95, 96] between the stuck grains h and the flowing grains ρ as follows:

$$\frac{\partial h(x, t)}{\partial t} = D_h \nabla^2 h(x, t) - T + \eta(x, t), \tag{11.10}$$

$$\frac{\partial \rho(x, t)}{\partial t} = D_\rho \nabla^2 \rho(x, t) + T, \tag{11.11}$$

$$T(h, \rho) = -v(\nabla h)_- - \lambda\rho(\nabla h)_+, \tag{11.12}$$

with $\eta(x, t)$ representing white noise as usual.

Here,

$$z_+ = z \quad \text{for} \quad z > 0,$$
$$= 0 \quad \text{otherwise;} \tag{11.13}$$
$$z_- = z \quad \text{for} \quad z < 0,$$
$$= 0 \quad \text{otherwise.} \tag{11.14}$$

The two terms in the transfer term T represent two different physical effects which we will discuss in turn. The first term represents the effect of tilt, in that it models the transfer of particles from the boundary layer at the 'stuck' interface to the flowing species whenever the local slope is steeper than some threshold (in this case zero, so that negative slopes are penalised). The second term is restorative in its effect, in that in the presence of 'dips' in the interface (regions where the slope is shallower, i.e. more positive than the zero threshold used in these equations), the flowing grains have a chance to resettle on the surface and replenish the boundary layer [69]. We notice that because one of the terms in T is independent of ρ we are no longer restricted to a coupling which exists only in the presence of flowing grains: i.e. this model is applicable to intermittent flows when ρ may or may not always exist on the surface. In the following we examine the effect of this interaction on the profiles of h and ρ respectively.

The complexity of the transfer term with its discontinuous functions precludes any attempts to solve this model analytically. Numerical solutions are presented and analysed in the subsections that follow.

11.5.1 Results for the single Fourier transforms

The single Fourier transforms $S_h(k, t = 0)$ (Fig. 11.12) and $S_h(x = 0, \omega)$ (Fig. 11.13) show power-law behaviour corresponding to

$$S_h(k, t = 0) \sim k^{-2.56 \pm 0.060},$$
$$S_h(x = 0, \omega) \sim \omega^{-1.68 \pm 0.011},$$

which implies that the roughness and growth exponents are given by, respectively, $\alpha_h = 0.78 \pm 0.030$ and $\beta_h = 0.34 \pm 0.005$. This suggests $z_h = \alpha_h/\beta_h \approx 2$. However, the small k limit of $S_h(k, t = 0)$ indicates a downward curvature and thus a deviation from the linear behaviour at higher k (Fig. 11.12). This curvature, which had also been observed in previous work [95], indicates a smaller roughness exponent α_h there, i.e. an asymptotic *smoothing*. In the light of current knowledge about anomalous ageing [220], where two-time correlation functions turn out to be crucial, we therefore turn to an investigation of the double Fourier transforms.

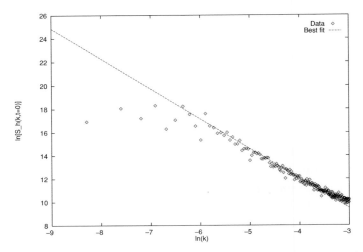

Fig. 11.12 Log–log plot of the single Fourier transform $S_h(k, t = 0)$ vs k for Case C. The slope of the fitted line is given by $-1 - 2\alpha_h = -2.56 \pm 0.060$.

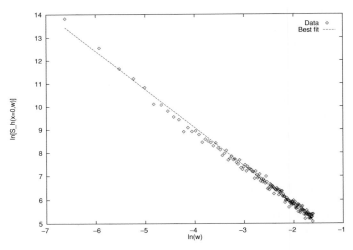

Fig. 11.13 Log–log plot of the single Fourier transform $S_h(x = 0, \omega)$ vs ω for Case C. The best fit has a slope of $-1 - 2\beta_h = -1.68 \pm 0.011$.

11.5.2 Results for the double Fourier transforms

The double Fourier transforms $S_h(k, \omega = 0)$ (Fig. 11.14) and $S_h(k = 0, \omega)$ (Fig. 11.15) show power-law behaviour corresponding to

$$S_h(k = 0, \omega) \sim \omega^{-1.80 \pm 0.007},$$

$$S_h(k, \omega = 0) \sim k^{-4.54 \pm 0.081} \quad \text{for large wavevectors,}$$

$$\sim \text{constant} \quad \text{for small wavevectors.}$$

The double Fourier transform $S_h(k = 0, \omega)$ shows the usual ω^{-2} behaviour [96].

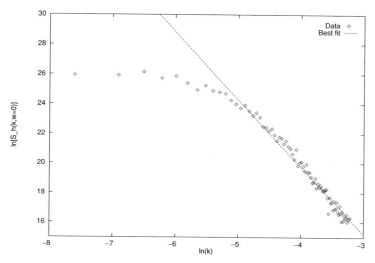

Fig. 11.14 Log–log plot of the double Fourier transform $S_h(k, \omega = 0)$ vs k obtained for Case C. The best fit for high wavevector has a slope of $-(1 + 2\alpha_h + z_h) = -4.54 \pm 0.081$. As $k \to 0$ we observe a crossover to a slope of zero.

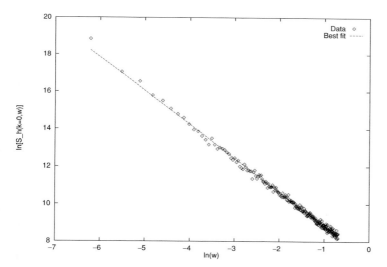

Fig. 11.15 Log–log plot of the double Fourier transform $S_h(k = 0, \omega)$ vs ω obtained for Case C. The best fitted line shown in the figure has a slope of $-(1 + 2\beta_h + 1/z_h) = -1.80 \pm 0.007$.

The structure factor $S_h(k, \omega = 0)$ signals a dramatic behaviour of the roughening exponent α_h, which crosses over from

- a value of 1.3 indicating anomalously large roughening at intermediate wavevectors, to
- a value of about -1 for small wavevectors indicating asymptotic hypersmoothing.

The anomalous roughening $\alpha_h \sim 1$ seen here is consistent with that observed via the single Fourier transform (Fig. 11.12) and suggests, via perturbative

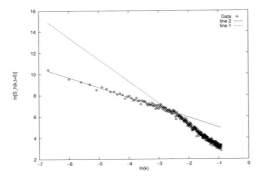

Fig. 11.16 Log–log plot of the single Fourier transform $S_h(k, t = 0)$ vs k obtained from the mean-field equations. The high k region is fitted with a line of slope $-1 - 2\alpha_h = -2.05 \pm 0.017$. The low k region is fitted with a line of slope $-1 - 2\alpha_h = -0.93 \pm 0.024$. Note the crossover from $\alpha_h = 0.5$ at large k to zero at small k.

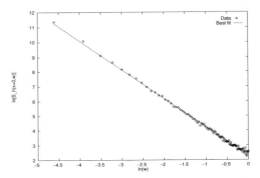

Fig. 11.17 Log–log plot of the single Fourier transform $S_h(x = 0, \omega)$ vs ω for the mean-field equations. The best fit has a slope of $-1 - 2\beta_h = -1.94 \pm 0.001$.

arguments [96] that $z_h = 1$. The anomalous smoothing obtained here ($\alpha_h \sim -1$) is also consistent with the downward curvature in the single Fourier transform $S_h(k, t = 0)$, as both imply a negative α_h; we mention also that the wavevector regime where this smoothing is manifested is almost identical in both Figs. 11.12 and 11.14.

The mean-field equations corresponding to Case C (which turn out to be identical to the so-called BCRE equations [214] have also been solved numerically [96]; from Fig. 11.16 and Fig. 11.17 we find that there is a crossover in $S_h(k, t = 0)$ (Fig. 11.16) from a diffusive behaviour ($z_h = 2$) at high wavevectors to a smoothing behaviour at low wavevectors, also in this case.

This behaviour is reflected in the results for Case C. At low frequencies the region of anomalous smoothing can be understood by comparison with the corresponding region in the mean-field equations which also manifest this. At large k, $S_h(k, t = 0)$ and $S_h(k, \omega = 0)$ indicate anomalous roughening with $\alpha_h \approx z_h \approx 1$, which is consistent with infrared divergence. However, as in Case A, $S_h(x = 0, \omega)$, there is also a strong presence of the the diffusive mechanism, $z_h = 2$ [96]. The presence of these *two* dynamical exponents ($z_h = 1$ and $z_h = 2$) in the problem suggests that the present model is an integrated version of the earlier two, reducing to their behaviour in different wavevector regimes, as is set out in more detail elsewhere [96].

11.6 Discussion

We have presented in the above a discussion of three models of sandpiles, all of which manifest asymptotic smoothing: Cases A and C manifest this in the species h of stuck grains, while Case B manifests this in the species ρ of flowing grains. We reiterate that the fundamental physical reason for this is the following: Cases A and C both contain couplings which are independent of the density ρ of flowing grains, and are thus applicable, for instance, to the dynamical regime of intermittent avalanching in sandpiles, when grains occasionally (but not always) flow across the 'bare' surface. In Case B, by contrast, the equations are coupled only when there is continuous avalanching, i.e. in the presence of a finite density ρ of flowing grains.

The analysis of Case A is straightforward, and was undertaken really only to explain features of the more complex Case C; that of Case B shows satisfactory agreement between perturbative analysis [96] and simulations. Anomalies persist, however, when such a comparison is made in Case C, because the discontinuous nature of the transfer term makes it analytically intractable. These are removed when the analysis includes a mean-field solution [96] which is able to reproduce the asymptotic smoothing observed.

We suggest therefore an experiment where the critical roughening exponents of a sandpile surface are measured in

(i) a rapidly rotated cylinder, in which the time between avalanches is much less than the avalanche duration. The results presented here predict that for small system sizes we will see only diffusive smoothing, but that for large enough systems, we will see extremely smooth surfaces.

(ii) a slowly rotated cylinder where the time between avalanches is much more than the avalanche duration. In this regime, the results of Case C make a fascinating prediction: anomalously large spatial roughening for moderate system sizes crossing over to an anomalously large spatial smoothing for large systems.

Finally, we make some speculations in this context concerning natural phenomena. The qualitative behaviour of blown sand dunes [5, 6, 224] is in accord with the results of Case B, because sand moves swiftly and virtually continuously across their surface in the presence of wind. By contrast, on the surface of a glacier, we might expect the sluggish motion of boulders to result in intermittent flow across the surface, making the results of Case C more applicable to this situation. It would be interesting to see if the predictions of anomalous roughening at moderate, and anomalous smoothing at large, length scales is applicable here.

11.7 A more complicated example: the formation of ripples

In this last section, we use the general formalism of the continuum equations first presented in [69] to look at a very different problem, that of Aeolian sand ripples. These are formed by the action of the wind on the sand bed in the desert and at the seashore. They have also recently been observed on Mars [225]. Aeolian ripples are a few centimetres in wavelength and their crests lie perpendicular to the prevailing wind direction. Bagnold (1941) made an influential early study [224]. He identified the importance of saltation, where sand grains are entrained by the wind, and whipped along the sand bed, colliding with it at high speed (of order ≤ 1m/s [226]), and causing other grains to jump out of and along the bed, thus sculpting ripples. The impact angle remains roughly constant at about $10°$–$16°$ despite the gusting of the wind. The stoss slope of the ripple lies in the range $8°$–$10°$ [227]. The lee slope is composed of a short straight section near the crest at an angle of about $30°$–$34°$ to the horizontal [224] [227] [228], followed by a longer and shallower concave section. The deposition of grains on the lee slope of sand dunes leads to oversteepening and avalanching [229] [230], which maintains the lee slope at an angle of around $30°$–$34°$ near the crest. A similar, if less dramatic, mechanism is likely to hold for ripples.

Numerical simulations of ripples and dunes based on tracking individual sand grains or on cellular automata have in the past few years yielded good qualitative agreement with observations [231]. However, these methods are computationally expensive. Continuum models provide a complementary approach allowing faster calculation of ripple evolution, and the possibility of obtaining certain information, such as the preferred ripple wavelength, analytically. Anderson [232] produced an analytical model of the initial generation of ripples from a flat bed. A one-species analytical continuum model was formulated by Hoyle and Woods [233]. Here, we present work [216] which extends the one-species model [233] to include relaxation effects, inspired by models of sandpiles [95, 96], in order to obtain more realistic ripple profiles and to predict the ripple wavelength and speed.

We consider two-dimensional sand ripples: in the spirit of the previous section [69, 95, 96] we assume that the effective surface of the ripple comprises a 'bare' surface defined by the local heights of clusters $h(x, t)$ sheathed by a thin layer of flowing mobile grains whose local density is $\rho(x, t)$, with x the horizontal coordinate and t time. The ripple evolves under the influence of two distinct types of mechanism. Firstly, the impact of a constant flux of saltating grains knocks grains out of the 'bare' surface, causing them to hop along the ripple surface and land in the layer of flowing grains. This is the underlying cause of ripple formation. Secondly, the ripples are subject to intracluster and intercluster granular relaxation mechanisms which result in a smoothing of the surface.

Following [233], we consider grains to be bounced out of the 'bare' surface by a constant incoming saltation flux, which impacts the sand bed at an angle β to the horizontal. These hopping or 'reptating' [234] grains subsequently land in the flowing layer. The saltating grains are highly energetic and continue in saltation upon rebounding from the sand bed. We assume that the number $N(x, t)$ of sand grains ejected per unit time, per unit length of the sand bed, is proportional to the component of the saltation flux perpendicular to the sand surface, giving

$$N(x, t) = J \sin(\alpha + \beta) = \frac{J (\sin \beta + h_x \cos \beta)}{(1 + h_x^2)^{1/2}}, \tag{11.15}$$

where $\alpha = \tan^{-1}(h_x)$, and J is a positive constant of proportionality. We assume that each sand grain ejected from the surface hops a horizontal distance a, with probability $p(a)$, and then lands in the flowing layer. We consider the flight of each sand grain to take place instantaneously, since ripples evolve on a much slower timescale than that of a hop. The hop length distribution $p(a)$ can be measured experimentally [234, 235], so we consider it to be given empirically. It is possible that $p(a)$ and hence the mean and variance of hop lengths could depend upon factors such as wind speed. The number $\delta n_o(x, t)$ of sand grains leaving the surface between positions x and $x + \delta x$ in time δt, where δx and δt are infinitesimal, is given by $\delta n_o(x, t) = N(x, t)\delta x \delta t$. The change δh in the surface height satisfies $\delta x \delta h(x, t) = -a_p \delta n_o(x, t)$, where a_p is the average cross-sectional area of a sand grain. In the limit $\delta t \to 0$ we find that the contribution to the evolution equation for $h(x, t)$ from hopping alone is

$$h_t = -a_p N(x, t) = -a_p J \frac{\sin \beta + h_x \cos \beta}{(1 + h_x^2)^{1/2}}. \tag{11.16}$$

There may be regions on the sand bed which are shielded from the incoming saltation flux by higher relief upwind. In these regions there will be no grains bounced out of the surface, and there will be no contribution to the h_t equation from hopping. The number $\delta n_i(x, t)$ of sand grains arriving on the layer of flowing grains between positions x and $x + \delta x$ in time δt is given by

$$\delta n_i(x, t) = \int_{-\infty}^{+\infty} p(a) N(x - a, t) da \, \delta x \, \delta t. \tag{11.17}$$

The change in depth of the flowing layer satisfies $\delta x \delta \rho(x, t) = a_p \delta n_i(x, t)$, and hence the contribution to the evolution equation for the flowing layer depth from hopping alone is

$$\rho_t = a_p J \int_{-\infty}^{+\infty} p(a) \frac{\sin \beta + h_x(x - a, t) \cos \beta}{\left(1 + h_x^2(x - a, t)\right)^{1/2}} da. \tag{11.18}$$

We incorporate diffusive motion [85] as well as processes governing the transfer between flowing grains and clusters, following [95, 96], leading to the equations [216]:

$$h_t = D_h h_{xx} - T(x, t) - a_p J \frac{\sin \beta + h_x \cos \beta}{\left(1 + h_x^2\right)^{1/2}}, \tag{11.19}$$

$$\rho_t = D_\rho \rho_{xx} + \chi(\rho h_x)_x + T(x, t)$$
$$+ a_p J \int_{-\infty}^{+\infty} p(a) \frac{\sin \beta + h_x(x - a, t) \cos \beta}{\left(1 + h_x^2(x - a, t)\right)^{1/2}} da, \tag{11.20}$$

where D_h, D_ρ and χ are positive constants and where $T(x, t)$, which represents the transfer terms, is given by

$$T(x, t) = -\kappa \rho h_{xx} + \lambda \rho(|h_x| - \tan \alpha), \tag{11.21}$$

for $0 \le |h_x| \le \tan \alpha$ and by

$$T(x, t) = -\kappa \rho h_{xx} + \frac{\nu(|h_x| - \tan \alpha)}{\left(\tan^2 \gamma - h_x^2\right)^{1/2}}, \tag{11.22}$$

for $\tan \alpha \le |h_x| < \tan \gamma$, with κ, λ and ν also positive constants. The term $D_h h_{xx}$ represents the diffusive rearrangement of clusters while the term $D_\rho \rho_{xx}$ represents the diffusion of the flowing grains. The flux-divergence term $\chi(\rho h_x)_x$ models the flow of surface grains under gravity. The current of grains is assumed proportional to the number of flowing grains and to their velocity, which in turn is proportional to the local slope to leading order [233]. The $-\kappa \rho h_{xx}$ term represents the inertial filling in of dips and knocking out of bumps on the 'bare' surface caused by rolling grains flowing over the top. The $\lambda \rho(|h_x| - \tan \alpha)$ term represents the tendency of flowing grains to stick onto the ripple surface at small slopes; it is meant to model the accumulation of slowly flowing grains at an obstacle. Clearly, for this to happen the obstacle must be stable, or else it would be knocked off by the oncoming grains, so that this term only comes into play for slopes less than $\tan \alpha$, where α is the *angle of repose*. The term $\nu(|h_x| - \tan \alpha)(\tan^2 \gamma - h_x^2)^{-1/2}$ represents tilt and avalanching; it comes into play only for slopes greater than $\tan \alpha$ and models the tendency of erstwhile stable clusters to shed grains into the flowing layer when tilted. This shedding of grains starts when the surface slope exceeds the angle of repose. For slopes approaching the angle γ, which is the *maximum angle of stability*, the rate of tilting out of grains becomes very large: an avalanche occurs. This term, among other things, is a novel representation of the well-known phenomena of *bistability* and *avalanching* in the context of continuum equations.

We renormalise the model equations, setting $x \to x_0\tilde{x}$, $t \to t_0\tilde{t}$, $a \to x_0\tilde{a}$, $\rho \to \rho_0\tilde{\rho}$, $h \to h_0\tilde{h}$, where $x_0 = D_h/a_pJ\cos\beta$, $t_0 = D_h/(a_pJ\cos\beta)^2$, $h_0 = D_h\tan\gamma/a_pJ\cos\beta$, $\rho_0 = a_pJ\sin\beta/\lambda\tan\alpha$, which gives for $0 \le h_x \le \tan\alpha/\tan\gamma$:

$$h_t = (1 + \hat{\kappa}\rho)h_{xx} - \rho\frac{\tan\beta}{\tan\alpha}\left(|h_x| - \frac{\tan\alpha}{\tan\gamma}\right) - f(x), \tag{11.23}$$

$$\rho_t = \frac{h_0}{\rho_0}\left\{-\hat{\kappa}\rho h_{xx} + \rho\frac{\tan\beta}{\tan\alpha}\left(|h_x| - \frac{\tan\alpha}{\tan\gamma}\right)\right\}$$
$$+ \frac{h_0}{\rho_0}\int_{-\infty}^{+\infty} p(a)f(x-a)da + \frac{D_\rho}{D_h}\rho_{xx} + \hat{\chi}(\rho h_x)_x, \tag{11.24}$$

and for $\tan\alpha/\tan\gamma \le h_x < 1$,

$$h_t = (1 + \hat{\kappa}\rho)h_{xx} - f(x) - \frac{\hat{v}(|h_x| - \tan\alpha/\tan\gamma)}{(1 - h_x^2)^{1/2}}, \tag{11.25}$$

$$\rho_t = \frac{h_0}{\rho_0}\left\{-\hat{\kappa}\rho h_{xx} + \frac{\hat{v}(|h_x| - \tan\alpha/\tan\gamma)}{(1 - h_x^2)^{1/2}}\right\}$$
$$+ \frac{h_0}{\rho_0}\int_{-\infty}^{+\infty} p(a)f(x-a)da + \frac{D_\rho}{D_h}\rho_{xx} + \hat{\chi}(\rho h_x)_x, \tag{11.26}$$

where the tildes have been dropped and where

$$f(x) = \left(h_x + \frac{\tan\beta}{\tan\gamma}\right)(1 + h_x^2\tan^2\gamma)^{-1/2}, \tag{11.27}$$

and $\kappa = \kappa\rho_0/D_h$, $\hat{v} = vt_0/h_0$, $\hat{\chi} = \chi h_0/D_h$. Wherever the sand bed is shielded from the saltation flux, the hopping term must be suppressed by removing the term $-f(x)$ in the h_t equation.

Close to onset of the instability that gives rise to sand ripples, the slopes of the sand bed will be small, since surface roughness is of small amplitude; hence the regime $0 \le \tan\alpha/\tan\gamma$ is relevant. There are no shielded regions at early times, since the slope of the bed does not exceed $\tan\beta$. Note that $h_x = 0$, $\rho = 1$ is a stationary solution of Eqs. (11.23) and (11.24). Setting $h = \hat{h}e^{\sigma t + ikx}$ and $\rho = 1 + \hat{\rho}e^{\sigma t + ikx}$, where $\hat{h} \ll 1$ and $\hat{\rho} \ll 1$ are constants, linearising, and Taylor-expanding the integrand gives a dispersion relation for σ in terms of k. The presence of the $|h_x|$ term means that strictly we are considering different solutions for sections of the ripple where $h_x > 0$ and sections where $h_x < 0$, but in fact the effect of the $|h_x|$

terms appears only as a contribution $\pm \tan\beta/\tan\alpha$ to the bracket $(1 \pm \tan\beta/\tan\alpha)$, where the $h_x > 0$ case takes the $+$ sign and the $h_x < 0$ case takes the $-$ sign. Since typically we have $\tan\beta/\tan\alpha \ll 1$, the two solutions will not be very different. One growth rate eigenvalue is given by $\sigma = -h_0 \tan\beta/\rho_0 \tan\gamma + O(k)$ and is the rate of relaxation of ρ to its equilibrium value of 1. To $O(k^4)$ the other eigenvalue is

$$\sigma = (\bar{a} - 1 - \hat{\chi}\rho_0/h_0)k^2 + iAk^3 + Bk^4, \tag{11.28}$$

where

$$A = -\frac{1}{2}\overline{a^2} + \frac{\rho_0 \tan\gamma}{h_0 \tan\beta}\left(1 + \hat{\chi}\frac{\rho_0}{h_0} - \bar{a} - \frac{D_\rho}{D_h}\right)\left(1 \pm \frac{\tan\beta}{\tan\alpha}\right), \tag{11.29}$$

$$B = -\frac{1}{6}\overline{a^3} - \frac{1}{2}\overline{a^2}\frac{\rho_0 \tan\gamma}{h_0 \tan\beta}\left(1 \pm \frac{\tan\beta}{\tan\alpha}\right)$$

$$- \frac{\rho_0 \tan\gamma}{h_0 \tan\beta}\left(\bar{a} - 1 - \hat{\chi}\frac{\rho_0}{h_0} + \frac{D_\rho}{D_h}\right)$$

$$\times \left\{\bar{a} + \hat{\kappa} + \frac{\rho_0 \tan\gamma}{h_0 \tan\beta}\left(1 \pm \frac{\tan\beta}{\tan\alpha}\right)^2\right\}, \tag{11.30}$$

and where $\overline{(.)}$ denotes $\int_{-\infty}^{+\infty}(.)p(a)da$. We have neglected higher order terms as we are looking for long wave modes where $|k|$ is small, since short waves are damped by the diffusion terms. Sand ripples grow if $\bar{a} > 1 + \hat{\chi}\rho_0/h_0$, which is equivalent to requiring that $\bar{a}a_pJ\cos\beta > D_h + \chi a_pJ\sin\beta/\lambda\tan\alpha$ holds in physical variables, giving a threshold saltation flux intensity for ripple growth. This is in agreement with the threshold found in [233]. Since B is negative ($\beta < \alpha$), the fastest growing mode has wavenumber $k^2 = -(\bar{a} - 1 - \hat{\chi}\rho_0/h_0)/2B$ with growth rate $\sigma = -(\bar{a} - 1 - \hat{\chi}\rho_0/h_0)^2/4B$. The allowed band of wavenumbers for growing modes is $0 < k^2 < -(\bar{a} - 1 - \hat{\chi}\rho_0/h_0)/B$. The wave speed is given by $c = -Ak^2 > 0$; it is higher for larger k^2, which implies that shorter waves move faster, as indeed was seen in the numerical simulations described below. The speed is higher for $h_x > 0$ than for $h_x < 0$, leading to wave steepening.

The renormalised model equations (11.23)–(11.26) were integrated numerically using compact finite differences [236] with periodic boundary conditions. The $-f(x)$ term in the h_t equation was suppressed in shielded regions. A normal distribution for the hop lengths with mean $\bar{a}$ and variance s^2 is used. In the run illustrated, the parameters $\bar{a} = 3.1$, $s = 0.1$, $D_\rho/D_h = 1.0$, $h_0/\rho_0 = 20.0$, $\hat{\chi} = 0.1$, $\hat{\upsilon} = 1.0$, $\hat{\kappa} = 0.1$, $\beta = 10°$, $\alpha = 30°$ and $\gamma = 35°$ have been chosen. The angles were chosen to agree with observational evidence, the ratio h_0/ρ_0 to ensure a thin layer of flowing grains, and the remaining parameters to allow ripple growth. The output was rescaled back into physical variables using $D_h = 1.0$ and $\lambda = 10.0$. The initial

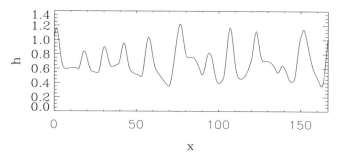

Fig. 11.18 The surface height h at time $t = 10.0\Delta t$.

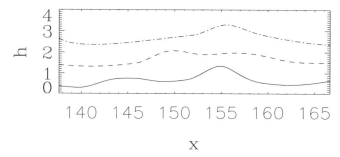

Fig. 11.19 A sequence of profiles showing a small ripple catching up and merging with a larger ripple. Sand is transferred from the larger to the smaller ripple until only the latter remains. The profiles are shown at times $t = 11.0\Delta t$ (solid line), $t = 12.0\Delta t$ (dashed line) and $t = 13.0\Delta t$ (dot-dash line). The later profiles are each offset by one additional unit in height.

conditions for the dimensionless variables were $h = 1.0 + 0.1\eta_h$, $\rho = 0.95 + 0.1\eta_\rho$, where η_h and η_ρ represent random noise generated by random variables on $[0, 1)$ in order to model surface roughness. In this case $B = -8.47$ and $A = -5.26$ (taking the minus sign in the brackets), giving a preferred wavenumber of $k = 0.352$, and a wave speed of $c = 5.26k^2$. The length of the integration domain was chosen to be ten times the linearly preferred wavelength.

Figure 11.18 shows the surface height at time $t = 10.0\Delta t$, where $\Delta t = 2.78$. Note the emergence of a *preferred wavelength*, with wavenumber $k \approx 0.457$ lying in the permitted band for growing modes predicted by the linear stability analysis and arrived at by a process of ripple merger. The wave speed close to onset was also measured and found to be $c \approx (4.13 \pm 0.39)k^2$, which is reasonably close to the predicted value. Ripple merger typically occurs when a small fast ripple catches up and merges with a larger slower ripple (Fig. 11.19), the leading ripple transferring sand to its pursuer until only the pursuer remains. Occasionally a small ripple emerges from the front of the new merged ripple and runs off ahead.

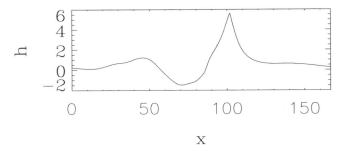

Fig. 11.20 The surface height h at time $t = 89\Delta t$, showing fully developed ripples.
Note the straight segments on the lee slopes close to the crests.

Figure 11.20 shows the surface height h at time $t = 89\Delta t$, with one shallow and
one fully developed ripple.

Note the long shallow stoss slopes, and the shorter steeper lee slopes with straight
sections near the crests and concave tails. The leftmost ripple has a maximum
stoss slope angle of 3.3°, and a maximum lee slope of 9.2°, whereas the more
fully developed rightmost ripple has a maximum stoss slope angle of 24.8° and
a maximum lee slope angle of 33.5°, which lies between the angle of repose and
the maximum angle of stability. The height to length ratio of the ripples is in the
range 1:8–1:22, which is in reasonable agreement with observations [227]. In the
long time limit, we would expect sand ripples to grow until the maximum lee slope
reaches an angle close to $\tan \gamma$. In reality, there is only a relatively shallow layer
of loose sand available for incorportion into ripples, and this together with the
maximum slope condition will determine the size of the fully-developed ripples.

This analytical continuum model for Aeolian sand ripples using a two-species
model [216] embodies intracluster and intercluster relaxation, in a description that
leads naturally to bistable behaviour at the angle of repose, and its cutoff at the
angle of maximal stability [72]. It is able to predict the preferred ripple wavelength,
the wave speed and the threshold saltation flux required for ripples to form. Also,
numerical simulations of these equations show the development of realistic ripple
profiles from initial surface roughness via growth and ripple merger.

11.8 Conclusions

In this chapter, we have demonstrated the power of a simple idea; set forth qualitatively now more than a decade ago [69], it embodies the notion that, locally, there
is a coupling between the instantaneous heights of a sandpile surface containing
'stuck' grains, and the instantaneous densities of mobile grains which move on the
surface. The basis of this idea has been set forth in the formal equations (11.4),
(11.7) and (11.12) with varying forms of the transfer term $T(h, \rho)$ to model specific

situations. While the use of these equations was, in their initial stages, to model relatively simple phenomena like the smoothing of post-avalanche surfaces, they have since been used with great success on a variety of problems, one of which, that of ripple formation [216], has been discussed in this chapter. With the burgeoning of the field of granular physics, each such field – ripple formation, dune formation – is now an area in itself, occasionally with real input from oceanographers and geophysicists. It is therefore all the more surprising that the theoretical basis for such complex phenomena could lie in a simple coupling between stuck and mobile grains – truly, it seems, that the world can indeed be 'seen in a grain of sand'.

12

Theory of rapid granular flows

Isaac Goldhirsch

Tel Aviv University, Israel

12.1 Introduction

The term 'rapid granular flows' is short for 'rapidly sheared granular flows' [194]. Indeed, the paradigm for a fluidised granular system had been for a long time a strongly sheared granular system, as in the classic experiments of Bagnold [5]. In recent years it seems that the main method of fluidisation in research laboratories is vertical vibration, see, e.g., [49, 237] and, at times, horizontal shaking of collections of grains, see, e.g., [238–240], or even electromagnetic fluidisation [241]. The fluidised state of a granular assembly is recently referred to as a 'granular gas', probably following the terminology introduced in [242]. Although most granular gases on earth are 'man made', there are naturally occurring granular gases, as part of snow and rock slides are fluidised. In outer space one finds interstellar dust and planetary rings (the latter being composed of ice particles).

In many cases, the grains comprising a granular gas are embedded in a fluid, hence technically they are part of a suspension. However, as noted by Bagnold, when the stress due to the grains sufficiently exceeds the fluid stress (the ratio of the two is known as the Bagnold number [5, 194]) one can ignore the effect of the ambient fluid (clearly, when the air is pumped out of a granular system, as in, e.g, [49], or when one considers celestial granular gases, one need not worry about the ambient fluid). Suspensions will not be considered in this chapter.

As the constituents of a granular gas collide, like in the classical model of a molecular gas, it is natural to borrow the terminology of the kinetic theory of gases [243–246] to describe them, and its methods to calculate 'equations of state' and 'constitutive relations'. Surprisingly, this has not always been the case: the old

This work has been partially supported by the United-States – Israel Binational Science Foundation (BSF) and the Israel Science Foundation (ISF).

literature contains criticism of the initial attempts to define a 'granular temperature' or similar entities which are taken over from the statistical mechanics of gases.

The similarity of a granular gas to a molecular gas should not be taken too literally. Granular collisions are inelastic and this fact alone has significant implications on the properties of granular gases, some of which are presented below. However, to the extent that the two can be considered to be similar, granular gases comprise a valuable model for studies of molecular gases; since they are composed of macroscopic particles they provide an opportunity to follow the path of each grain or, e.g., look 'inside' a shock wave by merely using a (fast) camera [247]. Of course, the study of granular gases does not need a justification based on an analogy with molecular gases.

The theoretical descriptions of granular gases are at least as varied as those of molecular gases, ranging from phenomenology, through mean free path theory, to the Boltzmann equation description, and its extensions to moderately dense gases [248]. Some classical many-body techniques, such as response theory [249], have also been applied to the study of granular gases [250, 251]. As this is not a review article, but rather an (somewhat biased) introduction to the field, we shall not describe the wealth of experimental and theoretical results concerning granular gases, many of which are quite recent [248]. The emphasis here is on theory with strong focus on results one can obtain from the pertinent Boltzmann equation. The analysis of the Boltzmann equation, properly modified to account for inelasticity, is not a straightforward extension of the theory of classical gases. In addition to the technical modifications of the Boltzmann equation and the Chapman–Enskog (CE) expansion, needed to study granular gases, one has to be aware of the limitations on the validity of the Boltzmann equation and the Chapman–Enskog expansion (beyond the obvious restriction to low densities for the 'regular' Boltzmann equation, and to moderate densities for the Enskog–Boltzmann equation [243, 245]), many of which are consequences of the lack of scale separation in granular gases [195]. These must be elucidated in detail in order to properly interpret the results of analyses of the Boltzmann equation, or apply them. The same holds for other methods of statistical mechanics, such as response theory. On the other hand one must keep in mind that some theories 'work' beyond their nominal domain of applicability; an example can be found in [252].

12.2 Qualitative considerations

As mentioned in the introduction, the central feature distinguishing granular gases from molecular gases (ignoring quantum effects) is the dissipative nature of grain collisions. One can draw several immediate conclusions from this property alone.

Consider the following idealisation, which is the granular 'equivalent' of a state of equilibrium, i.e., a granular gas of uniform macroscopic density and isotropic and uniform velocity distribution, centred around zero (i.e., the macroscopic velocity vanishes). Furthermore, for the sake of simplicity, ignore gravity. This state is known as the homogeneous cooling state (HCS). As the collisions are inelastic, the HCS cannot be stationary. The least one expects is that its kinetic energy decreases with time. Therefore the only stationary state of a granular gas is one corresponding to zero kinetic energy (or zero 'granular temperature', see more below). In order to remain at nonzero granular temperature a granular gas (whether in the HCS or not) must be supplied with energy, hence its state is always of nonequilibrium nature.

Interestingly, the HCS is not a stable state. It is unstable to clustering [242, 253, 254] and collapse [196, 255]. Forced granular gases exhibit similar instabilities (see below). These phenomena, and some of their consequences, are explained next.

12.2.1 Clustering

Consider a homogeneous cooling state first. Like every many body system, the HCS experiences fluctuations. Consider a fluctuation in the number density. In a domain in which the density is relatively large (without a change in the granular temperature) the rate of collisions is higher than in domains in which the density is relatively small (the collision rate is proportional to the square of the number density). Since the collisions are inelastic the granular temperature decreases at a faster rate in the dense domain than elsewhere in the system, hence the pressure in the dense domain decreases as well. The lower pressure in the dense regime causes a net flow of particles (or grains) from the surrounding more dilute domains, thus further increasing the density in the dense domains. This self-amplifying (and nonlinear) effect 'ends' when the low rate of particles escaping the resulting dense 'cluster' is balanced by the particles entering the cluster from its dilute surroundings. In due course clusters may merge in a 'coarsening' process, see, e.g., [256, 257], the result being (in a finite system) a state consisting of a single cluster containing most of the grains in the (finite) system. It thus follows that the HCS is unstable to the formation of clusters by the above 'collisional cooling' effect, and it does not remain homogeneous. In spite of this fact, the Boltzmann equation does have an HCS solution [258], which turns out to be useful for several purposes (see below).

A stability analysis of the HCS, using the granular hydrodynamic equations (see [242, 253, 254, 259, 260] and refs. therein) reveals that these equations are unstable, on a certain range of scales, to the formation of density inhomogeneities as well as shear waves. Sufficiently small granular systems do not develop clusters, but they become inhomogeneous and exhibit the above mentioned shear waves. However, for systems larger than a certain scale, the above nonlinear mechanism rapidly takes over and dominates the cluster creation process. The shear waves appear, e.g., in simulations with periodic boundary conditions [254, 256, 259]. Approximately half the system acquires a velocity in one direction and the other half moves in the opposite direction (the total momentum remaining zero, by momentum conservation). Thus, even in the absence of clustering, a HCS does not remain homogeneous.

The above arguments (with minor modifications) are relevant to forced and/or initially inhomogeneous systems as well. Consider, for instance, a granular gas confined by two parallel walls that move with equal speeds in opposite directions [261, 262] (the granular equivalent of a Couette flow; reference [261] is actually the first forced granular system in which clusters have been observed in a simulation). Next, imagine that the walls are allowed to change their velocities (still keeping them equal in size and opposite in direction) at some time. As the walls are the only source of energy in this system, the result of this change is an injection of energy into the system at the walls. This injection will cause the granular temperature to increase near the walls, the same holding for the pressure. The elevated pressure near the walls will move material towards the centre of the system, where its density will be higher. In the domain of elevated density the clustering mechanism will cause a further increase in the density, leading to a plug in the centre of the system.

The above 'method' for inducing a plug is not necessary to initiate clusters or plugs in a sheared (or any other) granular gas. A density fluctuation of sufficiently large size can increase and become a cluster by the collisional cooling mechanism. As a matter of fact, stability analysis of the hydrodynamic equations for the granular Couette system [262–266] reveals that this system is unstable to density fluctuations on certain scales. Therefore, clustering is always expected in a sheared system. Indeed, it has been observed in numerous experiments, e.g., [237, 267–269]. Interestingly, due to the rotational nature of shear flow, the clusters in such a flow are rotated and stretched by the flow. When two adjacent clusters are rotated in the same direction they are bound to collide with each other [262]. Such collisions may disperse the material in the clusters, but the above mentioned instability will cause new clusters to emerge, and so on. It therefore turns out that a 'stationary' and 'homogeneous' shear flow can be embedded with clusters that are born, destroyed and reborn; thus, this flow is neither stationary nor 'homogeneous' on the scales on which clusters can be resolved [261, 262].

The states of dense granular systems are known to be metastable [49, 262]. For instance, the ground state of a sandpile is one in which all grains reside on the floor. It turns out that most states of granular gases are metastable as well, see, e.g., [257, 270, 271], and that metastability can arise from the clustering phenomenon. Consider the simple shear flow again. Imagine that the initial condition for a sheared system is of a much higher granular temperature than that expected on the basis of steady-state solutions of the corresponding hydrodynamic equations. In this case the effect of shear on the system is (at least in the 'beginning') of secondary importance and the system will develop clusters much like in the HCS. At a later time the system will 'cool down' to essentially the expected average temperature, but its density distribution will remain similar to that of the HCS [262]. Due to cluster–cluster interactions, coarsening of the clusters is not expected in this case (in contrast to the HCS). Recall that a different initial condition for the same system leads to a plug flow. One may therefore conclude that the state of a granular gas does depend on history, rendering it metastable (and multi-stable), and that clustering is behind (at least some of) the mechanisms responsible for this property of granular matter.

Multistability is also observed in vibrated shallow granular beds [49, 272, 273], and numerous other granular systems. It is unclear whether these 'other' kinds of multi-stability are, or are not, related to clustering-like instabilities.

There is a significant difference between clusters and thermally induced density fluctuations of the kind that exist in every fluid. One of the key distinguishing features is that the granular temperature in the interior of clusters is lower than that in the ambient low density granular gas; clearly a molecular fluid does not spontaneously create long-lived structures whose temperature is different from the average temperature of the system or the local temperature (when temperature gradients are imposed). Furthermore, density fluctuations in molecular fluids are usually weak and they decay according to the Onsager hypothesis. This is not the case for granular clusters.

An interesting phenomenon related to clustering is the 'Maxwell demon effect' [274]. First published in a German teachers' journal cited in [274], the effect can be observed in the following experiment. A container is divided into two compartments by a vertical partition. A small opening in the partition allows grains to flow between the compartments. Grains are then symmetrically poured into the container, which is subsequently vertically vibrated. For sufficiently low values of the vibration frequency clustering commences in one of the compartments (lowering the pressure there), which then accumulates more mass flowing in from the other container, thus breaking the symmetry between the two compartments. Some interesting further experimental and theoretical studies of the Maxwell demon effect followed this discovery [275].

Clusters affect the stress in a granular system. A question of scientific as well as engineering importance is whether the value(s) of the average stress in a sheared granular system converge(s), as the system size is increased. Such a saturation is expected if, for sufficiently large systems, the cluster statistics does not depend on the system size any more. This question has been taken up in [276], see also [277] for implications concerning fluidised beds.

Although some arguments against a hydrodynamic description of clustering have been put forward [278], it is important to state that the clustering phenomenon is predicted by granular hydrodynamics, and therefore there is every reason to think of it as a hydrodynamic effect. In contrast, the collapse phenomenon discussed immediately below is not of hydrodynamic origin.

12.2.2 Collapse

As we all learned in high school physics (or should have), a ball hitting a floor with velocity v recedes with a velocity $e\,v$, where e is the coefficient of restitution. Although we know that e is velocity dependent [279], it is sufficient, for practical considerations, and certainly for the following explanation, to assume (as Newton did) that e is constant for given materials. An elementary calculation is then used to show that if the ball is dropped from rest at height h_0, its next maximal height is $e^2 h_0$, and the nth maximal height is $e^{2n} h_0$. The next calculation, though as trivial, is rarely taught in high schools. Denote by τ_n the time that elapses between the positions h_n and h_{n+1} of the ball. It is easy to show that $\tau_n = \tau_0\, e^n$ Since the sum of τ_n is finite (as $0 < e < 1$) it follows that an infinite number of collisions can occur in a finite time, during which the ball is brought to rest. Physical balls do not actually experience an infinite number of collisions, but when e is not too small the estimate for the total bouncing time is very good [280]. A similar process, now known as 'inelastic collapse' or 'collapse', may take place in many-grain systems [196, 255], leading (via a theoretically infinite number of collisions) to the emergence of strings of particles whose relative velocities vanish [196]. The collapse mechanism is a source of difficulties encountered in MD simulations since a very large number of events (collisions) occurs in a finite time while nothing much changes in the system. The 'collapse' process has been the subject of a number of studies which followed the pioneering work of [255], see, e.g., the review [278]. Clearly 'collapse' is a non-hydrodynamic phenomenon. In most three-dimensional *excited* granular gases there is no (saturation of the) collapse sequence because a particle external to the 'collapsing string' is essentially always available to break it up. Furthermore, the coefficient of restitution of real particles is velocity dependent and thus the 'collapse' stops when the relative velocities of the colliding particles are sufficiently small [279]. The above arguments notwithstanding, there is a report of collapse in a two-dimensional shear flow [281]. In MD simulations the collapse phenomenon is usually avoided by changing the collision law at low relative velocities from inelastic to elastic, thus mimicking real collisions. Another method [259] is to rotate the relative velocity of the colliding particles after the collision, so as to prevent the emergence of a (nearly) collinear string. Still another method is provided by the TC model whereby a finite collision time is allowed for [282]. When external forcing is stopped, any granular system collapses to a stationary state in which none of the particles moves any more.

Clustering can be a precursor to collapse as it creates conditions under which nearby particles can form strings or other shapes amenable to collapse. A one dimensional demonstration of this phenomenon can be found in [283].

12.2.3 Granular gases are mesoscopic

One of the important consequences of inelasticity is the lack of scale separation in granular gases [195]. Therefore one should be very careful in applying some of the standard methods of statistical mechanics (many of which are based on the existence of strong scale separation) to granular gases.

It is convenient to demonstrate the lack of scale separation in granular gases by considering a monodisperse granular gas, the collisions of whose constituents are characterised by a fixed coefficient of normal restitution, e. Assume the gas is (at least locally) sheared, i.e., its local flow field is given by $\mathbf{V} = \gamma y \hat{\mathbf{x}}$, where γ is the shear rate. In the absence of gravity, γ^{-1} provides the only 'input' variable that has dimensions of time. Let T denote the granular temperature, defined as the mean square of the fluctuating particle velocities. It is clear on the basis of dimensional considerations that $T \propto \gamma^2 \ell^2$, where ℓ is the mean free path (the only relevant microscopic length scale). Define the *degree of inelasticity*, ϵ, by $\epsilon \equiv 1 - e^2$. Clearly, T should be larger for a given value of γ the smaller ϵ is. In a steady sheared state without inelastic dissipation one expects T to diverge. Therefore, one may guess that $T = C\gamma^2 \ell^2 / \epsilon$. A mean field theoretical study yields the same result, as does a systematic kinetic theoretical analysis [284–288]). The value of C is about 1 in two dimensions and 3 in three dimensions.

Consider the change of the macroscopic velocity over a distance of a mean free path, in the spanwise, y, direction: $\gamma \ell$. A shear rate can be considered small if $\gamma \ell$ is small with respect to the thermal speed, $\sqrt{T}$. Employing the above expression for T one obtains: $\gamma \ell / \sqrt{T} = \sqrt{\epsilon} / \sqrt{C}$, i.e. the shear rate is not 'small' unless the system is nearly elastic (notice that for, e.g., $e = 0.9$, $\sqrt{\epsilon} = 0.44$). Thus, except for very low values of ϵ the shear rate is always 'large'. Incidentally, this also shows that the granular system is supersonic. Shock waves in granular systems have been reported, e.g., in [247, 289, 290]. This result also implies that the Chapman–Enskog (CE) expansion of kinetic theory (an expansion of the distribution function 'in powers of the gradients', one of which is the shear rate) may encounter difficulties; the reason is that the 'small parameter' of this expansion is truly the mean free path times the 'values of the gradients' of the hydrodynamic fields, or, in other words, the ratio of the mean free path and the scale on which the hydrodynamic fields change in space. Indeed, it is argued below that one needs to carry out this expansion beyond its lowest order (the Navier–Stokes order) and include at least the next (Burnett) order in the gradients. One of the results obtained from the

Burnett order is that the normal stress ('pressure') in granular gases is anisotropic (see also the next section). While the Burnett equations yield good results for steady states, they are dynamically ill posed. A resummation of the CE expansion has been proposed in [291]. The Burnett and higher orders are well defined in the framework of kinetic theory but they are 'not defined', i.e., divergent [292] in the more general framework of nonequilibrium statistical mechanics (i.e., at finite densities). This is taken to imply that higher orders in the gradient expansion may be non-analytic in the gradients [293], indicating non-locality.

Consider next the mean free time, τ, i.e. the ratio of the mean free path and the thermal speed: $\tau \equiv \ell/\sqrt{T}$. Clearly, τ is the microscopic timescale characterising any gas, and, as mentioned, γ^{-1} is a macroscopic timescale characterising a sheared system. The ratio $\tau/\gamma^{-1} = \tau\gamma$ is a measure of the temporal scale separation in a sheared system. Employing the above expression for the granular temperature one obtains $\tau\gamma = \sqrt{\epsilon}/\sqrt{C}$, typically an $O(1)$ quantity. It follows that (unless $\epsilon \ll 1$) there is no temporal scale separation in this system, *irrespective of its size or the size of the grains*. Consequently, one cannot a priori employ the assumption of 'fast local equilibration' and/or use local equilibrium as a zeroth order distribution function (e.g., for perturbatively solving the Boltzmann equation), unless the system is nearly elastic (in which case, scale separation is restored). The latter result sets a further restriction on the applicability of the hydrodynamic description: consider the stability of, e.g., a simply sheared granular system; since the 'input' time scale is $1/\gamma \approx \tau$, it is plausible (and it can be checked by direct calculations [262–266]) that some stability eigenvalues are of the order of τ^{-1}. When one of these eigenvalues corresponds to an unstable mode, as is the case in the above example, one is faced with the result that the equations of motion predict an instability on a scale which they do not resolve! It is possible that this observation is related to Kumaran's findings [294] that there are some inconsistencies between the stability spectrum obtained from granular hydrodynamics and that deduced directly from the Boltzmann equation.

In the realm of molecular fluids, when they are not under very strong thermal or velocity gradients, there is a range, or *plateau*, of scales, which are larger than the mean free path and far smaller than the scales characterising macroscopic gradients, and which can be used to define 'scale independent' densities (e.g. mass density) and fluxes (e.g. stresses, heat fluxes). Such *plateaus* are virtually nonexistent in systems in which scale separation is weak, and therefore these entities are expected to be scale dependent. By way of example, the 'eddy viscosity' in turbulent flows is a scale dependent (or resolution dependent) quantity, since in the inertial range of turbulence there is no scale separation. It can be shown [295] that due to this lack of scale separation in granular gases, the stresses and other entities measured by using the 'box division method' are strongly scale dependent. For instance, the velocity profile changes by a significant amount in a box whose dimensions exceed the mean free path, thus contributing to the 'velocity fluctuations'.

12.3 Kinetic theory

Kinetic theory has its roots in Maxwell's work on molecular gases, yet its main power stems from the existence of a fundamental equation, viz. the Boltzmann equation. Following Boltzmann's phenomenological and intuitive derivation of this equation, there have been a series of systematic derivations, most notably using the BBGKY hierarchy (and applying e.g., the Grad limit), see e.g. [243–246].

The classical derivations of the Boltzmann equation involve the assumption of 'molecular chaos' (originally named in German: Stosszahlansatz), namely that the positions and velocities of colliding molecules (more accurately, molecules about to collide) are uncorrelated. This assumption is not justified for dense gases, as molecules have a chance to recollide with each other, thereby becoming correlated. A model Boltzmann equation, which partially accounts for such a-priori correlations is known as the Enskog–Boltzmann equation [243, 245]. In some cases, e.g., for hard sphere models, the latter equation is known to produce good results [296] (compared to MD simulations). The Enskog–Boltzmann equation is not described below.

When one wishes to describe granular gases one needs to modify the Boltzmann equation to account for the inelasticity of the collisions [286, 287]. This can be easily done by a slight modification of the standard (e.g., phenomenological) derivation of the Boltzmann equation. Thus, the derivation of the Boltzmann equation for granular gases poses no serious technical problem. However, as mentioned, the justification of the assumption of molecular chaos for granular gases, even for low densities, is not as good as for molecular gases. To see this, consider the following simple model of a granular gas, namely a collection of monodisperse hard spheres, whose collisions are characterised by a constant coefficient of normal restitution. The binary collision between spheres labelled i and j results in the following velocity transformation:

$$\mathbf{v}_i = \mathbf{v}_i' - \frac{1+e}{2}(\hat{\mathbf{k}} \cdot \mathbf{v}_{ij}')\hat{\mathbf{k}}, \tag{12.1}$$

where $(\mathbf{v}_i', \mathbf{v}_j')$ are the precollisional velocities, $(\mathbf{v}_i, \mathbf{v}_j)$ are the corresponding post-collisional velocities, $\mathbf{v}_{ij}' \equiv \mathbf{v}_i' - \mathbf{v}_j'$, and $\hat{\mathbf{k}}$ is a unit vector pointing from the centre of sphere i to that of sphere j at the moment of contact. An important feature of this collision law is that the normal relative velocity of two colliding particles is reduced upon collision. This implies that the velocities of colliding particles become more correlated after they collide. Indeed, such correlations have been noted in MD

simulations [254, 297, 298]. In particular, since only grazing collisions involve a
minimal loss of relative velocity, the grains in a homogeneous cooling state show a
clear enhancement of grazing collisions [254] (a sign of correlation). This feature
is less pronounced in, e.g., shear flows [262, 297] but it is still measurable. As the
coefficient of restitution approaches unity, these correlations become smaller. This
implies (again) that the Boltzmann equation for granular gases should apply (at
best) to near-elastic collisions. The above mentioned lack of scale separation in
granular gases dictates that the standard method of obtaining constitutive relations
from the Boltzmann equation is limited to the case of near-elastic collisions as
well. Therefore this restriction applies to all kinetic and hydrodynamic theories of
granular gases ('hydrodynamic theories' are defined here as theories in which the
constitutive relations involve low order gradients of the fields, as they result from
appropriate gradient expansions).

The Boltzmann equation is an equation for the 'single particle distribution func-
tion', $f(\mathbf{v}, \mathbf{r}, t)$, which is the number density of particles having velocity $\mathbf{v}$ at a point
$\mathbf{r}$, at time t. Upon dividing f by the local number density, $n(\mathbf{r}, t)$, one obtains the
probability density for a particle to have a velocity $\mathbf{v}$ at point $\mathbf{r}$, at time t.

The Boltzmann equation for a monodisperse gas of hard spheres of diameter
d and unit mass, whose collisions are described by Eq. (12.1) is well established
[286, 287]. It reads:

$$\frac{\partial f}{\partial t} + \mathbf{v}_1 \cdot \nabla f = d^2 \int_{\hat{\mathbf{k}} \cdot \mathbf{v}_{12} > 0} d\mathbf{v}_2 d\hat{\mathbf{k}} (\hat{\mathbf{k}} \cdot \mathbf{v}_{12}) \left(\frac{1}{e^2} f(\mathbf{v}'_1) f(\mathbf{v}'_2) - f(\mathbf{v}_1) f(\mathbf{v}_2) \right),$$
(12.2)

where ∇ is a gradient with respect to the spatial coordinate $\mathbf{r}$. The unit vector $\hat{\mathbf{k}}$
points from the centre of particle '1' to the centre of particle '2'. The dependence of
f on the spatial coordinates and on time is not explicitly spelled out in Eq. (12.2), for
the sake of notational simplicity. Notice that in addition to the explicit dependence
of Eq. (12.2) on e, it also implicitly depends on e through the relation between the
postcollisional and precollisional velocities. The condition $\hat{\mathbf{k}} \cdot \mathbf{v}_{12} > 0$ represents
the fact that only particles whose relative velocity is such that they approach each
other can collide.

The basis physical idea underlying the Champan–Enskog method of solving the
Boltzmann equation is scale separation. It is assumed that the macroscopic fields
change sufficiently slowly on the time scale of a mean free time, and the spatial
scale of a mean free path, so that the system has a chance to basically locally
equilibrate (up to perturbative corrections, which are proportional to the Knudsen
number), the local equilibrium distribution depending on the values of the fields.
Since it is normally assumed that the only fields 'remembered' by the system are
the conserved fields (in some cases, such as liquid crystals, a non-conserved order

parameter may be 'remembered'); in other words the fields that determine the local distribution function are the densities of the conserved entities, i.e., the number density (or mass density), the energy density and the momentum density. In the case of granular gases, the (kinetic) energy density is not strictly conserved, but when the degree of inelastictiy is sufficiently small, it is justified to take it as an appropriate hydrodynamic field. Furthermore, since the kinetic energy density is an important characterisation of the state of a granular gas, it is rather clear that it should be included among the hydrodynamic fields. All of the aforementioned fields, the number density field, $n(\mathbf{r}, t)$, the macroscopic velocity field, $\mathbf{V}(\mathbf{r}, t)$ (which is the ratio of the momentum density field and the mass density field), and the granular temperature field, $T(\mathbf{r}, t)$ (which is related to the energy field in an obvious way; see more below) are moments of the single particle distribution. These quantities are given by:

$$n(\mathbf{r}, t) \equiv \int d\mathbf{v} f(\mathbf{v}, \mathbf{r}, t), \tag{12.3}$$

$$\mathbf{V}(\mathbf{r}, t) \equiv \frac{1}{n} \int d\mathbf{v} \, \mathbf{v} f(\mathbf{v}, \mathbf{r}, t) \tag{12.4}$$

and

$$T(\mathbf{r}, t) \equiv \frac{1}{n} \int d\mathbf{v} (\mathbf{v} - \mathbf{V})^2 f(\mathbf{v}, \mathbf{r}, t), \tag{12.5}$$

respectively; in the above $1/n$ denotes $1/n(\mathbf{r}, t)$. As mentioned, the mass, m, of a particle, is normalised to unity. The granular temperature, defined above (without the factor $1/3$ often used in the literature), is a measure of the squared fluctuating velocity. It is a priori unclear whether these fields are sufficient for a proper closure of the hydrodynamic equations of motion for granular gases, since one cannot naively extrapolate from the case of molecular gases, but this turns out to be the case (within the framework of the Chapman–Enskog expansion).

The equations of motion for the above defined macroscopic field variables, i.e., the corresponding continuum mechanics equations, can be formally derived by multiplying the Boltzmann equation, Eq. (12.2), by 1, $\mathbf{v}_1$ and v_1^2 respectively, and integrating over $\mathbf{v}_1$. A standard procedure (which employs the symmetry properties of the collision integral on the right-hand side of the Boltzmann equation) yields equations of motion for the hydrodynamic fields [285]:

$$\frac{Dn}{Dt} + n \frac{\partial V_i}{\partial r_i} = 0, \tag{12.6}$$

$$n \frac{DV_i}{Dt} + \frac{\partial P_{ij}}{\partial r_j} = 0, \tag{12.7}$$

$$n \frac{DT}{Dt} + 2 \frac{\partial V_i}{\partial r_j} P_{ij} + 2 \frac{\partial Q_j}{\partial r_j} = -n\Gamma, \tag{12.8}$$

where $\mathbf{u} \equiv \mathbf{v} - \mathbf{V}$ is the fluctuating velocity, $P_{ij} \equiv n\langle u_i u_j \rangle$ is the stress tensor, and $Q_j \equiv n\langle u^2 u_j \rangle/2$ is the heat flux vector, where $\langle\rangle$ denotes an average with respect to f. In addition, $D/Dt \equiv \partial/\partial t + \mathbf{V} \cdot \nabla$ is the material derivative, and Γ, which accounts for the energy loss in the (inelastic) collisions, is given by:

$$\Gamma \equiv \frac{\pi(1-e^2)d^2}{8n} \int d\mathbf{v}_1 d\mathbf{v}_2 v_{12}^3 f(\mathbf{v_1}) f(\mathbf{v_2}). \tag{12.9}$$

Equations (12.6)–(12.8) are *exact* consequences of the Boltzmann equation. They also comprise the equations of continuum mechanics, and thus their validity is very general [295]; in particular, they do not depend on the correctness or relevance of the Boltzmann equation. The specific expressions presented above for the stress field, the heat flux and the energy sink term are results of the Boltzmann equation (there are corrections to these expressions in the dense domain [295]). The microscopic details of the interparticle interactions affect the values of the averages $\langle u_i u_j \rangle$, $\langle u^2 u_i \rangle$ and Γ. As mentioned, a standard method for obtaining these quantities for molecular gases is the Chapman–Enskog expansion. It involves a perturbative solution of the Boltzmann equation in powers of the spatial gradients of the hydrodynamic fields (formally, in the Knudsen number, see below); the zeroth order solution yields the Euler equations, the first order gives rise to the Navier–Stokes equations, the second order begets the Burnett equations, etc. The Chapman–Enskog method is tailored for systems that have a stationary homogeneous (equilibrium) solution; the latter serves as a zeroth order solution of the expansion. We reiterate that the physical justification for the use of this zeroth order solution (for molecular gases) is that when there is sufficiently good scale separation and the gradients are sufficiently small (in the sense that the hydrodynamic fields change in a minute amount over the scale of a mean free path or during a mean free time), the system evolves towards local equilibrium everywhere, and the effects of the gradients in the fields are perturbations around the local equilibrium states. As mentioned, the scale separation in granular gases is not nearly as good as in typical molecular gases. Furthermore, since granular systems do not possess such equilibrium-like solutions, the Chapman–Enskog technique is not directly applicable to such systems. As shown below, one can extend the CE expansion method to the case of granular gases.

There are at present two systematic methods for extending the CE expansion to granular gases. Both require a zeroth order for the respective perturbation theories they develop. The method proposed in [288] is based on an expansion in the Knudsen number (gradients) around a local HCS. This method does not formally restrict the value of the degree of inelasticity, ϵ, to be small, hence, in principle, it is correct for all values of this parameter. However, as explained above, this can't be the case because of the lack of scale separation. This fact notwithstanding, the constitutive relations obtained this way are claimed to agree with DSMC simulations for values of e as low as 0.6 (recently this method has been extended by the Goldhirsch group to apply to all values of e.). The method proposed by the author and coworkers is presented below, following a brief description of less systematic approaches.

One of the methods that has been applied to the study of granular gases is the Grad expansion [299]. It is based on a substitution of a Maxwellian times a series of polynomials in the fluctuating velocity into the Boltzmann equation. The method leads to a set of nonlinear equations of motion for the coefficients of the polynomials. Scale separation (and truncation in the order of the polynomials) is then used to render the scheme manageable, and obtain a closure. This is not a systematic method, but it can be used to obtain constitutive relations for molecular as well as granular gases [300, 301]. Another approach to the study of constitutive relations is the use of simplified or model kinetic equations, such as the BGK equation [197, 302]. The advantage of this approach is its relative simplicity, and the possibility it affords to study, e.g., strongly nonlinear effects. However, one must alway remember that the BGK equation is an approximation.

The method developed by the author and coworkers is based on a different physical limit. The classical Chapman–Enskog expansion assumes the smallness of the Knudsen number, $K \equiv \ell/L$, where ℓ is the mean free path given by $\ell = 1/(\pi n d^2)$, and L is a macroscopic length scale, i.e., the length scale which is resolved by hydrodynamics, not necessarily the system size. Here we employ a second small parameter, the degree of inelasticity $\epsilon \equiv 1 - e^2$. Prior to explaining the meaning of this expansion, we need to dwell on some minor technicalities.

It is convenient to perform a rescaling of the Boltzmann equation, as follows: spatial gradients are rescaled as $\nabla \equiv \tilde{\nabla}/L$, the rescaled fluctuating velocity (in terms of the thermal speed) is $\tilde{\mathbf{u}} \equiv \sqrt{3/(2T)}(\mathbf{v} - \mathbf{V})$, and $f \equiv n (3/2T)^{3/2} \tilde{f}(\tilde{\mathbf{u}})$. In terms of the rescaled quantities, the Boltzmann equation assumes the form:

$$
\tilde{\mathbf{D}} \tilde{f} + \tilde{f} \tilde{\mathbf{D}} \left(\log n - \frac{3}{2} \log T \right)
$$

$$
= \frac{1}{\pi} \int_{\hat{\mathbf{k}} \cdot \tilde{\mathbf{u}}_{12} > 0} d\tilde{\mathbf{u}}_2 d\hat{\mathbf{k}} (\hat{\mathbf{k}} \cdot \tilde{\mathbf{u}}_{12}) \left(\frac{1}{e^2} \tilde{f}(\tilde{\mathbf{u}}_1') \tilde{f}(\tilde{\mathbf{u}}_2') - \tilde{f}(\tilde{\mathbf{u}}_1) \tilde{f}(\tilde{\mathbf{u}}_2) \right)
$$

$$
\equiv \tilde{\mathbf{B}}(\tilde{f}, \tilde{f}, e), \tag{12.10}
$$

where

$$
\tilde{\mathbf{D}} \equiv K \sqrt{\frac{3}{2T}} \left(L \frac{\partial}{\partial t} + \mathbf{v} \cdot \tilde{\nabla} \right). \tag{12.11}
$$

Notice that $\tilde{\mathbf{D}}$ is not a material derivative since the velocity $\mathbf{v}$ is not the hydrodynamic velocity but rather the particle's velocity.

Clearly, the double limit $\epsilon \to 0$ and $K \to 0$, with constant number density, corresponds to a homogeneous, elastically colliding collection of spheres for which the distribution function is Maxwellian. This limit is not singular. It is known that local equilibration occurs on a time scale of a few mean free times (i.e., following few collisions per particle) [244–246]; during such time a small degree of inelasticity has almost negligible effect. Hence, for (formally) $K \ll 1$ and $\epsilon \ll 1$, $\tilde{f}$ can be expressed as follows: $\tilde{f}(\tilde{\mathbf{u}}) = \tilde{f}_0(\tilde{u})(1 + \Phi)$ where $\tilde{f}_0(\tilde{u}) = e^{-\tilde{u}^2}/\pi^{3/2}$ and Φ is considered to be a 'small' perturbation. Employing the above form of $\tilde{f}$, and making use of $\tilde{u}^2 = 3(\mathbf{v} - \mathbf{V})^2/(2T)$, it follows that Eq. (12.10) can be transformed to:

$$(1 + \Phi)\left(\tilde{\mathbf{D}}\log n + 2\sqrt{\frac{3}{2T}}\tilde{u}_i\tilde{\mathbf{D}}V_i + \left(\tilde{u}^2 - \frac{3}{2}\right)\tilde{\mathbf{D}}\log T\right) + \tilde{\mathbf{D}}\Phi = \frac{1}{\tilde{f}_0}\tilde{\mathbf{B}}(\tilde{f}, \tilde{f}, e).$$

$$(12.12)$$

The following relations follow directly from Eqs. (12.6)–(12.8) and the definition of $\tilde{\mathbf{D}}$:

$$\tilde{\mathbf{D}}\log n = K\left(\tilde{u}_i\frac{\partial\log n}{\partial\tilde{r}_i} - \sqrt{\frac{3}{2T}}\frac{\partial V_i}{\partial\tilde{r}_i}\right), \qquad (12.13)$$

$$\tilde{\mathbf{D}}V_i = K\left(\tilde{u}_j\frac{\partial V_i}{\partial\tilde{r}_j} - \frac{1}{n}\sqrt{\frac{3}{2T}}\frac{\partial P_{ij}}{\partial\tilde{r}_j}\right) \qquad (12.14)$$

and

$$\tilde{\mathbf{D}}\log T = K\left(\tilde{u}_j\frac{\partial\log T}{\partial\tilde{r}_j} - \frac{2}{nT}\sqrt{\frac{3}{2T}}P_{ij}\frac{\partial V_i}{\partial\tilde{r}_j} - \frac{2}{nT}\sqrt{\frac{3}{2T}}\frac{\partial Q_j}{\partial\tilde{r}_j}\right) - \epsilon\tilde{\Gamma}, \quad (12.15)$$

where

$$\tilde{\Gamma} \equiv \frac{1}{12}\int d\tilde{\mathbf{u}}_1 d\tilde{\mathbf{u}}_2 \tilde{u}_{12}^3 \tilde{f}(\tilde{\mathbf{u}}_1)\tilde{f}(\tilde{\mathbf{u}}_2). \qquad (12.16)$$

Note that in the derivation of Eq. (12.15), one encounters a product $K \cdot \epsilon/K$, which turns a nominally $O(K)$ term to an $O(\epsilon)$ term. Next expand Φ in both small parameters, ϵ and K: $\Phi = \Phi_K + \Phi_\epsilon + \Phi_{KK} + \Phi_{K\epsilon} + \cdots$, where subscripts indicate the order of the corresponding terms in the small parameters, e.g., $\Phi_K = O(K)$. It is perhaps worthwhile mentioning that the $O(K\epsilon^n)$, for all $n \geq 0$, corrections to the single particle distribution function are considered to be of Navier–Stokes (or Chapman–Enskog) order, whereas the $O(K^2\epsilon^n)$ corrections are Burnett terms. In parallel to the expansion of Φ in the small parameters, the operation of $\tilde{\mathbf{D}}$ on any function of the field variables, ψ, can be formally expanded as the following sum: $\tilde{\mathbf{D}}\psi = \tilde{\mathbf{D}}_K\psi + \tilde{\mathbf{D}}_\epsilon\psi + \tilde{\mathbf{D}}_{KK}\psi + \tilde{\mathbf{D}}_{K\epsilon}\psi + \tilde{\mathbf{D}}_{\epsilon\epsilon}\psi + \cdots$, where, e.g., $\tilde{\mathbf{D}}_{K\epsilon}\psi$ is the

$O(K\epsilon)$ term in the expansion of $\tilde{\mathbf{D}}\psi$ in powers of K and ϵ. Since this expansion is well defined we shall refer to the symbols $\tilde{\mathbf{D}}_K$, $\tilde{\mathbf{D}}_\epsilon$ etc. as operators in their own right.

12.3.1 Some technical details and constitutive relations

Upon substituting $e = 1$ (or $\epsilon = 0$) in the right-hand side of Eq. (12.12) and retaining only $O(K)$ terms, one obtains:

$$\tilde{\mathbf{L}}(\Phi_K) = \tilde{\mathbf{D}}_K \log n + 2\sqrt{\frac{3}{2T}}\tilde{u}_i \tilde{\mathbf{D}}_K V_i + \left(\tilde{u}^2 - \frac{3}{2}\right)\tilde{\mathbf{D}}_K \log T, \qquad (12.17)$$

where $\tilde{\mathbf{L}}$ is the (standard) rescaled linearised Boltzmann operator [243–246] for elastically colliding hard spheres, given by:

$$\tilde{\mathbf{L}}(\Phi) \equiv \frac{1}{\pi^{5/2}}\int_{\hat{\mathbf{k}}\cdot\tilde{\mathbf{u}}_{12}>0} d\hat{\mathbf{k}}d\tilde{\mathbf{u}}_2(\hat{\mathbf{k}}\cdot\tilde{\mathbf{u}}_{12})e^{-\tilde{u}_2^2}(\Phi(\tilde{\mathbf{u}}_1') + \Phi(\tilde{\mathbf{u}}_2') - \Phi(\tilde{\mathbf{u}}_2) - \Phi(\tilde{\mathbf{u}}_1)).$$
$$(12.18)$$

The operation of $\tilde{\mathbf{D}}_K$ on the hydrodynamic fields can be read off Eqs. (12.13)–(12.15). One obtains:

$$\tilde{\mathbf{D}}_K \log n = K\left(\tilde{u}_i \frac{\partial \log n}{\partial \tilde{r}_i} - \sqrt{\frac{3}{2T}}\frac{\partial V_i}{\partial \tilde{r}_i}\right), \qquad (12.19)$$

$$\tilde{\mathbf{D}}_K V_i = K\left(\tilde{u}_j \frac{\partial V_i}{\partial \tilde{r}_j} - \frac{1}{2}\sqrt{\frac{2T}{3}}\frac{\partial \log n}{\partial \tilde{r}_i} - \frac{1}{2}\sqrt{\frac{2T}{3}}\frac{\partial \log T}{\partial \tilde{r}_i}\right) \qquad (12.20)$$

and

$$\tilde{\mathbf{D}}_K \log T = K\left(\tilde{u}_j \frac{\partial \log T}{\partial \tilde{r}_j} - \frac{2}{3}\sqrt{\frac{3}{2T}}\frac{\partial V_j}{\partial \tilde{r}_j}\right). \qquad (12.21)$$

In deriving Eqs. (12.20) and (12.21) use has been made of the (easy to check) facts that $P_{ij} = nT\delta_{ij}/3$ to zeroth order in K and ϵ, and $Q_i = O(K)$ (hence its spatial derivatives are of higher order in K). Substitution of Eqs. (12.19)–(12.21) in Eq. (12.17) results in:

$$\tilde{\mathbf{L}}(\Phi_K) = 2K\overline{\tilde{u}_i\tilde{u}_j}\sqrt{\frac{3}{2T}}\frac{\partial V_i}{\partial \tilde{r}_j} + K\left(\tilde{u}^2 - \frac{5}{2}\right)\tilde{u}_i\frac{\partial \log T}{\partial \tilde{r}_i}, \qquad (12.22)$$

where the overline denotes a symmetrised traceless tensor, i.e.,

$$\overline{A_{ij}} \equiv \frac{1}{2}(A_{ij} + A_{ji}) - \frac{1}{3}A_{kk}\delta_{ij}.$$

The linear inhomogeneous Eq. (12.22) is soluble only when its right-hand side is orthogonal to the (left) eigenfunctions of the operator $\tilde{L}$; this is known as the solubility condition or the 'Fredholm alternative'. This issue is further elaborated upon below.

Notice that Eq. (12.22) is identical to that obtained in the classical CE expansion to first order in spatial gradients (since $e = 1$ at this order). The isotropy and linearity of the operator $\tilde{L}$ [243–246] imply that the solution of Eq. (12.22) is of the form:

$$\Phi_K(\bar{\mathbf{u}}) = 2K\,\hat{\Phi}_v(\tilde{u})\overline{\tilde{u}_i\tilde{u}_j}\sqrt{\frac{3}{2T}\frac{\partial V_i}{\partial \tilde{r}_j}} + K\,\hat{\Phi}_c(\tilde{u})\left(\tilde{u}^2 - \frac{5}{2}\right)\tilde{u}_i\frac{\partial \log T}{\partial \tilde{r}_i}, \qquad (12.23)$$

where $\tilde{u} \equiv \parallel \bar{\mathbf{u}} \parallel$. It is common [243–246] to expand $\hat{\Phi}_v$, $\hat{\Phi}_c$ and similar functions in (truncated) series of Sonine polynomials. More accurate results can be obtained by expanding in sets of functions which obey the symmetry and asymptotic properties of the sought functions and solving for the appropriate coefficients by numerical means. For instance, it turns out [286] that the functions $\hat{\Phi}_v(\tilde{u})$ and $\hat{\Phi}_c(\tilde{u})$ are both formally even in $\tilde{u}$ and they are both proportional to $1/\tilde{u}$ at large values of $\tilde{u}$ (a property that cannot be obeyed by a truncated Sonine polynomial series). An appropriate complete set of functions, satisfying these properties, has been used [286] to numerically compute these two (and other) functions.

The local equilibrium distribution function, f_0, is defined in such a way that the hydrodynamic fields are *given* by its appropriate moments. It is perhaps important to iterate that the CE expansion is designed to find the distribution of fluctuating velocities when the macroscopic, or continuum, fields are given; in other words the local distribution function corresponds, by construction, to the true values of the macroscopic fields.[1] Thus, the corrections to the zeroth order term, i.e. the local equilibrium distribution function, should not change the values of the macroscopic fields. This can be translated to the following technical condition: the contribution of the correction, Φ, to the above mentioned moments should vanish, i.e. Φ *should be orthogonal (with respect to the weight function f_0) to the invariants of the (linearised) Boltzmann operator* (the eigenfunctions which correspond to zero eigenvalues), 1, $\bar{\mathbf{u}}$ and $\tilde{u}^2$, whose respective averages are the density, the velocity and the temperature field. The concept of orthogonality is employed here in the functional sense, i.e. two functions g and h are considered to be orthogonal with respect to the weight function f_0 if the integral over the product $f_0 \cdot g \cdot h$ over a predefined range (here: all values of the velocity) of integration vanishes. In the present context, the orthogonality of a pair of functions with respect to the equilibrium

---

[1] In a sense the CE expansion is a closure. One assumes the knowledge of the hydrodynamics fields in order to obtain the distribution function. The latter is employed to obtain the constitutive relations in terms of those fields. Once the constitutive relations are known, one can solve (in principle) the equations of motion, with given boundary conditions, to obtain the hydrodynamic fields.

distribution function means that the equilibrium average of their product vanishes. This orthogonality property should hold to all orders in perturbation theory [286]; it is also the reason the (generalised) CE expansion can be systematically carried out to all orders in the small parameters [286], as the same conditions are also the solubility conditions of the linearised Boltzmann equation, at each order in the expansion. Since the solution of equations of the type of Eq. (12.22) is determined up to the addition of an arbitrary combination of 1, $\tilde{\mathbf{u}}$ and $\tilde{u}^2$, which comprise the homogeneous solution of Eq. (12.22) (they are also the additive invariants of the collision, in the elastic limit), it is the above orthogonality property that determines the appropriate coefficients. The orthogonality of the function Φ_K to $\tilde{\mathbf{u}}$ leads to the condition (on the basis of Eq. (12.23)) $\int_0^\infty d\tilde{u}\, \tilde{u}^4 e^{-\tilde{u}^2} \hat{\Phi}_c(\tilde{u})(\tilde{u}^2 - 5/2) = 0$. The other orthogonality conditions are identically satisfied by the right-hand side of Eq. (12.23). The determination of $\hat{\Phi}_v$ does not require the application of the orthogonality conditions.

The contribution of Φ_K to the stress tensor reads

$$P_{ij}^K = \int d\mathbf{u}\, u_i u_j f_0(\tilde{u}) \Phi_K = K n M_v \frac{16}{15\sqrt{\pi}} \sqrt{\frac{2T}{3}} \frac{\partial V_i}{\partial \tilde{r}_j}, \tag{12.24}$$

where M_v is given by $M_v = \int_0^\infty dx\, x^6 \hat{\Phi}_v(x) e^{-x^2} \approx -1.3224$ (the integration employs the numerically determined function, $\hat{\Phi}_v$). Hence, one obtains

$$P_{ij}^K = -2\tilde{\mu}_0 n \ell \sqrt{T} \frac{\partial V_i}{\partial r_j}, \tag{12.25}$$

where $\tilde{\mu}_0 \approx 0.3249$. Similarly, the contribution of Φ_K to the heat flux is

$$Q_i^K = \frac{1}{2} \int d\mathbf{u}\, u^2 u_i f_0(\tilde{u}) \Phi_K = K n M_c \frac{2}{3\sqrt{\pi}} \left(\frac{2T}{3}\right)^{\frac{3}{2}} \frac{\partial \log T}{\partial \tilde{r}_i}, \tag{12.26}$$

where $M_c = \int_0^\infty dx\, x^6 \left(x^2 - 5/2\right) \hat{\Phi}_c(x) e^{-x^2} \approx -2.003$. One thus obtains

$$Q_i^K = -\tilde{\kappa}_0 n \ell \sqrt{T} \frac{\partial T}{\partial r_i}, \tag{12.27}$$

where $\tilde{\kappa}_0 \approx 0.4101$. These calculated values of the transport coefficients are in very close agreement with those calculated before for hard (smooth, elastic) spheres, see, e.g. [244]. Since, following Eq. (12.9), the energy sink term, Γ, has a prefactor $\epsilon \equiv 1 - e^2$, the function Φ_K should contribute an $O(\epsilon K)$ term to it. This term, $\Gamma_{K\epsilon}$, can be computed by exploiting the invariance of the double integral in Eq. (12.9) to the exchange $\mathbf{v}_1 \leftrightarrow \mathbf{v}_2$:

$$\Gamma_{K\epsilon} = \frac{\epsilon \pi d^2}{4n} \int d\mathbf{u}_1 d\mathbf{u}_2 u_{12}^3 f_0(\tilde{u}_1) f_0(\tilde{u}_2) \Phi_K(\mathbf{u}_1). \tag{12.28}$$

Clearly $\int d\mathbf{u}_2 u_{12}^3 f_0(\tilde{u}_2)$ is an isotropic function of u_1. Symmetry considerations and the orthogonality conditions imply that the integral in Eq. (12.28) vanishes, hence $\Gamma_{K\epsilon} = 0$.

The equation determining Φ_ϵ is obtained from Eq. (12.12) by expanding $\tilde{\mathbf{B}}(\tilde{f}, \tilde{f}, e)$ to first order in ϵ and retaining terms of $O(\epsilon)$. One obtains

$$\tilde{\mathbf{L}}(\Phi_\epsilon) = \tilde{\mathbf{D}}_\epsilon \log n + 2\sqrt{\frac{3}{2T}} \tilde{u}_i \tilde{\mathbf{D}}_\epsilon V_i + \left(\tilde{u}^2 - \frac{3}{2}\right) \tilde{\mathbf{D}}_\epsilon \log T$$

$$- \frac{\epsilon}{\pi^{\frac{5}{2}}} \int_{\hat{\mathbf{k}}\cdot\tilde{\mathbf{u}}_{12}>0} d\hat{\mathbf{k}} d\tilde{\mathbf{u}}_2 (\hat{\mathbf{k}} \cdot \tilde{\mathbf{u}}_{12}) \left(1 - \frac{1}{2}(\hat{\mathbf{k}} \cdot \tilde{\mathbf{u}}_{12})^2\right) e^{-\tilde{u}_2^2}. \quad (12.29)$$

The integral on the right-hand side of Eq. (12.29) is obtained by utilising the relation $\tilde{u}_1'^2 + \tilde{u}_2'^2 = \tilde{u}_1^2 + \tilde{u}_2^2 + \epsilon(\hat{\mathbf{k}} \cdot \tilde{\mathbf{u}}_{12})^2/2 + O(\epsilon^2)$. Clearly (cf. Eqs. (12.13)–(12.15)), $\tilde{\mathbf{D}}_\epsilon \log n = \tilde{\mathbf{D}}_\epsilon V_i = 0$ and $\tilde{\mathbf{D}}_\epsilon \log T = -\epsilon \tilde{\Gamma}_0$, where $\tilde{\Gamma}_0$ is the zeroth order term in the expansion of $\tilde{\Gamma}$ (obtained by substituting $\tilde{f}_0$ for $\tilde{f}$ in Eq. (12.29)); its value is: $\tilde{\Gamma}_0 = (2/3)\sqrt{2/\pi}$. Consequently, to $O(\epsilon)$, $\Gamma_\epsilon = (\epsilon/\ell)\sqrt{16/(27\pi)}T^{3/2}$. Upon carrying out the integral on the right-hand side of Eq. (12.29), the equation for Φ_ϵ assumes the form:

$$\tilde{\mathbf{L}}(\Phi_\epsilon) = -\epsilon \left[\sqrt{\frac{2}{\pi}} \left(\frac{2}{3}\tilde{u}^2 - 1\right) + \frac{3 - 2\tilde{u}^2}{8\sqrt{\pi}} e^{-\tilde{u}^2} + \frac{(5 + 4\tilde{u}^2 - 4\tilde{u}^4)\mathrm{erf}(\tilde{u})}{16\tilde{u}}\right].$$

$$(12.30)$$

The right-hand side of Eq. (12.30) is orthogonal to the invariants 1, $\tilde{\mathbf{u}}$ and $\tilde{u}^2$. The isotropy of $\tilde{\mathbf{L}}$ implies that the solution of Eq. (12.30) assumes the form $\Phi_\epsilon(\tilde{\mathbf{u}}) = \epsilon \hat{\Phi}_e(\tilde{u})$, where $\hat{\Phi}_e$ is a function of the *speed* $\tilde{u}$. It can also be shown [286] that $\hat{\Phi}_e(\tilde{u})$ is formally even with respect to $\tilde{u}$ and that it is asymptotically (for $\tilde{u} \gg 1$) proportional to $\tilde{u}^2 \log \tilde{u}$. An expansion of $\hat{\Phi}_e$ in a set of functions obeying these symmetry and asymptotic properties can be used [286] in order to obtain a numerical solution of Eq. (12.30); to this (inhomogeneous) solution one must add a combination of the invariants to render it orthogonal to the invariants.

It is straightforward to deduce from the isotropy of Φ_ϵ and its orthogonality to the invariants that *it does not contribute to the stress-tensor nor to the heat flux*. It only contributes a second order, in ϵ, term to Γ:

$$\Gamma_{\epsilon\epsilon} = \frac{\epsilon^2 \pi d^2}{4n} \int d\mathbf{u}_1 d\mathbf{u}_2 u_{12}^3 f_0(\tilde{u}_1) f_0(\tilde{u}_2) \hat{\Phi}_e(\tilde{u}_1). \quad (12.31)$$

The integrals over $\hat{\mathbf{u}}_1$ and $\hat{\mathbf{u}}_2$ in Eq. (12.31) are trivial. The remaining double integral over u_1 and u_2 can be evaluated by numerical means. The result is $\Gamma_{\epsilon\epsilon} \approx -0.0352\epsilon^2 nd^2 T^{3/2}$. Another possible contribution to Γ arises from the product of Φ_K and Φ_ϵ in the expansion of f, substituted in Eq. (12.16). This term can be shown to vanish by symmetry arguments.

The solutions at $O(K\epsilon)$ and $O(K^2)$ proceed along the same lines as the above derivations. They are not reproduced here [286].

The above method yields the following constitutive relations, presented below to second order in K and first order in ϵ. The heat flux assumes the form

$$
\begin{aligned}
Q_i = & -\tilde{\kappa} n\ell\sqrt{T}\frac{\partial T}{\partial r_i} - \tilde{\lambda}\ell\sqrt{T^3}\frac{\partial n}{\partial r_i} \\
& + \tilde{\theta}_1 n\ell^2\frac{\partial V_j}{\partial r_j}\frac{\partial T}{\partial r_i} + \tilde{\theta}_2 n\ell^2\left(\frac{2}{3}\frac{\partial}{\partial x_i}\left(T\frac{\partial V_j}{\partial r_j}\right) + 2\frac{\partial V_j}{\partial r_i}\frac{\partial T}{\partial r_j}\right) \\
& + \tilde{\theta}_3\ell^2\frac{\partial V_j}{\partial r_i}\frac{\partial(nT)}{\partial r_j} + \tilde{\theta}_4 n\ell^2 T\frac{\partial^2 V_j}{\partial r_i\partial r_j} + \tilde{\theta}_5 n\ell^2\frac{\partial V_j}{\partial r_i}\frac{\partial T}{\partial r_j},
\end{aligned} \tag{12.32}
$$

where $\tilde{\kappa} \approx 0.4101 + 0.1072\epsilon + O(\epsilon^2)$, $\tilde{\lambda} \approx 0.2110\epsilon + O(\epsilon^2)$ and the values of the $\tilde{T}_i$s are: $\tilde{\theta}_1 \approx 1.2291$, $\tilde{\theta}_2 \approx -0.6146$, $\tilde{\theta}_3 \approx -0.3262$, $\tilde{\theta}_4 \approx 0.2552$, $\tilde{\theta}_5 \approx 2.6555$.

Notice that the heat flux includes a 'non-Fourier' term at order $K\epsilon$, which is proportional to the density gradient and to ϵ. This term had been first discovered in [286], rediscovered shortly thereafter in [288], and further rediscovered on the basis of numerical simulations in [303]. The reason it was not discovered in the old kinetic approaches, see, e.g., [304–306], is that the latter were based on guesses of the form of the distribution function, and these guesses were not sufficiently close to the 'true' distribution function to obtain this term. A similar term was obtained [307] as a finite density correction to the heat flux (stemming from the Enskog correction). The present term is a pure Boltzmann level (i.e., it does not vanish in the low density limit) contribution. This term is of much importance in applications, e.g., in the determination of the temperature and density profile of a vertically vibrated granular system [308].

The stress-tensor reads

$$
\begin{aligned}
P_{ij} = & \frac{1}{3}nT\delta_{ij} - 2\tilde{\mu} n\ell\sqrt{T}\overline{\frac{\partial V_i}{\partial r_j}} \\
& + \tilde{\omega}_1 n\ell^2\overline{\frac{\partial V_k}{\partial r_k}\frac{\partial V_i}{\partial r_j}} - \tilde{\omega}_2 n\ell^2\left(\frac{1}{3}\overline{\frac{\partial}{\partial r_i}\left(\frac{1}{n}\frac{\partial(nT)}{\partial r_j}\right)}\overline{\frac{\partial V_i}{\partial r_k}\frac{\partial V_k}{\partial r_j}} + 2\overline{\frac{\partial V_i}{\partial r_k}\frac{\partial V_k}{\partial r_j}}\right) \\
& + \tilde{\omega}_3 n\ell^2\overline{\frac{\partial^2 T}{\partial r_i\partial r_j}} + \tilde{\omega}_4\frac{\ell^2}{T}\overline{\frac{\partial(nT)}{\partial r_i}\frac{\partial T}{\partial r_j}} + \tilde{\omega}_5\frac{n\ell^2}{T}\overline{\frac{\partial T}{\partial r_i}\frac{\partial T}{\partial r_j}} + \tilde{\omega}_6 n\ell^2\overline{\frac{\partial V_i}{\partial r_k}\frac{\partial V_k}{\partial r_j}},
\end{aligned} \tag{12.33}
$$

where $\tilde{\mu} \approx 0.3249 + 0.0576\epsilon + O(\epsilon^2)$ and the values of the $\tilde{\omega}_i$s are [286] $\tilde{\omega}_1 \approx 1.2845$, $\tilde{\omega}_2 \approx 0.6422$, $\tilde{\omega}_3 \approx 0.2552$, $\tilde{\omega}_4 \approx 0.0719$, $\tilde{\omega}_5 \approx 0.0231$, $\tilde{\omega}_6 \approx 2.3510$.

The inelastic dissipation term, Γ, reads, to $O(\epsilon K^2)$,

$$\Gamma = \frac{\tilde{\delta}}{\ell}T^{\frac{3}{2}} + \tilde{\rho}_1 \epsilon \ell \sqrt{T} \overline{\frac{\partial V_i}{\partial r_j} \frac{\partial V_i}{\partial r_j}} \tilde{\rho}_2 \frac{\epsilon \ell}{\sqrt{T}} \frac{\partial T}{\partial r_i} \frac{\partial T}{\partial r_i}$$

$$+ \tilde{\rho}_3 \frac{\epsilon \ell}{n\sqrt{T}} \frac{\partial(nT)}{\partial r_i} \frac{\partial T}{\partial r_i} + \tilde{\rho}_4 \epsilon \ell \sqrt{T} \frac{\partial^2 T}{\partial r_i \partial r_i}, \qquad (12.34)$$

where [286] $\tilde{\delta} \approx \sqrt{16/(27\pi)}\epsilon - 0.0112\epsilon^2$, $\tilde{\rho}_1 \approx 0.1338$, $\tilde{\rho}_2 \approx 0.2444$, $\tilde{\rho}_3 \approx -0.0834$ and $\tilde{\rho}_4 \approx 0.0692$. Notice that Γ is proportional to $1/\ell$, hence to non-vanishing leading order in K and ϵ (i.e. K^0 and ϵ^1) it is proportional to $1/\ell$; the next non-vanishing contribution to Γ is $O(K^2\epsilon)$ and it is proportional to ℓ. This property, which is specific to inelastic systems, indicates that (unlike in elastic systems) one cannot deduce the Knudsen orders of terms in the hydrodynamic equations by counting powers of ℓ; instead one must consider the appropriate order in the (gradient) expansion of f.

The normal stress difference (between P_{xx} and P_{yy}, normalised by their average), calculated from Eq. (12,33), equals 0.45 for $e = 0.8$ and 0.88 for $e = 0.6$, in good agreement with numerical results [48]: 0.42 and 0.86, respectively (for a volume fraction of $\nu = 0.025$). In general the normal stress difference in granular gases is $O(1)$.

The latter finding can be understood without the (rather cumbersome) technical details involved in the expansion. Consider again the case of simple shear: $\mathbf{V} = \gamma y \hat{\mathbf{x}}$. Straightforward tensorial and dimensional analysis reveals that to second order in the gradients the normal stresses, P_{xx} and P_{yy} should assume the form $P_{xx} = p(1 + c_1\gamma^2)/3$ and $P_{yy} = p(1 + c_2\gamma^2)/3$, where p is the pressure. Notice that the linear, Navier–Stokes order term does not contribute to P_{xx} or P_{yy} as $\partial V_x/\partial x = \partial V_y/\partial y = 0$. Since the above expressions in the brackets have to be dimensionless, $c_i\gamma^2 = d_i\gamma^2\ell^2/T$, where d_i are $O(1)$ dimensionless numbers (which depend on ϵ). For molecular gases (e.g. at STP) typical values of $\gamma^2\ell^2/T$ are of the order of 10^{-18}. As $T = C\gamma^2\ell^2/\epsilon$ for granular gases, the anisotropic correction to the pressure is typically $O(1)$ when $\epsilon = O(1)$. Again, it is the lack of scale separation in granular gases that elevates a rather negligible (Burnett) effect in molecular gases to the level of an $O(1)$ (and measurable, of course) term.

The above constitutive relations, though correct to Burnett order, are nominally insufficient for the description of steady states of rapid granular flows, in the following sense. Consider, e.g., the stress tensor P_{ij}. The leading order (elastic) viscous contribution is $O(K)$ and the leading inelastic correction is $O(\epsilon K)$. In a steady shear flow, at given T, $\epsilon \propto \gamma^2 = O(K^2)$, thus the leading inelastic correction is also $O(K^3)$, hence one needs to calculate the super-Burnett contributions (a rather frightening prospect) alongside the leading order inelastic corrections to retain consistency in the orders of K or ϵ. It is possible to develop a theory for steady states, which is systematic and consistent in the orders of the small parameters, see, e.g. [285, 300].

12.4 Boundary conditions

Equations of motion without corresponding boundary conditions are quite useless, unless all one wishes to study is, e.g., periodic solutions. The establishment of boundary conditions for gases is well known to be a hard problem, since it turns out that one needs to solve an integrodifferential equation (see below) which lacks a small parameter. In Maxwell's pioneering work on this problem, he assumed that the distribution function of the particles that impinge on a solid wall is Maxwellian. When a wall collision law ('accommodation function') is assumed or known, this approach enables one to obtain a closed form for the near-wall distribution function, hence the boundary conditions. A similar approach has been invoked in early studies of granular gases, see, e.g., [309]. A recent theory, in which a boundary condition has been derived for a specific state of a granular gas (the vertically vibrated gas) can be found in [310]. See also [311] for a numerical study of boundary conditions for a sheared flow. Since it is quite obvious (and correct) that the presence of a solid wall changes the distribution of the particles hitting the wall (because of recollisions), it is clear that the Maxwell approach, or its granular application, are at best low order approximations. Indeed, some (approximate) theories for boundary conditions of molecular gases go beyond the Maxwell level [312, 313].

In this section we present the essentials of (what we believe to be a first) a systematic method for deriving boundary conditions for granular gases (which is also relevant to molecular gases), see also [314]. The method is based on the well known fact that a gas locally equilibrates in a matter of about three collisions per particle. These collisions are represented by the right-hand side of the Boltzmann equation. By constructing a formalism for solving the Boltzmann equation near a solid boundary, in which each new order takes into account one more collision per particle than the previous order, we have been able to develop a systematic method for establishing boundary conditions for molecular and granular gases alike. It also turns out that this expansion converges very rapidly. The new method is outlined immediately below.

Consider a solid boundary situated at $z = 0$ and a (granular) gas occupying $z > 0$. The solution of the pertinent Boltzmann equation can be written as $f = f_{ce} + f_0 \Phi_w = f_0 (1 + \Phi_{ce} + \Phi_w)$, where Φ_w represents the effect of the wall and $\tilde{f}_0 (1 + \Phi_{ce})$ is the CE (or, in principle, an exact solution of the Boltzmann equation far enough from the boundary). The function Φ_w should vanish far away from the boundary (in practice, a few mean free paths away from it). Notice that the value of Φ_{ce} at the boundary is determined by extrapolating the CE solution to the boundary, since the CE solution itself is not valid near the boundary. It is convenient to choose a frame of reference in which the solid boundary is stationary (when the boundary moves at constant velocity this involves only a Galilean transformation; below we specialise to this case). Inside the domain influenced by the boundary, i.e. the *Knudsen layer*, the macroscopic flow field **V** (rescaled by the square root of the temperature field, T) is of the order of the Knudsen number (see also below), hence it is justified to expand the (extrapolated) CE solution f_{ce} in the Knudsen layer in powers of the rescaled velocity field.

The boundary conditions are conditions to be satisfied by the hydrodynamic fields extrapolated to the boundary, not by the true values of these fields there. The role of the boundary conditions is to ensure that the solutions of the hydrodynamic equations outside the Knudsen layer, whose width is a few mean free paths, are compatible with the kinetics near the boundary. In other words, the hydrodynamic equations provide 'outer' solutions that should match the 'inner' kinetic solution next to the boundary. Since Φ_w vanishes when the hydrodynamic fields are space independent it follows that Φ_w is $O(K)$.

The method for obtaining boundary conditions is outlined below at $O(\epsilon^0 K)$; results are presented to $O(\epsilon K)$. At order $\epsilon^0 K$ it is easy to see that it is sufficient to consider the following linearised Boltzmann equation:

$$v_z \frac{\partial \Phi_w}{\partial z} = \mathbf{L}(\Phi_w). \tag{12.35}$$

whose solubility conditions are

$$\int \mathrm{d}\mathbf{v}\, v_z \psi_i(\mathbf{v}) e^{-v^2} \Phi_w(z=0) = 0,$$

where ψ_i are the invariants of **L**. These relations require that Φ_w does not contribute to the mass, momentum and energy flux in the z direction. The latter fluxes are completely determined by the CE solution. The physical role of these requirements is to ensure the continuity of the fluxes, i.e. the values of the fluxes in the bulk, given by the CE expansion, should match the rate of the transfer of the corresponding moments to the boundary. Equation (12.35) is not easy to solve since it is a nontrivial integrodifferential equation.

As mentioned above, the basis of the method described below is the observation that initial distribution functions converge rapidly (for $\epsilon \ll 1$) to local equilibrium or equilibrium-like distributions in a matter of a few collisions. Indeed, as the results presented below indicate, this approach is justified since the contributions of multiple collisions with the wall to the transport coefficients are increasingly smaller. It is convenient to use the Fredholm [315] form of the linearised Boltzmann operator, $\mathbf{L}$: $\mathbf{L} = \mathbf{A} - q$, where

$$\mathbf{A}\Phi = \frac{1}{\pi^{\frac{3}{2}}} \int d\mathbf{v}' e^{-v'^2} \left(\frac{2}{R} e^{w^2} - R \right) \Phi(\mathbf{v}') \equiv \int d\mathbf{v}' K(\mathbf{v}, \mathbf{v}') \Phi(\mathbf{v}'), \qquad (12.36)$$

where $R = |\mathbf{v} - \mathbf{v}'|$, $w = \dfrac{\mathbf{v} \times \mathbf{v}'}{R}$

and

$$q(v) = \frac{1}{\sqrt{\pi}} \left(e^{-v^2} + \frac{\sqrt{\pi}}{2} \left(2v + \frac{1}{v} \right) \mathrm{erf}(v) \right)$$

is a positive definite function which depends on the speed v alone. The operator $\mathbf{A}$ includes the full 'gain term' of $\mathbf{L}$ and part of the 'loss term'. The second part of the 'loss term' is q. The function q essentially represents the rate of 'loss' of particles, having velocity $\mathbf{v}$, due to collisions, and it is trivially related to the (velocity dependent) mean free path. The operator $\mathbf{A}$ represents the *net* rate of 'creation' of particles with velocity $\mathbf{v}$ at a given point in space. Using Eq. (12.36) one can transform Eq. (12.35) as follows:

$$\frac{\partial}{\partial z} \left(e^{qz/v_z} \Phi_w \right) = \frac{1}{v_z} e^{qz/v_z} \mathbf{A}\Phi_w. \qquad (12.37)$$

The solution of Eq. (12.37) can be formally written as follows. When $v_z > 0$:

$$\Phi_w(z) = e^{-qz/v_z} \Phi_w(0) + \frac{1}{v_z} \int_0^z dz' e^{q(z'-z)/v_z} \mathbf{A}\Phi_w(z'), \qquad (12.38)$$

where only the dependence of Φ_w on z is explicitly spelled out. This equation should be interpreted as follows: particles having positive v_z are either particles that arrived from the boundary ($z = 0$) and proceeded upwards to z without any further collision, or particles that arrived from z', where they last encountered a collision, and moved up to z (note that $z' \le z$). When $v_z < 0$,

$$\Phi_w(z) = -\frac{1}{v_z} \int_z^\infty dz' e^{q(z'-z)/v_z} \mathbf{A}\Phi_w(z'). \qquad (12.39)$$

The interpretation of this result is similar to that for Eq. (12.38). Let $\mathbf{P}$ and $\mathbf{N}$ be projection operators on the $v_z \ge 0$ and $v_z < 0$ velocity subspaces, respectively (with $\mathbf{P} + \mathbf{N} = \mathbf{I}$). It follows that

$$\mathbf{P}\Phi_w = \mathbf{PGP}\Phi_w(0) + \mathbf{PQP}\Phi_w + \mathbf{PQN}\Phi_w$$

and

$$\mathbf{N}\Phi_w = \mathbf{NSP}\Phi_w + \mathbf{NSN}\Phi_w,$$

where

$$\mathbf{G}\phi = e^{-qz/v_z} \phi, \qquad (12.40)$$

$$\mathbf{Q}\phi = \frac{1}{v_z} \int_0^z dz' e^{q(z'-z)/v_z} \mathbf{A}\phi(z')$$

$$= \frac{1}{v_z} \int_0^z dz' \int d\mathbf{v}' e^{q(z'-z)/v_z} K(\mathbf{v}, \mathbf{v}')\phi(\mathbf{v}', z') \qquad (12.41)$$

and

$$\mathbf{S}\phi = -\frac{1}{v_z} \int_z^\infty dz' e^{q(z'-z)/v_z} \mathbf{A}\phi(z')$$

$$= -\frac{1}{v_z} \int_z^\infty dz' \int d\mathbf{v}' e^{q(z'-z)/v_z} K(\mathbf{v}, \mathbf{v}')\phi(\mathbf{v}', z'). \qquad (12.42)$$

The operator **NS** represents the events in which a particle whose velocity is $\mathbf{v}$, with $v_z < 0$, emerges from a collision at a point $z' > z$ and moves to the point z without further collision. Similarly, the operator **PQ** represents the events in which a particle whose velocity is $\mathbf{v}$, with $v_z > 0$, collides at $z' < z$ and proceeds, without further collision, to the point z. The operator **G** is the propagator corresponding to the motion of a particle from the boundary, $z = 0$, to z, without collision. Straightforward algebra yields:

$$\Phi_\mathrm{w} = (\mathbf{I} - \mathbf{C})^{-1}\mathbf{PGP}\Phi_\mathrm{w}(0) = (\mathbf{I} + \mathbf{C} + \mathbf{C}^2 + \cdots)\mathbf{PGP}\Phi_\mathrm{w}(0). \qquad (12.43)$$

where $\mathbf{C} \equiv \mathbf{NS} + \mathbf{PQ}$. The interpretation of Eq. (12.43) is rather simple: the function Φ_w is determined from its value at the boundary, $\Phi_\mathrm{w}(0)$, via successive processes of collisions and free motions. For example, the nth order term $\mathbf{C}^n\mathbf{PGP}\Phi_\mathrm{w}(0)$ is the contribution of the particles that come from the boundary (with positive z-component velocity), collide n times, following which their velocity is $\mathbf{v}$ and they move without collision to the point z. The practical implementation of these formal operators involves the calculation of integrals as per the definitions in Eqs. (12.40)–(12.42). The value of $\Phi_\mathrm{w}(0)$ can be obtained by using the accommodation function of the boundary [246, 314] and the solubility conditions. A similar formulation holds for the case of an inelastic gas.

Some results obtained using the above formulation are presented next. In the case of an inelastic gas and a diffusely (but elastically) reflecting boundary, the (slip) velocity parallel to the boundary is given by $V_x = \alpha \ell \partial V_x / \partial z$, where $\alpha \approx 0.728 + 0.130\epsilon$. When the boundary is characterised by a degree of inelasticity (for the normal part of the velocity), ϵ_w, and a thermal gradient is present in the z direction (as must be the case for an energy absorbing boundary), one obtains the following result: $V_z = -\zeta \ell \sqrt{T} \left(\dfrac{\partial \log n}{\partial z} + \dfrac{1}{2} \dfrac{\partial \log T}{\partial z} \right)$, where $\zeta \approx 0.044\epsilon$. This boundary condition for V_z is quite surprising. It implies that V_z does not necessarily vanish in the general case, unless a specific relation between the gradients of the number density and granular temperature is satisfied, hence mass conservation seems to be violated. This result pertains only to inelastically colliding systems as V_z is predicted to vanish for $\epsilon = 0$. The resolution of this 'paradox' can be found by noting (again) that in a steady sheared state the orders $K\epsilon$ (which is the order of the above expression for V_z) and K^3 are 'the same', hence a correction that is of super-Burnett order should be added to the above expression for V_z. When this is done the value of V_z vanishes in steady states, as it should. In the same situation, the boundary condition for the temperature, T, is $\epsilon_w T = \beta\ell\dfrac{\partial T}{\partial z} + \delta\ell\dfrac{T}{n}\dfrac{\partial n}{\partial z}$, where $\beta \approx 2.671 + 1.945\epsilon$ and $\delta \approx 3.810\epsilon$.

12.5 Weakly frictional granular gases

Most models of granular gases ignore the frictional interactions of the grains, even though all grains experience friction. Clearly, one of the reasons for this state of affairs is that it is not trivial to model frictional restitution; worse, it is hard to develop hydrodynamic equations of motion which account for friction (see, however, [306, 316–318]). In addition, it is noteworthy that frictionless models have encountered numerous successes in describing some aspects of the dynamics of granular gases [248]. This may indicate that tangential restitution may not be very important for the description of some properties of granular gases. However, it is known that some aspects of granular gas dynamics are strongly dependent on friction, see, e.g., the recent study of the effects of friction on granular patterns [319], and friction induced hysteresis [320]). Furthermore, it is known that friction enhances non-equipartition in the homogeneous cooling state, see, e.g., [321–325]. It is therefore curious that a rather small proportion of the granular literature is devoted to frictional granular hydrodynamics, see, e.g., [306, 327, 378, 483, 484].

The study of gases whose constituents experience frictional interactions started (in 1894!) in the realm of molecular gases [243, 329]. Applications include granular celestial systems [330, 331]. In previous kinetic theoretical based studies of granular hydrodynamics [306, 327, 328] it is assumed that the basic distribution function is Maxwellian in both the velocity and angular velocity (usually different rotational and translational temperatures are allowed for), and corrections due to gradients are identified (on the basis of symmetry). The assumed distribution function is substituted in the Enskog equations [2] [334], resulting in a closure for the constitutive relations. The above distributions with $\beta_0 \to 1$ and $e \to 1$ (see below) correspond to the rough elastic limit, which is appropriate for, e.g., rough molecules. Since (unlike in the molecular case) the basic model used for the description of granular gases is that of *smooth* particles (usually spheres), our goal here is to study the

---

[2] The Enskog equations are equations for averages of single particle properties, which can be derived from the Boltzmann equation, or by direct considerations. They should not be confused with the Enskog–Boltzmann equation, which is a modified Boltzmann equation, designed to account for finite density.

case of weak friction, as a perturbation around the smooth limit. Specifically, we consider, as in the above, a monodisperse system of spherical grains of mass $m = 1$, diameter d, and moment of inertia I (for homogeneous spheres, $I = 2 (d/2)^2 /5$), each. Denote the radius of gyration a grain by κ (with $\kappa \equiv 4I/d^2$). The velocity of particle 'i' is denoted by $\mathbf{v}_i$, and its angular momentum by ω_i. It is convenient to define a 'spin variable' $\mathbf{s}_i \equiv \dfrac{d}{2}\omega_i$, which has dimensions of speed.

Consider a binary collision between sphere '1' and sphere '2'. Let $\mathbf{k}$ be a unit vector pointing from the centre of sphere '2' to the centre of sphere '1'. The relative velocity of sphere '1' with respect to sphere '2', at the point of contact (when they are in contact), is $\mathbf{g}_{12} = \mathbf{v}_{12} + \mathbf{k} \times \mathbf{s}_{12}$, where $\mathbf{v}_{12} \equiv \mathbf{v}_1 - \mathbf{v}_2$, and $\mathbf{s}_{12} \equiv \mathbf{s}_1 + \mathbf{s}_2$. *In the following, precollisional entities are primed.* During a collision the normal component of the relative velocity changes according to $\mathbf{k} \cdot \mathbf{g}_{12} = -e\,\mathbf{k} \cdot \mathbf{g}'_{12}$, where e is the coefficient of normal restitution. The effect of the tangential impulse at a collision is modeled as in [323, 328], with slight notational differences. Let $\mathbf{J} = \mathbf{v}_1 - \mathbf{v}'_1$, its decomposition into normal and tangential components being

$$\mathbf{J} = A\mathbf{k} + B\frac{(\mathbf{k}\times(\mathbf{k} \times \mathbf{g}'_{12}))}{|\mathbf{k} \times \mathbf{g}'_{12}|}.$$

Let γ be the angle between $-\mathbf{k}$ and the relative velocity of the grains at the point of contact: $\cos \gamma \equiv -\mathbf{k} \cdot \mathbf{g}'_{12}/'g'_{12} = -\mathbf{k} \cdot \mathbf{v}'_{12}/'g'_{12}$, where $g'_{12} \equiv ||\mathbf{g}'_{12}||$; clearly, $0 \leq \gamma \leq \pi/2$. Define γ_0 such that if $\gamma > \gamma_0$ there is sliding (Coulomb friction) during the collision and $B = \mu A$, where μ is the friction coefficient, while if $\gamma \leq \gamma_0$ there is sticking (or the grain is 'rough') and $\mathbf{k} \times (\mathbf{k} \times \mathbf{g}_{12}) = -\beta_0 \mathbf{k} \times (\mathbf{k} \times \mathbf{g}'_{12})$, with $-1 \leq \beta_0 \leq 1$. In the case of sliding one obtains

$$\mathbf{k} \times (\mathbf{k} \times \mathbf{g}_{12}) = \left(1 - \frac{1+\kappa}{\kappa} (1+e) \mu \cot \gamma\right) \mathbf{k} \times (\mathbf{k} \times \mathbf{g}'_{12}). \qquad (12.44)$$

It follows that in both cases $\mathbf{k} \times (\mathbf{k} \times \mathbf{g}_{12}) = -\beta (\gamma) \mathbf{k} \times (\mathbf{k} \times \mathbf{g}'_{12})$, where (requiring $\beta (\gamma)$ to be a continuous function of γ)

$$\beta (\gamma) = \min \left\{\beta_0, -1 + \frac{1+\kappa}{\kappa} (1+e) \mu \cot \gamma\right\}, \qquad (12.45)$$

where

$$\cot \gamma_0 = \frac{\kappa}{1+\kappa} \frac{1+\beta_0}{1+e} \frac{1}{\mu}.$$

Using the conservation laws for the linear and angular momenta one obtains the transformation between the precollisional and postcollisional velocities and spins of a colliding pair of grains:

$$\mathbf{v}_i = \mathbf{v}'_i - \sigma_i \frac{1+e}{2}(\mathbf{k} \cdot \mathbf{g}'_{12})\mathbf{k} + \sigma_i \frac{\kappa}{2} \frac{1+\beta(\gamma)}{1+\kappa}\mathbf{k} \times (\mathbf{k} \times \mathbf{g}'_{12})$$

$$\mathbf{s}_i = \mathbf{s}'_i + \frac{1}{2}\frac{1+\beta(\gamma)}{1+\kappa}\mathbf{k} \times \mathbf{g}'_{12}, \qquad (12.46)$$

where $i = 1, 2$, $\sigma_1 = 1$ and $\sigma_2 = -1$. The Jacobian of this transformation is given by

$$J(\gamma) \equiv \frac{\partial\,(\mathbf{v}_1, \mathbf{v}_2, \mathbf{s}_1, \mathbf{s}_2)}{\partial(\mathbf{v}_1', \mathbf{v}_2', \mathbf{s}_1', \mathbf{s}_2')} = \begin{cases} e\beta_0^2, & \gamma < \gamma_0 \\ e\,|\beta(\gamma)|, & \gamma > \gamma_0. \end{cases} \tag{12.47}$$

Let $f(\mathbf{v}_1, \mathbf{s}_1, \mathbf{r}, t)(\equiv f_1)$ denote the single particle distribution (of the velocity and spin) function at point $\mathbf{r}$ and time t. The Boltzmann equation satisfied by f_1 is

$$\frac{\partial f_1}{\partial t} + \mathbf{v}_1 \cdot \nabla f_1 = B(f, f, \mathbf{v}_1, \mathbf{s}_1)$$

$$\equiv d^2 \int_{\mathbf{k}\cdot\mathbf{v}_{12}>0} d\mathbf{v}_2 d\mathbf{s}_2 d\mathbf{k}\,(\mathbf{k}\cdot\mathbf{v}_{12}) \left(\frac{1}{eJ(\gamma)} f_1' f_2' - f_1 f_2 \right). \tag{12.48}$$

As explained in the above, the basic premise of the Chapman–Enskog expansion is that the dependence of the single particle distribution function on space (i.e., $\mathbf{r}$) and time, t, can be replaced by a dependence on the 'slow' fields. In the case at hand, the momentum and the number of particles (or the total mass) are conserved, hence the momentum density and the particle number (or mass) density are hydrodynamic fields. The granular temperature corresponding to the translational degrees of freedom is nearly conserved in the near-elastic, near-frictionless case, and is therefore a hydrodynamic field. In the same *limit* the spin degrees of freedom are decoupled from the translational degrees of freedom and the number density corresponding to each value of the spin, $\mathbf{s}$, is conserved as well, hence the infinite set of spin dependent number densities should be taken as hydrodynamic fields (in other words, the system is considered to be a mixture whose components are indexed by the spin). All in all, the general set of hydrodynamic fields is given by

$$\mathbf{V}(\mathbf{r},t) \equiv \frac{1}{n} \int \mathbf{v} f\,(\mathbf{v}, \mathbf{s}, \mathbf{r},t)\,d\mathbf{v}d\mathbf{s},$$

$$T(\mathbf{r},t) \equiv \frac{1}{n} \int u^2 f\,(\mathbf{v}, \mathbf{s}, \mathbf{r},t)\,d\mathbf{v}d\mathbf{s}$$

and

$$\rho\,(\mathbf{s}, \mathbf{r},t) \equiv \frac{1}{n} \int f\,(\mathbf{v}, \mathbf{s}, \mathbf{r},t)\,d\mathbf{v},$$

where $\mathbf{u}$ is the peculiar velocity $\mathbf{u} = \mathbf{v} - \mathbf{V}(\mathbf{r},t)$. Another field of interest is the velocity field corresponding to particles of spin $\mathbf{s}$. It is defined by

$$\mathbf{V}(\mathbf{s}, \mathbf{r},t) = \frac{1}{n\rho\,(\mathbf{s}, \mathbf{r},t)} \int \mathbf{v} f\,(\mathbf{v}, \mathbf{s}, \mathbf{r},t)\,d\mathbf{v},$$

The latter field is not a hydrodynamic field, as it does not correspond to a conserved entity in the smooth elastic limit, and indeed, like any such field it is enslaved to the slow fields, i.e., expressible in terms of these fields (see below). The resulting expression qualifies as an additional constitutive relation. A similar procedure to that used in the frictionless case now produces the continuum equations of motion from the (moments of) Boltzmann equation, the result being [483, 484]

$$n \frac{DV_\alpha}{Dt} + \frac{\partial}{\partial r_\beta} P_{\alpha\beta} = 0$$

$$n \frac{DT}{Dt} + 2 \frac{\partial V_\alpha}{\partial r_\beta} P_{\alpha\beta} + 2 \frac{\partial Q_\alpha}{\partial r_\alpha} = -2\Gamma$$

$$n \frac{D\rho\,(s)}{Dt} = \int B\,(f, f, \mathbf{v}, \mathbf{s})\,d\mathbf{v} - \nabla \cdot (n\rho\,(s)\,, \delta \mathbf{V}\,(s))\,, \quad (12.49)$$

where D/Dt is the material derivative, and the summation convention is assumed. Also, $\delta \mathbf{V}\,(s, \mathbf{r}, t) = \mathbf{V}\,(s, \mathbf{r}, t) - \mathbf{V}\,(\mathbf{r}, t)$ is the relative velocity between the particles of spin s and the velocity field. The stress tensor is given by

$$P_{\alpha\beta} = \int u_\alpha u_\beta f\,d\mathbf{u}d\mathbf{s}.$$

The heat flux vector is given by

$$Q_\alpha = \frac{1}{2} \int u_\alpha u^2 f\,d\mathbf{u}d\mathbf{s},$$

and the energy sink term is

$$\Gamma = -\int d\mathbf{v}_1 d\mathbf{s}_1 \frac{v_1^2}{2} B(f, f, \mathbf{v}_1, \mathbf{s}_1).$$

Consider Eq. (12.45). When $\beta_0 = -1$, it follows that $\beta\,(\gamma) = -1$ for all allowed values of γ_0, and the collision is always 'smooth'. Energy is conserved if, in addition, $e = 1$. Next, consider the near-smooth case, for which $\beta_0 \approx -1$. Clearly, following Eq. (12.45), $\beta(\gamma) = \beta_0$ unless $\mu \cot(\gamma)$ is sufficiently small, i.e., either μ or $\cot(\gamma)$ is small. When μ is taken to be $O(1)$, $\beta(\gamma) = -1 + (1 + \kappa)(1 + e)\mu \cot(\gamma)\kappa$ for γ close to $\pi/2$, i.e., for near grazing collisions. On the other hand, when μ is very small, this equality can hold for almost all values of γ (except for γ near zero). This implies that the small parameters corresponding to friction can be taken to be either $\epsilon_3 \equiv 1 - \beta_0^2$ and μ, or just ϵ_3, assuming μ to be $O(1)$. With the first choice, both the Coulomb and the non-Coulombic friction are small and the transition angle between them is finite, whereas the second choice assumes an $O(1)$ Coulomb friction (and the transition angle is near grazing). These two choices of expansion near the smooth limit give different expansions. Here we choose to specialise to the case of finite Coulomb friction. In this case, the small parameters are: the degree of normal inelasticity, $\epsilon \equiv 1 - e^2$, the small parameter corresponding to the friction, ϵ_3, and the Knudsen number, $K \equiv \ell/L$, where $\ell = 1/(\pi n d^2)$ is the mean free path (n being the number density) and L is a macroscopic scale.

The zeroth order distribution function, f_0 (which solves the Boltzmann equation for $\epsilon = \epsilon_3 = K = 0$), is given by $f_0(\mathbf{u}, \mathbf{s}) = f_M(\mathbf{u}) \rho(\mathbf{s})$, where $f_M(\mathbf{u}) = n(3/2\pi T)^{3/2} e^{-3u^2/(2T)}$. The full single particle distribution function $f(\mathbf{v}, \mathbf{s})$ can be written as $f(\mathbf{v}, \mathbf{s}) = f_0(\mathbf{u}, \mathbf{s})(1 + \phi(\mathbf{u}, \mathbf{s}))$. As in the frictionless case, it is assumed that ϕ can be expanded in the small parameters:

$$\phi = K\phi_K + \epsilon\phi_\epsilon + \epsilon_3\phi_3 + K\epsilon\phi_{K\epsilon} + K\epsilon_3\phi_{K3} + \epsilon\epsilon_3\phi_{\epsilon3} + \cdots.$$

Since $f_0(\mathbf{u}, \mathbf{s})$ satisfies $\int d\mathbf{v} f_0(\mathbf{u}, \mathbf{s}) = n\rho(\mathbf{s})$, and has (by construction) the correct temperature, (spin dependent) number density and average velocity dependence, it follows that $\int f_0(\mathbf{u}, \mathbf{s})\phi(\mathbf{u}, \mathbf{s}) d\mathbf{v} = 0$, for any value of $\mathbf{s}$, and $\int f_0(\mathbf{u}, \mathbf{s}) u^2 \phi(\mathbf{u}, \mathbf{s}) d\mathbf{v} d\mathbf{s} = 0$. Also $\delta\mathbf{V}(\mathbf{s}) = \dfrac{1}{n\rho(\mathbf{s})} \int f_0(\mathbf{u}, \mathbf{s})\phi(\mathbf{u}, \mathbf{s})\mathbf{v}\, d\mathbf{v}$.

The perturbative solution of the Boltzmann equation, in powers of K, ϵ and ϵ_3, involves the repeated solution of $\tilde{L}\phi_{\text{given order}} = R_{\text{same order}}$, where $R_{\text{same order}}$ depends on previous orders, and the linearised Boltzmann operator, $\tilde{L}$, is given by $\tilde{L}\phi \equiv d^2 \int_{\mathbf{k}\cdot\mathbf{v}_{12}>0} d\mathbf{v}_2 d\mathbf{s}_2 d\mathbf{k}(\mathbf{k}\cdot\mathbf{v}_{12}) f_0(\phi'_1 + \phi'_2 - \phi_1 - \phi_2)$. The algebra involved in these inversions (which requires the use of rather high orders in expansions in Sonine polynomials; convergence is obtained at sixth order in the polynomials) is rather heavy. We have devised a computationally-aided method, using a symbolic program, to overcome this difficulty.

The application of the Chapman–Enskog expansion yields constitutive relations [483, 484] for the stress tensor, the heat flux, the energy sink, the field $\delta\mathbf{V}$ and the term $\int B d\mathbf{v}$. We have developed those constitutive relations to second (Burnett) order in the small parameters: as the results are rather cumbersome they are not present here. Instead, we present some of the consequences of the analysis of the Boltzmann equation for nearly frictional granular gases.

Consider the homogeneous cooling state of a nearly frictionless granular gas. In this case the hydrodynamic equations reduce to:

$$\frac{dT}{d\tau} = -2\alpha T + 2\beta T_{\text{rot}} + 4\beta S^2,$$

$$\frac{dT_{\text{rot}}}{d\tau} = \nu T - 2\eta T_{\text{rot}} + \xi S^2,$$

$$\frac{dS^2}{d\tau} = -\delta S^2, \tag{12.50}$$

where $\alpha, \beta, \nu, \eta, \xi$ and δ are numerical prefactors that depend on the parameters that define the collision process,

$$T_{\mathrm{rot}}(\mathbf{r}, t) \equiv \int (\mathbf{s} - \mathbf{S}(\mathbf{r}, t))^2 \, \rho(\mathbf{s}, \mathbf{r}, t) \, d\mathbf{s}$$

is the rotational temperature (representing the spin fluctuations), and

$$\mathbf{S}(\mathbf{r}, t) \equiv \int \mathbf{S}\rho(\mathbf{s}, \mathbf{r}, t) d\mathbf{s}$$

is the spin density field. The parameter τ represents the accumulated number of collisions per particle (a measure of time), given by

$$\tau \equiv \int_0^t \frac{T^{1/2}(t')}{\ell} \, dt'.$$

The use of this measure of time renders the equations of motion linear.

The three eigenvalues of this system of equations are negative. The long-time decay rate of T and T_{rot} follows Haff's law, where

$$\lim_{t \to \infty} \frac{S^2}{T} = 0$$

and

$$\lim_{t \to \infty} \frac{T_{\mathrm{rot}}}{T} = r = \frac{1}{2\beta} \left(\alpha - \eta + \sqrt{(\alpha - \eta)^2 + 2\nu\beta} \right).$$

Since S^2 decays to zero faster than T and T_{rot} it is justified to consider the asymptotic time dynamics for the case $\mathbf{S} = 0$. In this case one obtains from an analysis of the Boltzmann equation that the tail of the spin distribution function is of the form of a power law, and that the entire distribution is strongly non-Maxwellian. Interestingly, in the case of simple shear flow one can obtain an exact form for the distribution function:

$$\rho(\zeta) = A_0 (B + C\zeta^2)^{-\frac{A}{2C}}, \tag{12.51}$$

where $\zeta \equiv s/\sqrt{T}$ is the normalised spin, and A_0, B, C and A are numerical constants that depend on the collision parameters.

In summary, we have outlined an approach to obtaining hydrodynamic equations for near-frictionless granular gases. The essence of the approach is to regard such gases as mixtures of particles identified by their respective spins. The hydrodynamic equations explicitly couple the spin and translational degrees of freedom. The rotational energy distribution is highly non-Maxwellian in both the HCS and simple shear flow. As only the dilute case is considered here, the stress tensor (being the average of a symmetric entity) is symmetric, i.e., one does not obtain a micropolar theory [332].

12.6 Conclusion

An introduction to the kinetic theory of granular gases has been presented. Following a qualitative description of consequences of the inelasticity of the collisions, it has been shown that a generalised Chapman–Enskog expansion applied to the pertinent Boltzmann equation can produce a hydrodynamic description of granular gases. This description is limited to near elastic collisions, as scale separation in strongly inelastic granular gases is weak or nonexistent. The lack of scale separation in granular gases, which stems from the inelastic nature of the collisions, is a fundamental property of these gases, and it is the root cause of many important properties thereof, including the promotion of negligible effects in molecular fluids to strong effects in granular fluids, the normal stress difference being a prominent example (so are the long time tails [333]). An outline of a novel method for the derivation of boundary conditions for molecular and granular gases has been presented; in the latter case, the lack of scale separation is of importance in both the derivation and interpretation of the results. As we have not employed the Enskog correction in the above, our results are, strictly speaking, limited to dilute systems. It is possible to derive Green–Kubo relations for granular gases [250, 251] using methods that are similar to those employed in the derivation of these relations for molecular gases, thus enabling the extension of the constitutive relations to moderate densities in a systematic fashion.

It is perhaps important to contradistinguish the approach presented here from the approach employed in the pioneering works on granular kinetics. Those studies did not employ the pertinent Boltzmann equation, but rather the Enskog equations [334] satisfied by the low moments of the distribution function (i.e., the conservation equations, modified to account for inelastic energy losses). In order to obtain a closure, a form of the single particle distribution function was *guessed* and substituted in the Enskog equations. The results were therefore dependent on the quality of the guess; evidently no-one seems to have guessed a form for the distribution function that would yield the non-Fourier part of the heat flux. The choice whether to account for non-Maxwellian corrections was at the discretion of the modeller, and therefore some papers took into account such corrections whereas others didn't. The modern systematic approaches do not suffer from those deficiencies, and this is one of their great advantages.

Numerous other studies of granular gases have not been described here. For instance, a Cahn–Hilliard model has been proposed as a description of the (diffusive) large scale dynamics of granular systems [335]. A number of Landau–Ginzburg models (or 'amplitude equations') have been employed for the modelling of the patterns observed in vibrated granular beds, see, e.g. [271]. The latter class of models is not directly based on hydrodynamics but they seem to capture the phase diagram reported in [49]. In this respect it is interesting to note that a model that is claimed to be almost devoid of physics (essentially a recursion relation) has been successful in mapping out the above phase diagram [336]. While the precise reasons for the successes of this class of models are at present unclear, it seems obvious that some features of granular flows are sufficiently 'universal' to be captured by generic models.

Many of the pioneering kinetic models for rapid granular flows employed the Enskog corrected Boltzmann equation. As mentioned, this equation captures some of the corrections needed to describe moderately dense granular flows, but not all. There is a variety of engineering models for the constitutive relations and boundary conditions for granular flows, see, e.g. [265]. Several heuristically derived hydrodynamic models have been recently proposed to describe dense flows [337–339]. A more systematic method is offered in [250, 251] where the statistical mechanical response theory is modified to apply to inelastic collisions. The result is a set of Green–Kubo formulae relating transport coefficients with (integrals of) flux correlations in the HCS. The practical implications of these results are yet to be studied. There are two disadvantages of the way the response method is implemented: although it is formally correct for all physically allowed densities, it employs a gradient expansion and therefore does not allow for non-analytic dependences of e.g. the stress on the rate of strain. Furthermore, it uses the grand canonical variant of the HCS distribution which, as mentioned, is unstable. A ring level kinetic theory (which goes 'one step' beyond the Boltzmann description) has been studied in [340]. A study of the properties of the dense HCS has been presented in [341].

Many flow patterns observed in experiments are captured by granular hydrodynamics. This is perhaps not surprising in itself, but the fact that some of the experimental systems are only 'a few grains deep' [252] and still exhibit agreement with granular hydrodynamics is puzzling. It seems that these equations provide useful information on scales which are clearly beyond the range of validity implied by their derivation. In this sense, as well as due to the lack of scale separation, one could state that granular gases stretch statistical mechanical methods (and hydrodynamics) to their limit [342], and therefore provide valuable insights to the nature and underlying physics of these methods.

One of the problems studied in the field of granular matter is the degree of non-equipartition in granular gases, see, e.g. [321, 325, 343, 344]. In a way, the fact that a far-from-equilibrium system does not exhibit equipartition is hardly surprising. As a matter of fact, even a driven molecular system does not obey equipartition, as can be easily checked using the single particle distribution function one obtains from, e.g., the Chapman–Enskog expansion. What is probably more surprising is that such a system can obey approximate equipartition. However, the degree of non-equipartition, and the nontrivial velocity distribution functions characterising many granular systems [258, 345–348] are of interest and potential applicability to the derivation of improved constitutive relation.

There are many open questions remaining in this field. First, much has not yet been studied in a systematic way, e.g., polydisperse granular gases [349, 483, 484] and non-spherical grains (see, however, [350, 351]). The studies of the consequences of friction are still incomplete. The dense regime is far from being understood, as is the interface [352] between a granular gas and a granular solid (as in chute flows, for instance). The full implications of the lack of scale separation are still not fully elucidated. For instance, does one need (at least in principle) nonlocal and non-Markovian equations of motion or their replacement by a larger number of fields? Another important problem is to study the interface between hydrodynamic domains and Knudsen domains, a case in point being a vibrated granular system with no lid (the upper part of the gas is clearly in the Knudsen regime). This chapter has also not discussed the interesting problems concerning the description of patterns in granular flows, and the intriguing properties of granular gases in outer space, where gravity plays an important role. These and many other interesting properties of granular gases will undoubtedly be the subjects of future investigations.

13

The thermodynamics of granular materials

Sam Edwards and Raphael Blumenfeld

University of Cambridge

13.1 Introduction

Many granular and particulate systems have been studied in the literature and there is a wide range of parameters and physical states that they support [21, 106, 28, 353]. Here we confine ourselves to jammed ensembles of perfectly hard particles. There are extensive studies in the literature of suspensions of particles in liquids or gases using various methods, including Stokes or Einstein fluid mechanics and Boltzmann or Enskog gas mechanics. These, however, are not jammed and we therefore discuss them no further. This chapter is not intended as a comprehensive review but rather as an interim report on the work that has been done by us to date.

The simplest material for a general jammed system is that of hard and rough particles, ideally perfectly hard and infinitely rough. To a lesser extent it is also useful to study perfectly hard but smooth particles. The former is easily available in nature, for example sand, salt, etc., and we prefer to focus on this case. Nevertheless, the discussion can be readily extended to systems of particles of finite rigidity, as has been shown recently [354]. In jammed systems particles touch their neighbours at points, which have to be either predicted or observed. At these contact points the particles exert on one another forces that must obey Newton's laws. In general, determination of the structure and the forces requires prior knowledge about the history of formation of the jammed system. For example, if grains of sand are poured from a narrow orifice onto a plane they will form a conical sand pile which is known to have a minimum of pressure under the apex [355]. If, however, the sand grains are poured uniformly into a right cylinder standing on a plane the cylinder will fill at approximately a uniform rate, producing a relatively flat surface and a uniform pressure on the plane. If one starts pouring the sand from a narrow

We acknowledge discussions with Professor R. C. Ball and Dr D. V. Grinev.

Granular Physics, ed. Anita Mehta. Published by Cambridge University Press. © A. Mehta 2007.

orifice into a cylinder and changes to a uniform source when the edges of the pile reaches the cylinder walls then the original sand pile will be buried eventually by the uniform deposition and the pressure on the plane is some mixture of the two earlier pictures. Therefore, just given a cylinder full up to a certain level by sand is *insufficient* to determine the pressure at the bottom. Without knowledge of the formation history only a detailed tomography of the individual grains can help the investigator. This is usually the situation in the systems relevant to soil mechanics and to civil engineering.

But there is another situation which brings the problem into the realm of physics. In this set-up the cylinder of sand is prepared in such a way that there is an analogue of equilibrium statistical mechanics which opens the door to *ab initio* calculations of configurations and forces. Suppose the cylinder of sand is shaken with an amplitude A and a frequency ω, each shake being sufficient to break the jamming conditions and reinstate the grains for the next shake. The sand will then occupy a volume V which is a function of A and ω, $V(A, \omega)$. Changing A to A' and ω to ω' one will get a new volume $V' = V(A', \omega')$. If we now return to A and ω we will again find that the volume is $V(A, \omega)$. This suggests that, in analogy with the microcanonical ensemble in thermodynamics, the sand will possess an entropy which is the logarithm of the number of ways the N grains of sand will fit into the volume V, that is, the conventional expression for the entropy,

$$S(E, V, N) = \log \int \delta(E - \mathcal{H})\mathrm{d}\{\text{all degrees of freedom}\}, \qquad (13.1)$$

is replaced by

$$S(V, N) = \log \int \delta(V - \mathcal{W})\Theta\mathrm{d}\{\text{all degrees of freedom}\}, \qquad (13.2)$$

where $\mathcal{W}$ is a function of the structural characteristics of the grains that gives the volume for any arbitrary configuration of grains and Θ is the condition that all grains are touching their neighbours in such a way that the system is in mechanical equilibrium. If Eq. (13.2) is accepted (its derivation is given below) then one can pass to the canonical ensemble replacing the conventional expressions on the left by those on the right;

$$T = \frac{\partial E}{\partial S} \quad \leftrightarrow \quad X = \frac{\partial V}{\partial S}, \qquad (13.3)$$

$$F = E - TS \quad \leftrightarrow \quad Y = V - XS. \qquad (13.4)$$

In these, X is named the compactivity of the system, since $X = 0$ corresponds to maximum density and $X = \infty$ is where the condition of mechanical equilibrium fails due to a topology that cannot support the intergranular forces.

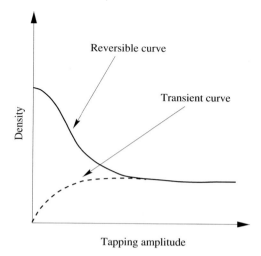

Fig. 13.1 A sketch of the density of granular matter in a vessel after being shaken at amplitude A. Adapted from [172, 173, 356].

Detailed studies of the density of shaken granular systems as a function of the number of 'tappings' and the force of a tap were first given by the Chicago group [172, 173, 356] and fit in with the above theoretical arguments.

13.2 Statistical mechanics

Consider a cylinder containing granular material whose base is a diaphragm that can oscillate with frequency ω and amplitude A. Suppose one vibrates the system for a long time. When the vibration is turned off the granular material occupies a volume $V_0 = V(A, \omega)$. Repeating the process with ω_1 and A_1 gives a volume $V_1 = V_1(A_1, \omega_1)$. Returning now to ω and A, it has been found that the system returns to $V(A, \omega)$. This is surely what one would expect, nevertheless the experiment, done firstly by the Chicago group [172, 173, 356], is new. A different version of this experiment has also been carried out in our department [357]: powdered graphite, after first being assembled, has a low density, as found by measuring its conductivity. But as it is shaken and allowed to come to rest again it exhibits a higher conductivity. Upon cycling the load applied to the powder one reaches, and moves along, the reversible curve shown in Fig. 13.1. By using a simple effective medium approximation [358] it is possible to estimate the mean coordination number as a function of the coordination. We shall see later that the mean coordination number is a parameter that plays a central role in the behaviour of granular materials.

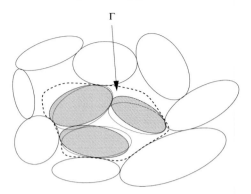

Fig. 13.2 An example of two states of a granular system that differ only by the positions of three particles confined to within a region, Γ.

The first rigorous theory of statistical mechanics came when Boltzmann derived his equation and proved that it describes a system whose entropy increases until equilibrium is achieved with the Boltzmann distribution. He needed a physical specification, that of a low density gas where he could assume only two body collisions, and a hypothesis, the Stosszahlansatz, that memory of a collision was not passed from one collision to another. The question is can we do the same for a powder?

Assuming that the grains are incompressible, a physical condition is that all grains are immobile when an infinitesimal test force is applied to a grain, namely, there are no 'rattlers' which carry no stress at all. A system is jammed when all grains have enough contacts and friction such that there is a finite threshold that a force has to exceed for motion to initiate. The hypothesis we need is that when the external force, say from a diaphragm, propagates stress through the system, then for a particular A and ω there exist bounded regions where motion results which rearranges the grains. We assume that outside these regions no rearrangement takes place. An example is illustrated in Fig. 13.2, where the region Γ consists of three particles that can rearrange in several configurations, of which two are sketched. Given the equation characterising the boundary of Γ and the configuration of the grains inside it, there must exist a function $\mathcal{W}_\Gamma$ that gives the volume of Γ in terms of variables which describe the local geometric structure and the boundary grains. Since the system is shaken reversibly then under the shake $\mathcal{W}_\Gamma$ remains the same, $\mathcal{W}_\Gamma = \mathcal{W}_{\Gamma'}$, and for the entire system

$$\sum \mathcal{W}_\Gamma = \sum \mathcal{W}_{\Gamma'}. \tag{13.5}$$

We can now construct a Boltzmann equation. There must be a probability f of finding any configuration with a specification of positions and orientations. Under a shake

$$\frac{d\mathcal{P}}{dt} = \int K(\Gamma, \Gamma') \left(\Pi_\Gamma f^\Gamma - \Pi_{\Gamma'} f^{\Gamma'} \right) d\{\text{all degrees of freedom}\}, \qquad (13.6)$$

where $\mathcal{P}$ consists of the probabilities f^Γ of finding particular configurations of grains inside regions Γ and their boundary specifications. The kernel function K contains all the information on the contacts between grains and the constraints on the forces expressed via δ-functions.

Now we are at the same situation as Boltzmann, for the steady state will depend only on $\delta(\mathcal{W}_\Gamma - \mathcal{W}_{\Gamma'})$ and the jamming specification. This is the analogue of the conservation of kinetic energy of two particles under collision in conventional statistical mechanics. Equation (13.2) means that the probability f which satisfies (13.2) is

$$f^\Gamma = e^{Y/X - \mathcal{W}_\Gamma/X} \Theta, \qquad (13.7)$$

where Θ specifies the jamming conditions and $e^{Y/X}$ is the normalisation. We can go further and deduce the entropy of the powder by

$$S = - \int f \log f \, d\{\text{all degrees of freedom}\}, \qquad (13.8)$$

where we have dropped, for convenience, the indices Γ and Γ'. From (13.4) we can derive, using symmetry arguments in the same way that Boltzmann did,

$$\frac{dS}{dt} = \int K \, \Pi f \left(\frac{\Pi f}{\Pi f'} - 1 \right) \log \frac{\Pi f}{\Pi f'} d\{\text{all degrees of freedom}\}. \qquad (13.9)$$

Since K and Πf are positive definite, as is $(x-1)\log x$ for $x > 0$, then

$$\frac{dS}{dt} > 0 \quad \text{until} \quad f = e^{(Y-\mathcal{W}_\Gamma)/X}. \qquad (13.10)$$

The Boltzmann approach leads naturally to the canonical ensemble, but the result (13.4) was first put forward for the microcanonical ensemble [17, 359, 360],

$$S = \log \int \delta(V - \mathcal{W})\Theta d\{\text{all degrees of freedom}\}, \qquad (13.11)$$

where now $\mathcal{W}$ is the complete volume function and Θ the complete jamming condition. This form is the analogue of

$$S = \log \int \delta(E - \mathcal{H})d\{\text{all degrees of freedom}\},$$

and the usual result

$$F = E - TS$$

becomes

$$Y = V - XS. \tag{13.12}$$

Similarily, the analogue of the temperature $T = \partial E / \partial S$ is now the compactivity

$$X = \frac{\partial V}{\partial S}. \tag{13.13}$$

This discussion, which has been presented for perfectly hard grains, can be readily extended to the analysis of grains that have internal energy. This leads to

$$S = \int \int \delta(E - \mathcal{H})\delta(V - \mathcal{W})\Theta d\{\text{all degrees of freedom}\}, \tag{13.14}$$

and we obtain

$$e^{S-E(\partial S/\partial E)_{V,N}-V(\partial S/\partial V)_{E,N}}$$
$$= \int e^{-\mathcal{H}(\partial S/\partial E)-\mathcal{W}(\partial S/\partial V)}\Theta d\{\text{all degrees of freedom}\} \tag{13.15}$$

or

$$e^{S-E/T-V/X} = \int e^{-\mathcal{H}/T-\mathcal{W}/X}\Theta d\{\text{all degrees of freedom}\}. \tag{13.16}$$

The Gibbs relation

$$S - E\left(\frac{\partial S}{\partial E}\right)_{V,N} - V\left(\frac{\partial S}{\partial V}\right)_{E,N} = S - E/T - PV = -G \tag{13.17}$$

identifies the inverse of the compactivity as

$$\frac{1}{X} = \left(\frac{\partial E}{\partial V}\right)_{S,N}\left(\frac{\partial S}{\partial E}\right)_{V,N} = -\frac{P}{T} \quad \text{as } T \to 0. \tag{13.18}$$

We regard this relation, however, as a curious formal analogue rather than a useful formula. Although, in general, entropies due to internal thermal effects and configurational rearrangements mix, the two can be readily separated (i.e. a heap of hot sand will have many of the characteristics of a heap of cold sand) and we can write

$$S = S_{\text{th}} + S_{\text{conf}}. \tag{13.19}$$

It is interesting to note that confirmation of this 'thermodynamics' of granular systems by numerical simulations has used the mixed, rather than the purely configurational, approach [361]. One can go further to the Grand canonical ensemble

$$\Omega = S - E\left(\frac{\partial S}{\partial E}\right)_{V,N} - V\left(\frac{\partial S}{\partial V}\right)_{E,N} - N\left(\frac{\partial S}{\partial N}\right)_{E,V}$$
$$= S - E/T - V/X - N\mu/T. \tag{13.20}$$

Since there can be many different kinds of grains, the last term should really be a sum over N_i and μ_i, but we have not looked into such systems yet.

If the system is subject to an external stress on its surface, P_{ij}, then one can be even more general and notice that S becomes $S(V, N, P_{ij})$ and (now discarding E and keeping N fixed)

$$\Omega = S - V\left(\frac{\partial S}{\partial V}\right)_{P_{ij}} - P_{ij}\left(\frac{\partial S}{\partial P_{ij}}\right)_V ,\qquad (13.21)$$

leading to a distribution

$$e^{-S+(V-W)\frac{\partial S}{\partial V}+(P_{ij}-\Pi_{ij})\frac{\partial S}{\partial P_{ij}}} ,\qquad (13.22)$$

where the simplest case only involves the external pressure P_{kk}, and Π_{kk} is related to the total force moment $\sum_{\text{grains}} f_i r_i / V_{\text{grain}}$. This latter form is briefly discussed below. Having named $\frac{\partial V}{\partial S}$ the compactivity, we name the quantity $\partial p/\partial S$, where p is the scalar pressure, *angoricity*. Note that in general the angoricity is the analogue of a *tensorial temperature*, $\partial P_{ij}/\partial S$.

Formula (13.11) was presented many years ago [17, 359, 360] but did not find wide acceptance. This was partly due to a lingering scepticism and partly due to the nonexistence of an exact way to characterise the analogue of a Hamiltonian, the volume function $\mathcal{W}$. Both these problems have been resolved. First, numerical simulations have appeared that validated the formalism [362]. The second development involved the discovery of an exact volume function both in two dimensions [363, 364] and in three dimensions. Nevertheless, to our minds, the validity of this approach was already implicit in the experiment in Refs. [172, 173, 356].

13.3 Volume functions and forces in granular systems

We have seen above that, provided a mechanism for changing configurations can be found, such as tapping and vibrational agitation, a reversible curve can be achieved. This implies that a statistical mechanical approach can be applied to this set of states in powders and that the probability distribution is governed by

$$e^{(Y-\mathcal{W})/X}\Theta.\qquad (13.23)$$

This is already enough for a simple theory of miscibility [17, 359, 360] and indeed any application of the conventional thermodynamic function $\exp(-(F - \mathcal{H})/k_B T)$ will have an analogue for granular systems. However, these systems also enjoy several new problems that have no equivalent in conventional thermal

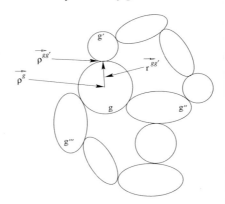

Fig. 13.3 A particle g in contact with three neighbours g', g'' and g'''. $\vec{\rho}^{gg'}$ is the position vector of the contact between g and g'; $\vec{\rho}^{g}$ is the centroid of the contact points; $\vec{r}^{gg'}$ points from the centroid to the contact point between g and g'; $\vec{R}^{gg'} = \vec{r}^{gg'} - \vec{r}^{g'g} = -\vec{R}^{g'g}$; $\vec{S}^{gg'} = \vec{r}^{gg'} + \vec{r}^{g'g} = \vec{S}^{g'g}$.

systems. One such problem concerns the distribution of forces and stresses within the granular packing. Many of the most interesting issues concerning force transmission in, e.g., heaps of particles, lie outside the above framework, for the force exerted by a sand pile on its base depends sensitively on how it was created. Nevertheless, there are quite a few problems that can be tackled with the analytical tools we have already.

The simplest case is probably that of perfectly hard and rough particles ('perfect' must be understood to not fully apply when the material is assembled, but once it has consolidated we can restrict ourselves to the application of forces below the yield limit). In the following we consider particles of arbitrary shapes and sizes. Presuming that the material is in mechanical equilibrium, force and torque balance must be satisfied. Let us consider a part of the material sketched in Fig. 13.3. We assume for simplicity that no two neighbouring particles contact at more than one point. This assumption is not essential to our discussion but it leads, as we shall see in the following, to the conclusion that in two dimensions the material is in isostatic mechanical equilibrium when the average coordination number per grain is exactly three. Figure 13.3 shows a particular grain g in contact with three neighbours, g', g'' and g'''. The contact point between, say, grains g and g' is $\vec{\rho}^{gg'}$ and each grain is assigned a centroid,

$$\vec{\rho}^{g} = \frac{1}{z_g} \sum_{g'} \vec{\rho}^{gg'}, \qquad (13.24)$$

that is defined to be the mean of the positions of all its z_g contacts. The vector

$$\vec{r}^{gg'} = \vec{\rho}^{gg'} - \vec{\rho}^{g} \qquad (13.25)$$

points from the centroid of grain g to the point of its contact with grain g'. The grains g and g' also exert a force on one another through the contact, and let $\vec{f}^{gg'}$ be the force that g exerts on g'. For later use we also define the vectors

$$\vec{R}^{gg'} = \vec{r}^{gg'} - \vec{r}^{g'g} = -\vec{R}^{g'g} \tag{13.26}$$

and

$$\vec{S}^{gg'} = \vec{r}^{gg'} + \vec{r}^{g'g} = \vec{S}^{g'g}. \tag{13.27}$$

Balance of forces and torque moments gives

$$\sum_{g'} \vec{f}^{gg'} = \vec{G}^g, \tag{13.28}$$

$$\sum_{g'} \vec{f}^{gg'} \times \vec{r}^{gg'} = 0, \tag{13.29}$$

where $\vec{G}^g$ is the external force acting on grain g. Newton's third law requires that at each contact

$$\vec{f}^{gg'} + \vec{f}^{g'g} = 0 . \tag{13.30}$$

Various useful tensors can be generated using these vectors:

$$\hat{\mathcal{E}}_{ij}^g = \sum_{g'} R_i^{gg'} R_j^{gg'},$$

$$\hat{\mathcal{F}}_{ij}^g = \sum_{g'} f_i^{gg'} f_j^{gg'},$$

$$S_{ij}^g = \frac{1}{2} \sum_{g'} \left(f_i^{gg'} r_j^{gg'} + f_j^{gg'} r_i^{gg'} \right). \tag{13.31}$$

The latter is sometimes known as the Love stress tensor. Other 'fabric tensors' that have appeared already in the literature can also be defined from these quantities, e.g. $\sum_{g'} r_i^{gg'} r_j^{gg'}$. We will show first that a simple theory of granular systems can be expressed in terms of these tensors. However, it does not yield a complete description. A new geometric characterisation has been formulated, which makes it possible to construct an exact microscopic theory of two-dimensional systems, and this will be described below.

In three dimensions the 3×3 tensor $\hat{\mathcal{E}}_{ij}^g$ has three Euler angles of orientation and three eigenvalues, $\lambda_1^2, \lambda_2^2, \lambda_3^2$, whose combinations have direct physical interpretations:

$$\langle \sum_i \hat{\mathcal{E}}_{ii} \rangle = \langle \lambda_i^2 \rangle = 3 \times \text{(the average radius squared)}, \tag{13.32}$$

$$\langle \lambda_1^2 \lambda_2^2 \lambda_3^2 \sum_i \lambda_i^{-2} \rangle = 3 \times \text{(the average cross section)} \tag{13.33}$$

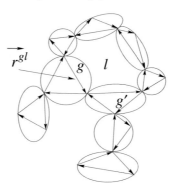

Fig. 13.4 The circulation of loops of $\vec{r}$-vectors around grains, e.g. g, is in the clockwise direction and around voids, e.g. l, in the anticlockwise direction.

and

$$\langle \lambda_1^2 \lambda_2^2 \lambda_3^2 \rangle = \{\text{the average volume squared}\}. \tag{13.34}$$

The total volume is approximately

$$V \approx \frac{1}{2} \sum_g \sqrt{\det(\hat{\mathcal{E}}^g)} \,. \tag{13.35}$$

Thus, from $\hat{\mathcal{E}}^g$ we can produce a first approximation to the volume function $\mathcal{W}$ of the entire system

$$\mathcal{W} = \sum_g \mathcal{W}^g = \frac{1}{2} \sum_g \sqrt{\det(\hat{\mathcal{E}}^g)} \,. \tag{13.36}$$

More recently Ball and Blumenfeld [363] have found an exact form for $\mathcal{W}$ in two dimensions, using a new geometric tensor that characterises differently the local microstructure around grains. This geometric tensor is constructed as follows. For lack of sufficient symbols we shall use in what follows $\vec{R}$ and $\vec{r}$ again but these should not be confused with the quantities defined in Eqs. (13.25) and (13.26). First, connect all the contact points around grain g by vectors $\vec{r}^{gl}$ that circulate clockwise, as shown in Fig. 13.4. The choice of this direction is not essential but it is important that these vectors circulate in the same direction around *all* grains. The vectors $\vec{r}^{gl}$ form a network that spans the system which we term the *contact network*. In two dimensions the grains form closed loops that enclose voids and around these loops the vectors $\vec{r}^{gl}$ circulate in the anticlockwise direction. Each $\vec{r}^{gl}$ is uniquely identified by the grain g that it belongs to and the void loop l that it encircles. Next, define the centroid of loop l as the mean position vector of all the contact points around it:

$$\vec{\rho}^{l} = \frac{1}{z_l} \sum_{g,g' \in \partial l} \vec{\rho}^{gg'}, \tag{13.37}$$

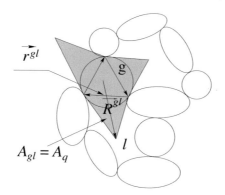

Fig. 13.5 The geometric variables around grain g. The $\vec{r}$-network connects the contact points around grains; for example, $\vec{r}^{gl}$ is a vector connecting two neighbouring contact points around grain g which are on the boundary of void loop l. The vector $\vec{R}^{gl}$ extends from the centroid of grain g to the centroid of void loop l. The pair $\vec{r}^{gl} - \vec{R}^{gl}$ forms a quadrilateral q that is the elementary unit of the structure – the quadron. The area of the quadron is termed $A_{gl} = A_q$.

where z_l is the number of grains around the loop and the sum is over the grains that surround it, ∂l. Finally, define a vector, $\vec{R}^{gl}$, that extends from the centroid of grain g to the centroid of void loop l (see Fig. 13.5),

$$\vec{R}^{gl} = \vec{\rho}^l - \vec{\rho}^g. \tag{13.38}$$

The vectors $\vec{R}^{gl}$ also form a network that spans the system and this network is the dual of the contact network. The Ball–Blumenfeld basic geometric tensor is expressed in terms of the outer product of these vectors:

$$\hat{C}^g_{ij} = \sum_l r^{gl}_i R^{gl}_j, \tag{13.39}$$

where i, j stand for x, y and the sum runs over all the loops that surround grain g. The antisymmetric part of each of the terms in the sum (13.39) can be written as

$$A\left(\vec{r}^{gl} \vec{R}^{gl}\right) \equiv A_{gl}\hat{\epsilon}, \tag{13.40}$$

where $\hat{\epsilon} = \begin{pmatrix} 0 & 1 \\ -1 & 0 \end{pmatrix}$ is the unit antisymmetric tensor corresponding to $\frac{\pi}{2}$-rotation in the plane. The prefactor A_{gl} is exactly the area of the quadrilateral of which the vectors $\vec{r}^{gl}$ and $\vec{R}^{gl}$ are the diagonals (see Fig. 13.5). A key observation is that the areas of the quadrilaterals tile the entire system without holes and with no overlaps, as long as there are no non-convex loops, which are unstable when the system is loaded only through its external boundaries. By summing these areas over the quadrilaterals that surround grain g we obtain the area associated with this grain,

$$A_g = \sum_l A_{gl}. \tag{13.41}$$

A summation of this quantity over all grains gives the area of the entire two-dimensional system. Thus, the volume function in two dimensions is exactly

$$\mathcal{W} = \sum_g A_g = \sum_{g,l} A_{gl}. \tag{13.42}$$

Note that we could index each quadrilateral by q and sum over all q directly instead of over the grains g and the void loops l. This indicates that the basic building blocks of the system are not the grains, as one would initially expect. Rather, each grain can be regarded as composed of z_g internal elements, the quadrilaterals, and these are the fundamental quasi-particles (or excitations, in the language of conventional statistical mechanics) of the system. In two-dimensions isostatic, or marginally rigid, systems have on average three quadrilaterals per grain and we term these elementary quasi-particles 'quadrons'.

To make use of this identification it is necessary to determine the distribution of areas A_q in any given system. This information, combined with the behaviour of the density of states Θ (which, as in conventional thermodynamic systems, is expected generically to vary as a power law), will make it possible to deduce the compactivity of the system X by fitting it to an exponential form. Alternatively, it makes it possible to estimate the density of states analytically and proceed to calculate the partition function

$$Z = \int \dots \int e^{-\beta \sum_{q=1}^{3N} A_q} \Theta(\{A_q\}) \mathrm{d}A_q \tag{13.43}$$

as a function of the compactivity.

The volume function (13.42) also makes it possible to identify a compact phase space of degrees of freedom, the vectors $\vec{r}^{gl} = \vec{r}^q$. There are altogether $3N$ such vectors, on average three per grain. These, however, are not all uncorrelated due to the constraints imposed by the topology of the structure. Basically, we need to determine how many *independent* degrees of freedom there are. A very significant advantage of the exact volume function (13.42) is that it enables us to pinpoint the correlations amongst these vectors. The key to this lies in the observation that the topological constraints that give rise to the correlations originate from the *irreducible loops* in the structure. The irreducible loops are the fundamental loops of which all other loops can be composed, as shown in Fig. 13.6. There are two types of irreducible loops: grain loops, which consist of the vectors $\vec{r}^q$ connecting the contacts around individual grains, and void loops, which consist of vectors $\vec{r}^q$ circulating around individual voids. There are N of the former, one per grain, and M of the latter, giving altogether $N + M$ dependent vectors. To determine the number

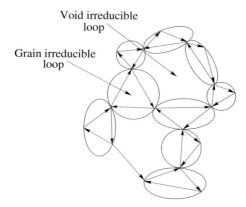

Void irreducible
loop

Grain irreducible
loop

Fig. 13.6 The irreducible loops consist of two types: loops circulating grains (in the clockwise direction) and loops circulating voids (in the anticlockwise direction). All other loops in the networks can be decomposed into combinations of the irreducible loops.

of voids M we can employ Euler's relation [365] between edges, cells and vertices, combined with the fact that the mean coordination number in two dimensions is exactly three per grain for these systems. This gives that there are on average six grains around a void. This means that there are two grains per void loop and therefore that $M = N/2$. Thus, $N + M = 3N/2$ and of the $3N$ vectors $\vec{r}^q$ only half are independent. With two degrees of freedom per vector, this gives that the phase space is $3N$-dimensional. Turning attention back to the quadrons, this argument leads to a surprising coincidence: the number of independent degrees of freedom is the same as the number of quadrons! This suggests that in two dimensions one can get rid of the function $\Theta(\{A_q\})$ in the partition function (13.43) by integrating over quadrons rather than over the independent vectors,

$$Z = \int P(\{A_q\})\Pi_{q=1}^{3N}e^{-\beta A_q}\mathrm{d}A_q, \tag{13.44}$$

where $P(\{A_q\})$ is the correlated probability density of the quadron areas. Various conventional statistical mechanical methods can be used to evaluate this probability density (e.g., cluster expansion). The simplest approximation would be to assume that the quadrons are independent, that is,

$$P(\{A_q\}) = \Pi_{q=1}^{3N}P_q(A_q), \tag{13.45}$$

where $P_q(A_q)$ is the probability density of the area of the q quadron. This form, which resembles the treatment of the density of states in conventional statistical mechanics, makes it simpler both to evaluate the partition function and to appreciate the implicit approximations that have been used in the literature when the areas of the grains, rather than the quadrons, were taken as the fundamental particles of the system.

At present we know of no first-principles theory that gives the form of $P_q(A_q)$ for any system. Therefore, to make progress, we are required to make assumptions on this form. One simple approximation is to assume that the area of any quadron is chosen from a uniform distribution of average A_0 that lies between a maximum value $A_0 + \Delta$ and a minimal value $A_0 - \Delta > 0$,

$$P(B_q) = \begin{cases} \dfrac{1}{2\Delta} & \text{if } A_0 - \Delta < B_q < A_0 + \Delta; \\ 0 & \text{otherwise.} \end{cases} \qquad (13.46)$$

In this case the partition function is straightforward to compute;

$$Z = \left(\frac{e^{-2\beta A_0} \sinh(\beta\Delta)}{\beta\Delta} \right)^{3N} \qquad (13.47)$$

and the mean total area of the system and its mean fluctuations are

$$\langle A_{\text{system}} \rangle = \frac{3N}{2} \left(2A_0 + \frac{1}{\beta} - \Delta\coth(\beta\Delta) \right), \qquad (13.48)$$

$$\langle \delta A^2_{\text{system}} \rangle = \frac{3N}{2} \left(\frac{1}{\beta^2} - \frac{\Delta^2}{\sinh(\beta\Delta)^2} \right). \qquad (13.49)$$

A plot of the mean area and the fluctuations as a function of the compactivity X are shown in Fig. 13.7. Recalling that the compactivity increases monotonically with the external agitation of the system (the tapping in the Chicago experiment), one can invert the plot of the mean area into a plot of the mean density (see Fig. 13.7c). The plot is in a reasonable agreement with the experimental results of Refs. [172, 173, 356].

13.4 The stress field

As mentioned already, in a packing of perfectly hard grains there cannot be any strain involved in the determination of the stress. This observation extends to systems of grains that are not infinitely rigid [354], as long as the condition that the mean coordination number is three in two dimensions and four in three dimensions. The system is then isostatic or statically determinate. The conditions for frameworks to be statically determinate have been worked out already by Maxwell [366] and Levy [367]. Since the forces in isostatic states are statically determinable, stress–strain relations are *redundant* and therefore the equations for the stress field have to be closed by another type of constitutive information. The new set of equations and their analysis are called isostaticity theory. In the following we shall make use the symbol $\hat{\sigma}$ for the stress tensor and σ_{ij} for its components. The global balance

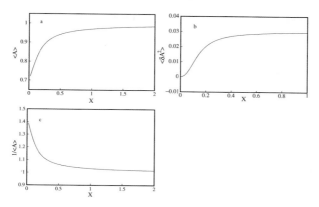

Fig. 13.7 (a) The mean area per quadron as a function of the compactivity for the density of states (the distribution of quadron areas) modelled by (13.46); (b) the mean area fluctuations per quadron as a function of compactivity; (c) the mean density per quadron as a function of compactivity. The plots agree reasonably well with the experimental results of Refs. [172, 173, 356].

equations for the stress field in isostatic states,

$$\frac{\partial \sigma_{ij}}{\partial x_i} = g_j \quad \text{(force balance)},$$

$$\sigma_{ij} = \sigma_{ji} \quad \text{(torque balance)}, \tag{13.50}$$

are the force and torque balance equations for $d(d+1)/2$ variables in d dimensions. The stress tensor has d^2 components to be determined and therefore further $d(d-1)/2$ equations are missing, one in 2D and three in 3D. These equations must depend on the geometry of the contacts and we derive them below. We shall give below two derivations. The first, after Edwards and Grinev [368], applies only to systems on the reversible curve. Systems not on this curve, such as sandpiles, depend on the history of preparation. In sandpiles the missing equations were postulated using two main hypotheses, 'fixed principal axis' and 'oriented stress linearity', each resulting in a different equation [369, 370]. The second derivation presented here, after Ball and Blumenfeld [363], applies to generally quenched systems.

13.4.1 First approach

Although the stress is a macroscopic variable it can be defined on the level of one grain by using the force moment, which on averaging over the volume becomes the stress,

$$S_{ij}^g = \frac{1}{2} \sum_{g'} \left(f_i^{gg'} r_j^{gg'} + f_j^{gg'} r_i^{gg'} \right), \tag{13.51}$$

where Newton's equations apply. We need the probability $\mathcal{P}(\hat{\sigma}^g)$ and, more usefully, its Fourier transform

$$\mathcal{P}(\zeta^g) = \int e^{i \sum_{ij} \zeta^g_{ij} \sigma^g_{ij}} \Pi_{ij} d\sigma^g_{ij} .$$ (13.52)

Using Newton's equations we are led to [368]

$$\mathcal{P}(\hat{\sigma}) = \int d\zeta \int d\eta e^{i \sum \eta^g F^g_{\text{ext}}} \Pi_{gg'} \delta \left(\zeta^g r^{gg'} - \zeta^{g'} r^{g'g} - \eta^g + \eta^{g'} \right),$$ (13.53)

where F_{ext} is the external force on grain g. To solve the set of equations

$$\zeta^g_{ij} r^{gg'}_j - \zeta^{g'}_{ij} r^{g'g}_j = \eta^g_i - \eta^{g'}_i$$ (13.54)

we can rearrange the terms:

$$\left(\zeta^g + \zeta^{g'} \right) \left(r^{gg'} - r^{g'g} \right) + \left(\zeta^g - \zeta^{g'} \right) \left(r^{gg'} + r^{g'g} \right) = \eta^g - \eta^{g'}.$$ (13.55)

Thus, to make a simple start, assume that $\zeta^g \approx \zeta^{g'}$, which gives

$$\zeta^g = \frac{\partial \eta^g}{\partial r} R^{gg'}.$$ (13.56)

This equation has the structure

$$\vec{a} \cdot \vec{u} = b,$$

whose general solution for $\vec{u}$ is

$$\vec{u} = b \frac{\vec{a}}{a^2} + (\vec{a} \times \vec{c}),$$ (13.57)

where $\vec{c}$ is an arbitrary vector perpendicular to $\vec{a}$ and $\vec{u}$. Thus, we expect

$$\zeta = \frac{\partial \eta}{\partial r} + \zeta^*,$$ (13.58)

where

$$\zeta^{*g}_{ij} r^{gg'}_i = \zeta^{*g'}_{ij} r^{g'g}_i.$$ (13.59)

Using the force balance conditions in Eq. (13.50) gives

$$\frac{\partial \sigma_{ij}}{\partial x_j} = f^g_i + \text{(small corrections)},$$ (13.60)

whilst the second term can be further reduced by noting that the correct number of missing equations appears when (13.59) is summed over g' (i.e. there are still redundancies in (13.59) which complicate the analysis). From the condition

$$\sum_{g'} \zeta_{ij}^{*g'} r_j^{g'g} = 0 \tag{13.61}$$

one finds that σ must have the structure

$$\sigma_{ij}^g = \frac{1}{2} \sum (\phi_i^{g'} r_j^{gg'} + \phi_j^{g'} r_i^{gg'}), \tag{13.62}$$

where the elimination of the parameter ϕ leads to the missing equations [?, 368]. Let us try a first approximation using the ansatz

$$\phi_i^{g'} = ar_i^{g'g} + br_i^{gg'}. \tag{13.63}$$

This gives the following averages:

$$\langle \sigma_{ij}^g \rangle = \frac{1}{2} \sum_{g'} \left(2a \langle r_i^{gg'} r_j^{gg'} \rangle + b \langle r_i^{gg'} r_j^{g'g} + r_j^{gg'} r_i^{g'g} \rangle \right), \tag{13.64}$$

leading to the simple form

$$\begin{vmatrix} \sigma_{xx} & C_{xx} + D_{xx} & E_{xx} + D_{xx} \\ \sigma_{xy} & C_{xy} + D_{xy} & E_{xy} + D_{xy} \\ \sigma_{yy} & C_{yy} + D_{yy} & E_{yy} + D_{yy} \end{vmatrix} = 0, \tag{13.65}$$

where

$$C_{ij} = \sum_{g'} R_i^{gg'} R_j^{gg'}; \quad D_{ij} = \sum_{g'} R_i^{gg'} S_j^{gg'}; \quad E_{ij} = \sum_{g'} S_i^{gg'} S_j^{gg'}.$$

The crudest approximation one can use is $C_{yy} = C_{xx} = a^2$, $C_{xy} \approx 0$, $E_{ij} \approx 0$, $D_{xx} = -D_{yy} = \sin\theta$, $D_{yx} = \cos\theta$, where θ is the angle between $\vec{R}$ and $\vec{S}$. This gives

$$\sigma_{xx} - \sigma_{yy} = 2\tan\theta\,\sigma_{xy}, \tag{13.66}$$

where θ varies in isotropic conditions within $-\pi < \theta < \pi$. From this expression one is prompted to define

$$\psi = \frac{\sigma_{xy}}{\sigma_{xx} - \sigma_{yy}}, \tag{13.67}$$

which, for θ random, has a probability density

$$P(\psi) = \frac{\psi}{\pi(1 + \psi^2)}. \tag{13.68}$$

Similarly, the probability density of ψ^{-1} is

$$P(\psi^{-1}) = \frac{\psi^{-1}}{\pi(1 + \psi^{-2})} \, . \tag{13.69}$$

In sandpiles, which are not isotropic, the form (13.66) has been successfully used to predict stresses throughout the pile, in particular reproducing the minimum under the apex [355].

A central difficulty of attempts to quantify granular physics is that these systems are very disordered. This means that, even when the existence of a statistical mechanical formalism is established, one must resort to approximations because a detailed statistical analysis is at present too difficult. Therefore, one should aim at this stage only for general laws. An example of such an approach is a recent exact formulation leading to an understanding of the onset of plasticity from a marginally rigid state of granular matter [371].

13.4.2 Second approach – coarse-graining a microscopic theory

To derive the stress transmission equations on the granular level in two dimensions Ball and Blumenfeld [363] have followed a different, more rigorous, approach. Consider a two dimensional granular packing, part of which is shown in Fig. 13.4 together with its contact network, which has been defined above. The vectors $\vec{r}$ connecting the contacts around grains circulate clockwise and therefore they form loops that circulate around the voids in the anticlockwise direction. Ball and Blumenfeld defined for every such void loop a force $\vec{f}_l$ located at the centre of the loop, which is defined as the mean position of the contacts that surround the loop. In two dimensions every contact, say between g and g', sits on the boundary between two void loops, say l and l'. The grain g' exerts a force $\vec{f}_{gg'}$ on g and this force is parametrised in terms of the loop forces as follows:

$$\vec{f}_{gg'} = \vec{f}_l - \vec{f}_{l'} \, . \tag{13.70}$$

The sign convention adopted in this expression is that if the vector $\vec{r}_{g'l}$ points towards grain g then $\vec{f}_l$ is preceded by a positive sign, and if not, a negative sign. The analysis of these forces, rather than the original contact forces, has several advantages.

(1) By definition, the force exerted by grain g on g' is $\vec{f}_{g'g} = \vec{f}_{l'} - \vec{f}_l = -\vec{f}_{gg'}$. Therefore, the loop forces *satisfy Newton's third law of action and reaction*.
(2) By writing the net force on grain g as a sum over its contact forces and representing the latter in terms of the loop forces, using (13.70), we observe that every loop force appears in the sum exactly twice, once with a positive and once with a negative sign. Thus, the net force on grain g vanishes identically, which means that the field of loop forces *automatically satisfies the force balance conditions on all the grains*. For consistency, the loop forces, just like the contact forces, must be uniquely determinate. If the system

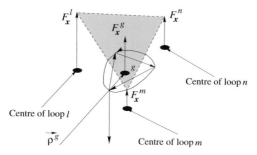

Fig. 13.8 The piecewise linear continuation of the field of loop forces. Around a grain construct a polygonal surface (a triangular example is shown) whose corners are at $(\rho_x^l, \rho_y^l, f_i^l)$. The union of all such polygons is the continuous function F_i.

consists of $N(\gg 1)$ grains then there are $3N$ balance equations in total, two of force and one of torque moment per grain. From Euler's theorem on the topology of edges, vertices and cells in the plane, we have that there are on average six grains around a loop. Combining this with the fact that the mean coordination number is three it is immediate to deduce that there are $N/2$ voids in the packing. This gives that there are $N/2$ loop forces and therefore N unknowns to determine. This is to be contrasted with the $3N$ unknown components of the contact forces in the original force field. Since the loop forces already satisfy the force balance conditions then the loop forces can be only determined through the torque balance equations. Of these there are exactly N, one per grain. Thus, there are exactly the same number of torque balance equations as there are unknowns and the loop forces are indeed uniquely determinate.

(3) The ratio of contact forces to loop forces is 3:1 and so the field of loop forces is three times more sparse then the original field of contact forces. Thus, a side effect of the parametrisation is that the field of loop forces is a coarse-grained version of the field of contact forces.

In terms of the loop forces and the vectors $\vec{r}$ the force moment around grain g is

$$S_{ij}^g = \sum_l r_i^{lg} f_j^l, \tag{13.71}$$

where the sum is over the loops that surround grain g. As mentioned above, the stress is the force moment normalised by a suitable area, and it is natural to use for this the grain area that consists of the areas of its quadrons, $A_g = \sum_q A_q$. Definition (13.71) is still discrete and we now need to pass to the continuum. This is done by a piecewise linear interpolation. We shall not describe this interpolation in detail, but an example of the interpolation of one component of the loop forces is shown in Fig. 13.8. Let us call the resulting continuous force field $\vec{F}$. In terms of $\vec{F}$ the original forces are

$$\vec{f}_l = \vec{f}_g + \vec{R} \cdot \nabla F(\vec{x}_g). \tag{13.72}$$

Substituting this form into (13.71), the expression for the force moment becomes

$$S_{ij}^g = \sum_l r_i^{lg} \left[\sum_k \vec{R}_k \partial_k F_j \right], \qquad (13.73)$$

which can be written as

$$S_{ij}^g = \sum_l C_{ik} \partial_k F_j. \qquad (13.74)$$

The geometric tensor $\hat{C}$ is exactly the tensor defined in (13.39) which gave rise to the exact volume function in the entropic analysis. Thus, the tensor $\hat{C}$ provides a natural characterisation of the geometry for the purpose of both the stress description and the entropy. Recalling that the stress tensor $\hat{\sigma}$ is the force moment properly normalised by area of a grain and integrating over a small region containing few grains, Ball and Blumenfeld were able to show that the imposition of torque balance gives *two* equations. One is

$$\langle \sigma_{ij} \rangle = \langle \sigma_{ji} \rangle,$$

which corresponds to the global torque balance condition (13.50) for the mean stress tensor. The other condition is new,

$$\left[\hat{P} \hat{\epsilon}^{-1} \hat{\sigma} \right]_{ij} = \left[\hat{P} \hat{\epsilon}^{-1} \hat{\sigma} \right]_{ji}, \qquad (13.75)$$

where $\hat{P} = \frac{1}{2} (\hat{C} + \hat{C}^T)$ is the symmetric part of $\hat{C}$. Equation (13.75) can be written explicitly as

$$p_{xx}\sigma_{yy} + p_{yy}\sigma_{xx} - 2p_{xy}\sigma_{xy} = 0, \qquad (13.76)$$

and is a manifestation of the local torque balance condition. It gives a first-principles microscopic relation between the stress field and local geometric properties of the (generally disordered) microstructure, as characterised by the geometric tensor $\hat{P}$. Relation (13.76) is the missing constitutive equation for general disordered systems in two dimensions.

It is instructive to consider the statistical properties of $\hat{P}$. It has been shown [363] that the volume average of all its components, p_{ij}, vanishes identically *regardless* of the geometrical characteristics. This means that Eq. (13.76) couples in fact between the stress field and *fluctuations* in the structural characteristics. The vanishing of the components p_{ij} under volume averaging presents a problem – it undermines conventional coarse-graining methods, which rely on finite-valued mean constitutive properties. This rather unique coarse-graining problem could, in principle, severely limit the applicability of the theory. Fortunately, this difficulty has been resolved by the development of a specialised procedure, based on concepts from frustration and antiferromagnetism [372]. An important result of that procedure is that the volume average of the coarse-grained p_{ij} became *finite*. This has made it possible to treat Eq. (13.76) as macroscopic, opening the way to macroscopic calculations.

The field equations (13.50) and (13.76) couple the stress components σ_{ij}. In a next development it was found that the equations can be resolved to yield explicit decoupled equations for each of these variables [373]. By making the assumption that for macroscopic scales the gradients of the fluctuations p_{ij} become small compared with the gradients of the stress field, general solutions have been obtained to the decoupled equations. The key to the analysis of the equations is a local linear transformation,

$$\begin{pmatrix} u \\ v \end{pmatrix} = \mathcal{M}\left(\{p_{ij}\}\right) \begin{pmatrix} x \\ y \end{pmatrix},$$

(13.77)

in terms of which the equations for the stress components become

$$\left(\frac{\partial^2}{\partial u^2} - \frac{\partial^2}{\partial v^2} \right) \sigma_{ij} = f_{ij}.$$

(13.78)

In this expression f_{ij} are source terms that depend only on the gradients of the external loading and on the local geometry. The differential operator on the left-hand side is *hyperbolic*, confirming earlier suggestions [369, 370] to this effect. Equation (13.78) is quite elegant in that all the stress components follow the same equation but with different source terms f_{ij}. The general solutions give rise to force chains that propagate through the isostatic granular medium [373], in good agreement with experimental observations [374, 375, 376]. The form (13.78) also made it straightforward to derive the Green function, resolving a long-standing debate in the literature. It has been further shown recently by one of us that introducing corrections due to the gradients of the constitutive parameters p_{ij} does not affect the force chains, leading to the conclusion that, at least in two dimensions, the force chains are a generic solution of the general stress field equations and that they follow the characteristic curves $\eta = u + v$ and $\zeta = u - v$.

The above analysis has been limited to planar systems. A promising extension of the theory to three dimensions is under construction at present, using insight from the two-dimensional case. However, it is unclear at this stage whether the three-dimensional theory would also give rise to force chain solutions. The experimental status is also not entirely clear. Detailed measurements by Brujic [377] in emulsions, where the colloidal particles exert force on one another, show no sign of force chains within the bulk. This may indicate either that force chains are not as prominent in three dimensions as they are in two or that they are negligible for very soft particles. Alternatively, this may point to inherent differences between systems of particles with very little, or no, friction, which emulsions are, and systems of particles with friction between them. Intriguingly, it has been suggested in [363] on the grounds of theoretical stress analysis that even in two dimensions there may exist an inherent difference between systems of smooth and rough grains, but how such a difference could manifest in the elimination of force chains, if it does, is unclear at present.

13.5 Force distribution

There is a distinct difference between the problem of force statistics and the distribution in configurations space. The packing of grains on the reversible branch is governed by the function $e^{-W/X} \Theta$. The forces have little effect on this distribution, but they are fully dependent on the configurations and the external loading. For a given structural configuration, the problem is the following. Given external, i.e. surface, loading and internal, i.e. body, forces, what is the solution of the balance conditions (13.28) and (13.29)?

We first recall that for infinitely rigid grains there can only be a statically determinate solution if the mean number of contacts between grains takes a particular value, z_c. In d dimensions this value is $d + 1$ for rough grains, $d(d + 1)$ for smooth arbitrary grains, and $2d$ for smooth spheres. These values are obtained upon requiring that in mechanical equilibrium the number of force and torque balance conditions should be equal to the number of unknown force components. Here we discuss only the case of rough grains and we shall assume that this condition is always obeyed. Several objections can be (and have been) raised to this assumption: one is that real systems are never fully rigid and another is that the number of contacts in real granular packings is usually higher than z_c, undermining the determinacy condition. The first objection has been shown recently to be misguided: the infinite rigidity is not a necessary condition and packings of compliant grains that satisfy the coordination number condition are also isostatic [354]. The second objection presents an interesting challenge. It is true that in most real granular packings $z > z_c$ even if the grains are very hard. In fact, a recent experiment has identified that upon consolidation of a granular pile the coordination number can only approach z_c as the consolidating grains approach a particular density and therefore that there exists a marginally rigid state where $z = z_c$ [378]. The question, however, is how much above z_c does z need to be for the isostatic behaviour (and for the isostaticity theory discussed above) to become irrelevant. Clearly, this cannot take place abruptly – adding one contact in an otherwise isostatic macroscopic system would not suddenly change the stress field across the entire system from a solution of the isostaticity equations to a solution of elasticity theory. Rather, the change must

be gradual. A preliminary discussion of the manner that the change might occur
has been presented recently [373] with the conclusion that the isostatic theory only
deteriorates gradually as the number of extra contacts increases. This new understanding, while exciting, is outside the scope of this presentation. To simplify the
following discussion, and without loss of much generality, we shall assume that all
grains have exactly $d + 1$ contacts. The generalisation to more general distributions
is straightforward although more tedious.

In a macroscopic system the forces will have a distribution that, for a homogeneous powder under external pressure, may be roughly like a hydrostatic pressure
p: $\vec{f} \approx -p\vec{n}$, where $\vec{n}$ is a coarse-grained normal, or $f \approx p$. The distribution of f
has been observed in several experiments [159, 379] to be exponential and several
theoretical models have been proposed to explain it. The derivation of an exact
equation is quite difficult if one attempts to take into account random shaped grains
or even a random distribution of grain contacts. Therefore, the present discussion
will start generally and then we shall simplify the internal structure of the equation
to allow analytic solutions.

Let us use the tensor $\hat{\mathcal{F}}^g$ and in particular base the analysis on $\mathrm{Tr}[\hat{\mathcal{F}}^g]$ which will
relate to the average

$$\langle f \rangle = \frac{1}{2}\sqrt{\mathrm{Tr}[\hat{\mathcal{F}}]}\,. \tag{13.79}$$

The tensor $\hat{\mathcal{F}}^g$ is related to the touching grains g_1, g_2, g_3, g_4 via Newton's equations
and therefore its probability density satisfies

$$\mathcal{P}(F^g) = \int \delta\left(F^g - \sqrt{\mathrm{Tr}[\hat{\mathcal{F}}^g]}\right) \Pi_i \left[\mathcal{P}(F^{g_i})\mathrm{d}F^{g_i}\right] \Phi([\mathcal{F}])\,, \tag{13.80}$$

where Φ is a set of δ-functions that fix the relationships of the variables F^{g_i}. We can
then use the crude approximation that F^g is linearly related to F^{g_i} with coefficients
λ^{g_i},

$$\mathcal{P}(F) = \int_0^1 \cdots \int_0^1 \delta\left(F - \sum_{g'=1}^d \lambda^{g'} F^{g'}\right) \Phi\left(\left[\lambda^{g'}\right]\right) \Pi_{g'} \left[\mathcal{P}\left(F^{g'}\right)\mathrm{d}F^{g'}\right]\,. \tag{13.81}$$

The Fourier transform of this expression gives

$$\mathcal{P}(K) = \int \cdots \int \Phi\left(\left[\lambda^{g'}\right]\right) \Pi_{g'} \left[\mathcal{P}\left(\lambda^{g'}K\right)\mathrm{d}\lambda^{g'}\right]\,. \tag{13.82}$$

The limits 0, 1 originate in the definition of F above. The simplest approximation
is to put $\Phi =$ constant, which gives

$$\frac{\mathrm{d}Q}{\mathrm{d}K} = \frac{Q^d}{K^d}\,, \tag{13.83}$$

where $\mathcal{P}(K) = dQ/dK$. Solving for Q and differentiating we obtain

$$\mathcal{P}(K) = \frac{1}{(1 + cK^{d-1})^{d/(d-1)}} . \tag{13.84}$$

In two dimensions the inverse transform gives

$$\mathcal{P}(F) = \frac{\text{constant}}{p^2} F e^{-F/p}, \tag{13.85}$$

whilst in three dimensions we can use fractional derivatives to express the Fourier transform

$$\mathcal{P}(F) = \frac{1}{2\pi} \int_{-\infty}^{\infty} \frac{e^{-iKF} \, dK}{(1 + p^2 K^2)^{3/2}} \tag{13.86}$$

as a sum over Bessel functions. For small K (large F) we get that

$$\mathcal{P}(F) \approx \frac{1}{(2/3)!} \left(\frac{F}{p}\right)^{2/3} e^{-F/p}. \tag{13.87}$$

Different approximations for the function Φ may alter the power 2/3 but the exponential structure is robust. Different experiments by Liu [158] and Brujic [377] and analyses [159, 379], all agree with the exponential form. One can always question the error involved in the approximation of the function Φ as a constant, but whether it is worth the effort of obtaining a more accurate expression given the randomness of grain shapes or even the topology of the contact network is debatable at this stage. One thing seems clear from the experimental evidence – $\mathcal{P}(F)$ rises from a very small value, possibly zero, peaks and then decreases exponentially towards large forces, and this feature is captured by this simple model.

The canonical distribution of stresses mentioned above offers an alternative to the direct self-consistent approach for it predicts

$$e^{\frac{-1}{V_{\text{grain}}} \sum f_i r_i \frac{\partial S}{\partial P}}, \tag{13.88}$$

where $f_i r_i$ is the diagonal element of the force moment and we are treating only the pressure part of the stress for simplicity. If $\vec{r}$ is roughly parallel to $\vec{f}$ the distribution (13.88) behaves as $e^{-c|\vec{f}|}$. This brief derivation needs to be refined by the fact that the neighbouring $\vec{f}$s are related and Newton's laws apply. This is a new approach which is currently being explored.

14

Static properties of granular materials

Philippe Claudin

CNRS, Paris

In this chapter, we are interested in static pilings of cohesionless grains. For example, we would like to be able to describe how forces or stresses are distributed in these systems. As a matter of fact, this is not a simple issue as, for instance, two apparently identical sandpiles but prepared in different ways can show rather contrasted bottom pressure profiles.

The aim is to be as complementary as possible to the existing books on granular media. There are indeed numerous ones which deal with Janssen's model for silos, Mohr–Coulomb yield criterion or elasto-plasticity of granular media or soils, see e.g. [21, 443, 405, 418, 482]. We shall then sum up only the basics of that part of the literature and spend more time with a review of the more recent experiments, simulations and modellings performed and developed in the last decade. This chapter is divided into two main sections. The first one is devoted to microscopic results, concerning in particular the statistical distribution of contact forces and orientations, while, in the second part, more macroscopic aspects are treated with stress profile measures and distribution. Finally, let us remark that, although the number of papers related to this field is very large, we have tried to cite a restricted number of articles, excluding in particular references written in another language than English, as well as conference proceedings or reviews difficult to access.

14.1 Statics at the grain scale

14.1.1 Static solutions

Equilibrium conditions

Let us consider a single grain in a granular piling at rest. As depicted in Fig. 14.1, this grain, labelled (i), is in contact with its neighbours (k). As suggested by this

I wish to thank Jean-Philippe Bouchaud, Chay Goldenberg, Isaac Goldhirsh and Jacco Snoeijer for essential discussions and great help with the writing of the manuscript. I am also grateful to the authors whose figures are reproduced in this chapter.

Granular Physics, ed. Anita Mehta. Published by Cambridge University Press. © A. Mehta 2007.

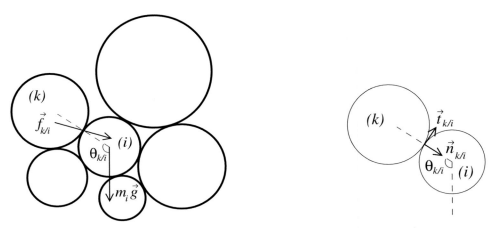

Fig. 14.1 Left: the grain labelled (i) is submitted to its own weight $m_i \vec{g}$ plus the forces $\vec{f}_{k/i}$ from its (here five) different neighbours (k). $\theta_{k/i}$ denotes the contact angle between the grains (i) and (k). Because of the intergranular friction, the orientation of the contact force may deviate from this angle. Right: normal $\vec{n}_{k/i}$ and tangential $\vec{t}_{k/i}$ contact unit vectors.

figure, we shall, except where otherwise stated, restrict, for simplicity, the following discussion to two-dimensional packings of polydisperse circular beads. The study of more realistic systems (polyhedral grains, for example) requires indeed more complicated notation, but does not involve any fundamentaly different physics, and the conclusions that will be drawn with these simple packings are in fact very generic.

At the scale of the grain, the relevant quantities are the different contact forces $\vec{f}_{k/i}$ exerted on this grain, and the corresponding contact angles $\theta_{k/i}$. Note that, for cohesionless granular materials as considered here, only compression can be supported. This is called the 'unilaterality' of the contacts. It means that the forces $\vec{f}_{k/i}$ are borne by vectors which point *to* the grain (i). Due to the action–reaction principle, we have of course $\vec{f}_{i/k} = -\vec{f}_{k/i}$. Likewise, $\theta_{i/k} = \theta_{k/i} + \pi$. If the grains are perfectly smooth, these forces are along the contact direction. However, for a finite intergranular friction coefficient $\mu_g \equiv \tan \phi_g$, the orientation of $\vec{f}_{k/i}$ may deviate from this angle by $\pm \phi_g$ at most. A contact between a grain and one of the walls of the system is not different from a contact between two grains, albeit a possible different friction coefficient μ_w. For the usual case of a packing of grains under gravity, grains are also subjected to their own weight $m_i \vec{g}$.

The conditions of static equilibrium are simply the balance equations for the forces and torques. More precisely, if the grain (i) has N_i neighbours in contact, these equations read

$$\sum_{k=1}^{N_i} \vec{f}_{k/i} + m_i \vec{g} = \vec{0}, \tag{14.1}$$

$$\sum_{k=1}^{N_i} \vec{f}_{k/i} \times \vec{n}_{k/i} = \vec{0}, \tag{14.2}$$

where $\vec{n}_{k/i}$ is unit vector in the direction of $\theta_{k/i}$. We can choose this unit vector to point inward – see Fig. 14.1. Likewise, $\vec{t}_{k/i}$ is the unit vector perpendicular to the contact direction. The condition of unilaterality for cohesionless grains can then be simply expressed by the fact that normal forces are positive:

$$\vec{f}_{k/i} \cdot \vec{n}_{k/i} > 0. \tag{14.3}$$

Finally, none of the contacts must be sliding. Defining normal and tangential contact forces as $N_{k/i} = \vec{f}_{k/i} \cdot \vec{n}_{k/i}$ and $T_{k/i} = \vec{f}_{k/i} \cdot \vec{t}_{k/i}$, the Coulomb friction condition can then be written as

$$|T_{k/i}| \le \mu_g N_{k/i}. \tag{14.4}$$

Multiplicity of static solutions

If N_g denotes the number of grains in the packing, equations and conditions (14.2–14.4) must be satified for each $i = 1, N_g$. For a given piling of grains and a given set of boundary conditions, the unkowns are the contact forces. The usual situation is that the total number of these forces is significantly greater than the total number of equations. The additional conditions are inequalities that partly reduce the space of admissible solutions, but the multiplicity of the solutions that is left is still very large. As a simple illustration, it is obvious that since the number of equations is fixed by N_g, an increasing number of contacts per grain will lead to a larger number of undetermined contact forces. In summary, the list of the position of all the grains and contacts is in general not sufficient to determine the precise state of a static packing of grains submitted to some given external load. This has sometimes been called the 'stress indeterminacy'.

There are, however, cases where the contact forces are uniquely determined by the configuration of the piling. This happens when the number of unknown forces exactly equals the number of equilibrium equations. Such situations are called *isostatic*. They may seem to be specific to rather particular configurations, but in fact it has been shown by Roux [459] and Moukarzel [438] that generic assemblies of polydisperse frictionless and rigid beads are *exactly* isostatic. For instance, this is the case in two dimensions when beads have four contacts on average, which gives two unknown contact forces per grain that are then determined

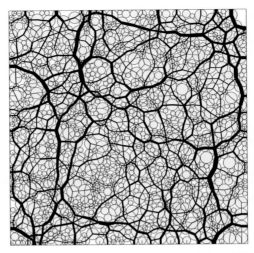

Fig. 14.2 Example of a granular system at rest obtained by Radjai *et al.* [454, 456] in a 'contact dynamics' simulation. The black lines represent the amplitude of the contact forces – the thicker the line, the larger the force. The force spatial distribution is rather inhomogeneous and shows so-called 'force chains'.

by the two force balance equations – the torque balance is automatically verified for perfectly smooth beads, and so is the sliding Coulomb condition, but of course the unilaterality must be checked. Such systems show some particular behaviours, like a strong 'fragility' under incremental loading [397], but have also many features that are very similar to those of more usual frictional bead packings (see below) and thus can be convieniently used to investigate the small and large scale properties of granular materials.

In real experiments or in standard numerical simulations run with molecular dynamics (MD) or contact dynamics (CD) for example, a definite final static state is of course reached from any given initial configuration. An example of the output of such a simulation is shown in Fig. 14.2. The force spatial distribution is rather inhomogeneous and shows so-called 'force chains', which can be also observed in experiments on photoelastic grains [401, 408]. The choice of one specific solution among all possible ones is then resolved by the dynamics of the grains before they come to rest and/or the elasticity of the contacts. In MD simulations, for instance, these contacts are treated as (possibly nonlinear) springs that give a force directly related to the slight overlap of the grains.

As a conclusion, for given boundary conditions (geometry, external load), but for different initial configurations of the grains (positions, velocities), the final static packing (positions, contacts, forces) will be different. The implicit hypothesis is that all these final states are statistically equivalent and can be used to compute averaged quantities or statistical distribution functions. The description of these averaged quantities (e.g. the stress tensor) at a larger scale is the subject of the second part of the chapter. In the following subsections, we shall rather study the probability distribution of the contact forces f and orientations θ.

14.1.2 Force probability distribution

A picture like Fig. 14.2 shows that the forces applied on a grain can be very different from point to point. Some grains belong indeed to chain-like structures that carry most of the external load, while others stay in between these chains and hardly support any stress. Many pieces of work have been devoted to the study of the probability distribution of the forces between grains. We shall start with experimental results, and then turn to the numerical ones.

The first reference experiment has been published by Liu *et al.* in [158], together with a simple scalar model that will be presented below. The sketch of the set-up of this experiment is shown on the left of Fig. 14.3: a carbon paper is placed at the bottom of a cylinder filled with glass beads. The granular material is compressed from the top. After the compression, the black spots left by the beads on the paper are analysed. Their size can be calibrated versus the intensity of the forces that were pushing on these beads. The experiment is repeated several times, and a force histogram can be obtained. On the right part of Fig. 14.3 is plotted the probability distribution function P of the forces f after they have been normalised by their mean value. The semi-log plots cleary show that the decay of P is exponential. This means that measuring a force which is twice or three times the mean value is quite frequent, or at least not that rare. This feature is very robust and does not depend on the place where the measurements were performed [159]. More surprisingly, it is also insensitive to the value of the friction coefficient between the grains [386]. Finally, the way the packing was initially built up seems to be unimportant too [386]: ordered HCP pilings and disordered amorphous packings have the same $P(f)$. This last result in fact suggests that a very weak amount of local geometrical disorder may be sufficient to generate a large variability of the forces at the contact level.

This carbon paper technique is pretty astute. However, it is not very well adapted to get a precise measure of small forces and needs a high confining pressure. Other experiments have been performed using different probes, such as that of Løvoll *et al.* [430] where the grains are compressed by their own weight only. Their results are plotted in Fig. 14.4. Again, forces have been normalised by their mean value. Besides the exponential decay of $P(f)$ at large force, they got almost a plateau distribution for small f. The same behaviour has been reported by Tsoungui *et al.* [474] on two-dimensional systems, and by Brockbank *et al.* [388]. At last, similar features have been shown with softer grains, either sheared in Couette cells [416, 417], or under moderate compression [161, 435].

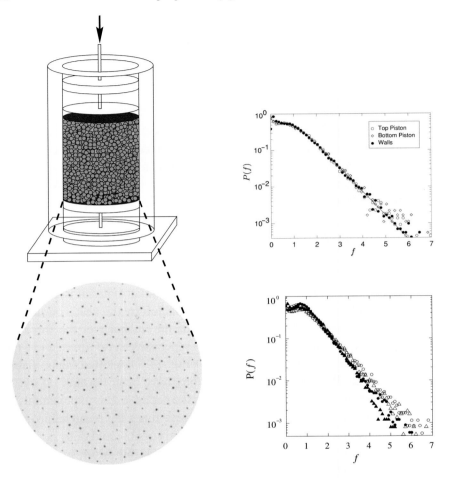

Fig. 14.3 Left: sketch of the carbon paper experimental set-up. The forces felt by the grains at the bottom of the cell are measured by the size of the black spot left on the paper below the grains. Right: force distribution function $P(f)$. The forces have been normalised by their mean value. This distribution is very robust and follows the same exponential curve, independent of the place where the measurements were performed (top), and of the ordering of the packing or the friction coefficient between beads. (bottom): smooth amorphous piling of glass beads ($\circ$), smooth HCP ($\bullet$), rough amorphous ($\triangle$) and rough HCP ($\blacktriangle$). These pictures are from Mueth *et al.* [159], and Blair *et al.* [386].

Numerical simulations have been another way to address the issue of the force probability distribution in granular systems. The work already cited of Radjai *et al.* [453, 454] gives the function $P(f)$ plotted in Fig. 14.5. Similar simulations [470, 461, 419, 422, 444, 384], a recent ensemble approach [464–466], as well as studies of frictionless rigid beads [379, 473], and of sheared granular systems [380], lead to

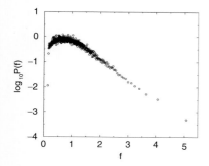 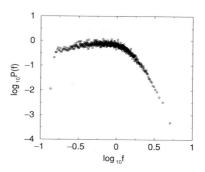

Fig. 14.4 Force distribution function measured by Løvoll *et al.* with an electronic pressure probe [430]. The left semi-log plot shows the exponential fall-off of $P(f)$ at large forces, while one can see the almost flat behaviour of the distribution at small f on the right.

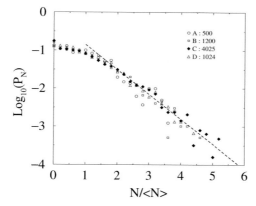

Fig. 14.5 Distribution function of the normal forces computed from simulations by Radjai *et al.* [454] such as the one displayed in Fig. 14.2. This distribution is independent of the number of grains in the sample. The behaviour of P at small forces is again almost flat. The distribution of tangential forces is very similar.

very similar results. As a broad statement, one can say that almost all experimental and numerical data can be reasonably well fitted with a force probability distribution of the form

$$P(f) \propto \begin{cases} (f/\bar{f})^{\alpha}, & \text{for} \quad f < \bar{f}, \\ e^{-\beta f/\bar{f}}, & \text{for} \quad f > \bar{f}, \end{cases} \tag{14.5}$$

where $\bar{f}$ is the mean value of the contact forces. In fact, some of the $P(f)$ plots of the above cited papers show a large force falloff slightly faster than an exponential – e.g. with a Gaussian cutoff – and the fine nature of the large f tail is certainly still a matter of discussion. Besides, interesting comparisons with supercooled liquids near the glass transition or random spring networks can be found in Refs.[447, 446, 412].

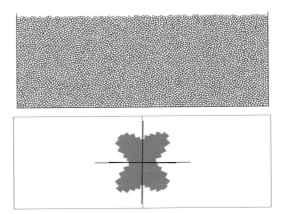

Fig. 14.6 Polar representation of the contact orientation distribution obtained in a numerical simulation of a granular layer prepared by a uniform 'rain' of grains [455]. Four lobes are clearly visible.

The coefficient β is always between 1 and 2. α stays very close to 0, but is sometimes found positive as in the experiments shown in Fig. 14.4, or negative as in Radjai's simulations. More important is the question whether the function P vanishes at small f or remains finite. This may be related to boundary effects [430, 464, 465], and will be discussed further at the end of the subsection on the q-model.

In conclusion, forces in granular materials vary much from a contact between two grains and the next, and therefore exhibit a rather wide probability distribution. This function $P(f)$ is almost flat at forces smaller than the mean force, which means that these small forces are very frequent. The exponential tail of $P(f)$ at large f leads to a typical width of the distribution which is quite large and in fact of the order of the mean force itself.

14.1.3 Texture and force networks

After the study of the probability distribution of the contact forces, another interesting microscopic quantity is the statistical orientation of these contacts $Q(\theta)$. As a matter of fact, getting an isotropic angular distribution in numerical simulations, for example, requires a very careful procedure. In general, the gravity or the external stresses applied to a granular assembly rather create some clear anisotropy in the contact orientation.

An example of such an anisotropy is shown in Fig. 14.6, which is extracted from the numerical work of Radjai *et al.* [455]. In this two-dimensional simulation, a layer of grains is created from a line source, i.e. a uniform 'rain' of grains. The gravity makes these grains fall and confines them into a rather compact packing. The probability distribution Q of the contact orientation θ between two grains is

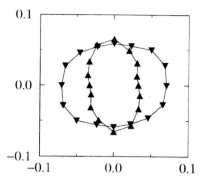

Fig. 14.7 Angular histograms of the orientation of the contact forces computed from simulations of Radjai *et al.* [454, 456] such as the one displayed in Fig. 14.2. The large forces (▲) are preferentially oriented along the main external stress which is vertical, while the small ones (▼) are distributed in a more isotropic way.

plotted in a polar representation. This distribution clearly shows four lobes. This means that vertical and horizontal contacts are less numerous than diagonal ones. This feature has been also reported in experiments [389].

As suggested by the analysis of the force distribution $P(f)$, it may be useful to distinguish between 'strong' and 'weak' contacts that carry a force larger or smaller than the average, and plot separated angular histograms $Q(\theta)$. This has been done by Radjai *et al.* in [454, 456], see Fig. 14.7. In this work, the system of grains confined in a rectangular box has been submitted to a vertical load which was larger than the horizontal one. As a result, large forces are preferentially oriented along the main external stress, while the small ones are distributed in a more isotropic way. Besides, they have shown more precisely that although the strong force network represents less than $\sim 40\%$ of the contacts, it supports all the external shear load.

In summary, by contrast to the force probability distribution $P(f)$, the angular histogram of contact orientation $Q(\theta)$ of a granular packing is very sensitive to the way this system was prepared. This function is then a good representation of its internal structure, or its so-called 'texture'. A good empirical fit of these polar histograms can be obtained by a Fourier modes expansion, i.e. with a function of the form

$$Q(\theta) = \frac{1}{2\pi} (1 + a \cos 2\theta + b \cos 4\theta). \tag{14.6}$$

Profiles of this function are shown in Fig. 14.8. People have tried to built several tensors that encode this microscopic information. The simplest texture tensor is probably

$$\varphi_{\alpha\beta} = \langle n_\alpha n_\beta \rangle, \tag{14.7}$$

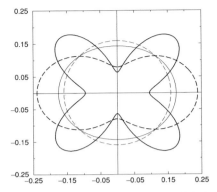

Fig. 14.8 Polar plot of the function defined by Eq. (14.6). The angle θ is taken here with respect to the vertical direction. The thin dashed line is the isotropic case $a = b = 0$. The thin solid line is for $a = -0.1$ and $b = 0$. The bold dashed line is again for $b = 0$ but $a = -0.5$. Note the qualitative change of the curve from an ellipse-like shape to a 'peanut-like' one when $|a| > 1/5$. A four-lobes profile is obtained with finite values of b: here the bold solid line is for $a = -0.1$ and $b = -0.5$.

where n_α is the αth component of the contact unit vector $\vec{n}$. The brackets represent an ensemble average over the contacts. In the case of $Q(\theta)$ of Expression (14.6), the principal directions of $\varphi_{\alpha\beta}$ are the vertical and horizontal axis, and the eigenvalues read $1/2 \pm a/4$, independent of b and of any additional higher order Fourier mode. Note that these principal directions may not coincide with those for which contacts are most (or least) frequent. If they should become so, more complicated texture tensors must be introduced.

A last interesting property of the angular distribution is a kind of 'signature' of its past history. Suppose, for example, that a layer of grain is prepared with a rain under gravity and shows a $Q(\theta)$ like the one in Fig. 14.6. Now, when this layer is gently sheared, say, to the right, the top right and bottom left lobes of $Q(\theta)$ will progressively shrink. When an eventual ellipse-like angular histogram is achieved, it will mean that all the initial preparation has been forgotten. We shall see in the next section the importance of the preparation procedure in the measure of the macroscopic stress tensor profiles.

14.1.4 The q-model

Presentation of the model

In order to understand the exponential distribution of contact forces in a granular system, a very simple stochastic model has been introduced by Liu *et al.* [158, 398]. They consider a packing of grains under gravity. The first strong simplification of

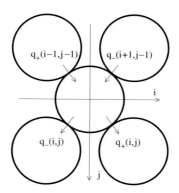

Fig. 14.9 Scheme of the q-model with $N = 2$ neighbours. The $q_{\pm}$s are independent random variables, except for the weight conservation constraint $q_{+}(i, j) + q_{-}(i, j) = 1$.

this model is to deal with a scalar quantity, the 'weight' w of the grains. The second
step is to describe how each of these grains receives some weight from its upper
neighbours, and distributes fractions of its own w to its lower ones. Such a point
of view works well with an ordered enough packing where one can identify grain
layers with upper and lower contacts. For example, one can assume that the grains
reside on the nodes of a two-dimensional lattice. We denote by $q_k(i, j)$ the fraction
of the weight that the grain labelled with the two integers (i, j) transmits to its kth
lower neighbour. Because real granular packings are disordered, the qs are taken as
independent random variables. They encode in a global phenomenological way all
the geometrical irregularities of the piling, the variations of friction mobilisation at
the contacts, and so on. To ensure weight conservation, they must, however, verify
$\sum_{k=1,N} q_k(i, j) = 1$, where N is the number of lower (or upper) neighbours. The
nice trick of this approach is thus to mix together a regular connection network
between the grains and random transmission coefficients. The random variables q
gave the name of the model.

In the following we shall focus for simplicity on the case of $N = 2$ neighbours, as
depicted in Fig. 14.9. In this case, the grain (i, j) has two transmission coefficients
q_+ and $q_- = 1 - q_+$. The case $q_+ = q_- = 1/2$ would correspond to a completely
ordered situation. In practice, they are distributed according to some distribution
function $\rho(q)$. We shall see below that the choice of this function is crucial for the
behaviour of the force distribution function $P(w)$. The simplest case is to consider
a uniform distribution between 0 and 1, for which $\rho(q) = 1$.

In this framework, the equations of static equilibrium reduce to the balance of
the vertical component of the forces, which reads

$$w(i, j+1) = w_0 + q_+(i-1, j)w(i-1, j) + q_-(i+1, j)w(i+1, j), \quad (14.8)$$

where w_0 is the weight of a single grain. For any given set of all the $q_{\pm}(i, j)$, the weights $w(i, j)$ can be computed, layer after layer, with this equation everywhere starting from the top surface $j = 0$. Because the $q_{\pm}$ are random, w fluctuates from point to point. The relevant quantity to look at is then the force distribution function $P(w)$.

Force distribution and the exponential tail

Coppersmith *et al.* [398] have shown that, in the limit of a very deep system, the weights of two neighbouring sites become independent for any generic function $\rho(q)$. Then $P(w)$ obeys the following mean-field equation for $j \to \infty$:

$$P_{j+1}(w) = \int_0^1 dq_1 dq_2 \rho(q_1) \rho(q_2)$$
$$\times \int_0^\infty dw_1 dw_2 P_j(w_1) P_j(w_2) \delta[w - (w_1 q_1 + w_2 q_2 + w_0)]. \quad (14.9)$$

For $\rho(q) = 1$, the expression of the stationary solution P^* is given by

$$P^*(w) = \frac{w}{\overline{w}^2} \exp -\frac{w}{\overline{w}}, \quad (14.10)$$

where $2\overline{w} = j w_0$ is the average weight. In the more general situation of N neighbours, P^* is instead a Gamma distribution of parameter N: its small w behaviour is w^{N-1}, while the large w tail is exponential.

This behaviour for P^* at small w is not specific to the choice $\rho(q) = 1$. For example, the condition for the local weight w to be small is that all the N qs reaching this site are themselves small; the phase space volume for this is proportional to w^{N-1}, if the distribution $\rho(q)$ is finite and regular around $q = 0$. This scaling is in fact very general and is also found in the ensemble approach of Snoeijer *et al.* [464, 465]. If instead $\rho(q) \propto q^{\gamma-1}$ when q is small, one expects $P^*(w)$ to behave for small w as $w^{-\alpha}$, with $\alpha = 1 - N\gamma < 0$. Similarly, the exponential tail at large w is sensitive to the behaviour of $\rho(q)$ around $q = 1$. In particular, if the maximum value of q is $q_M < 1$, one can study the large w behaviour of $P^*(w)$ by taking the Laplace transform of Eq. (14.9). One finds in that case that $P^*(w)$ decays *faster* that an exponential:

$$\log P^*(w) \underset{w \to \infty}{\propto} -w^b \quad \text{with} \quad b = \frac{\log N}{\log q_M N}. \quad (14.11)$$

Note that $b = 1$ whenever $q_M = 1$, and that $b \to \infty$ when $q_M = 1/N$: this last case corresponds to an ordered packing with no fluctuations. In this sense, the exponential tail of $P^*(w)$ in the q-model is *not* universal but requires the possibility that one of the q can be arbitrarily close to 1. This implies that all other qs originating from that point are close to zero, i.e. that there is a nonzero probability that one grain is entirely bearing on one of its downward neighbours. This is what could be called 'arching' in this context.

Finally, a qualitatively different behaviour is obtained if the qs can only take the values 0 and 1. The stationary force distribution at large depth is then a power law $P^*(w) \propto w^{-\alpha}$, with $\alpha = 4/3$ for $N = 2$. This law is truncated for large w as soon as values for q different from 0 and 1 are permitted. A generalisation of the q-model allowing for arching was suggested in [395], which dynamically generates some sites where $q_+ = 1$ and $q_- = 0$ (or vice versa).

The exponential behaviour of $P^*(w)$ at large w, in comparison to the experimental and numerical data of the previous subsections, is probably the main success of the q-model and made it popular. Note that this model underestimates the proportion of small forces, as $P^*(w) \to 0$ when $w \to 0$. However, it is not clear whether contact forces correspond to w or to qw. As a matter of fact, the probability distribution of the latter quantity is also exponential but finite at small qw for uniform qs. More generally, Snoeijer *et al.* have shown that the measure of the distribution of bulk contact forces, or of forces on a boundary of the system (e.g. a wall) which comes from a sum over several contacts, do not have the same behaviour at small f, see [464, 465].

Besides, the q-model suffers from other serious flaws. Indeed, due to its scalar nature, it neglects all the contribution of the horizontal forces, and therefore excludes shearing or proper arching effects. Another point is that Eq. (14.8) is equivalent at large scales to a diffusion equation, the vertical axis being the equivalent of the time. For the stresses in a silo or in response to a localised overload, this leads to a scaling behaviour that is not one of those observed experimentally – see next section. Several vectorial generalisations of the q-model have been proposed [467, 396, 445, 403], which also give a force distribution function $P(w)$ with an exponential (or slightly shrinked exponential) tail. In other studies, correlations have been taken into account, see e.g. [463].

Let us finish by mentioning another interesting type of approach for the description of the force probability distribution $P(f)$. This approach is based on the postulate that the statistics of a disordered grain packing is well encoded by an entropy of the type $S \propto \int df\, P \ln P$. If one maximizes this function under the constraints that $P(f)$ is normalised and that the overall stress is constant, one gets explicit expressions for $P(f)$ which have exponential tails [419, 422, 444, 384].

14.2 Large-scale properties

In this second section of the chapter, we would like to present large scale properties of static granular pilings. As a matter of fact, in many experiments stresses are measured at a rather 'macroscopic' scale, e.g. with captors in contact with typically hundreds of grains. We start with a review of such experiments performed in different situations (geometry of the pile, the silo or the uniform layer) and related numerical simulations. We switch after this to theoretical considerations, with firstly the question of change of scale (how to go from the contact forces to a continuum stress field) and secondly a brief description of the modellings and approaches introduced to interpret these experimental data.

14.2.1 Stress measurements in static pilings

We now present recent data concerning the measurement of the stresses in granular systems at rest. We shall start with one of the most studied case, namely the silo geometry, which is sometimes called the Janssen's experiment in reference to a paper published by this German engineer in 1895. Of course, the literature on this subject is particulary large, especially because of the applications of such a geometry in industrial processes. As already emphasised in the introduction of this chapter, we will not review all the existing papers, but only give here the basis of the screening effects that are observed in silos. Another simple geometry is that of the conical pile. As a matter of fact, the description of the pressure profile under a sandpile is probably one of the issues that has been at the origin of the interest of many physicists for granular materials [359]. The last point of this subsection will be dedicated to the study of the stress response function of a layer of grains. This situation is in some way a more elementary and fundamental configuration which contains in fact all the challenging difficulties of these systems – history dependency, anisotropy and so on.

Silos

The principle of a typical experiment in silos is sketched on the left of Fig. 14.10. Consider a column filled with a certain mass of grains M_{fill}. The question is to know what is the weight felt by the bottom plate of this silo. The experiments have shown that this weight corresponds to an apparent mass M_{app} which is only a fraction of M_{fill}. In other words, the lateral walls of the silo support a substantial part of the total mass of the grains.

More precisely, one can measure M_{app} as a function of M_{fill}. The corresponding plot is shown on the top right of Fig. 14.10. The curve grows and saturates to some value M_{sat} when M_{fill} becomes large enough. In this case, large enough means that the silo must be filled up to a height of the order of few times its diameter. Pouring more grains than this, or even adding an overload Q on the top of the grains will hardly affect the apparent mass at the bottom. The top of the silo is 'screened' by the walls and the bottom feels only what is just above it. For silos of smaller aspect

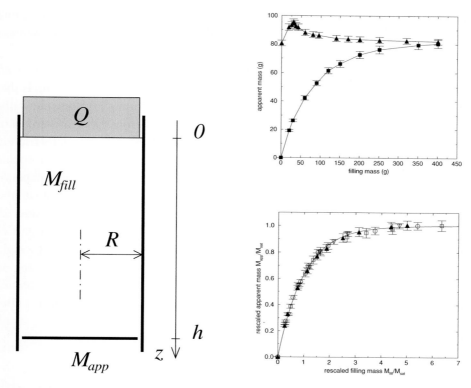

Fig. 14.10 Left: sketch of the silo experiment used by Vanel *et al.* [475] and Ovarlez *et al.* [452]. A mass M_{fill} is poured into the column, and the apparent mass M_{app} is measured at the bottom. An overload Q can be added on the top before the measure. The experimental protocol ensures that the friction is fully mobilised at the walls. Right: apparent mass vs filling mass without (■) and with (▲) an overload of 80.5 g (top), here for a medium-rough 38 mm column. These curves saturate to some well defined value M_{sat}. An 'overshoot' is observed when the grains are overloaded. Each data point has been obtained from a different run of controlled density. When rescaled by the saturation mass, the different unoverloaded data collapse onto a single curve (bottom): loose packing in the medium-rough 38 mm diameter column (▲), and dense packing in the rough (○) and smooth (■) 38 mm columns, dense packing in the medium-rough 80 mm column (▽). This master curve is very well fitted by Janssen's prediction (line). These two graphs are from [452].

ratios, however, the effect of the overload can be clearly seen, as it leads to an 'overshoot' of the saturation value.

The value of M_{sat} depends on the precise preparation procedure of the column, as well as the roughness of the walls. As expected, a larger friction coefficient between the grains and the walls gives a smaller M_{sat}. Likewise, a denser packing of grains also makes M_{sat} decrease. For columns of different sizes, the saturation mass scales like R^3. Interestingly, when rescaled by M_{sat}, all the unoverloaded screening curves collapse onto a single curve, see Fig. 14.10 (bottom right). In the presence of a finite overload, the rescaled maximum amplitude of the overshoot is found to increase with the wall friction or the density.

The data presented here in Fig. 14.10 have been obtained by Ovarlez *et al.* [452], see also [475]. A very important feature of their experimental set-up is that they make sure to have a wall friction fully mobilised uniformly all along the walls, which is done by a tiny displacement of the whole piling before the measurement of M_{app}. As a matter of fact, the screening effects discussed above are crucially friction dependent, and one should be aware that a less controlled experimental protocole can lead to rather different results. Similar data can be found in many other papers – see [427, 478, 471] for instance, or [425] for the corresponding numerical simulations – albeit that the effect of the additional overload Q is generally not considered.

In summary, the weight measured below a granular column is only a part of the total weight of the grains in that column. More precisely, this apparent weight progressively saturates to a value corresponding to the grains in the bottom region of the column, i.e. up to a height of the order of its diameter. The rest is screened or supported by the walls. As a consequence, an overload on the top surface does not affect the apparent weight at the bottom if the silo is tall enough. This overload, however, produces an interesting overshoot effect in small columns.

Sandpiles

Let us turn now to the pile geometry. Grains are poured on a rigid flat plate. They spontaneously form a conical pile at the angle of repose ϕ of the material. Can we predict what the pressure profile below this pile is? A naive guess would be that the pressure would simply be proportional to the local thickness of the pile. However, careful experiments of Vanel *et al.* [375] have shown that the shape of this profile strongly depends on the way the pile was built. The sketch of the experimental set-up is depicted in Fig. 14.11: the pressure p under a pile of height h is measured with a capacitive gauge at a horizontal distance r from the centre of the pile. As evidenced in the graphs of Fig. 14.12, the profile shows a minimum, or a 'dip' around $r = 0$ if the pile has been grown from a hopper, i.e. a point source. By contrast, $p(r)$ has a slight 'hump' when measured on a pile built by successive horizontal layers, i.e. from a distributed 'rain'. A very similar behaviour – with perhaps a less pronounced dip – is found in wedges [375], or in numerical calculations and simulations of two-dimensional heaps [376, 429, 431, 436, 437, 448].

The data presented in this figure have also been collected from the papers of Šmíd and Novosad [462] and Brockbank *et al.* [388]. What is interesting is that all these data have been obtained on piles of various heights, with rather different measurement techniques, and that they can be collapsed onto the same master curve. To do so, they have been rescaled by the height of the piles and the density

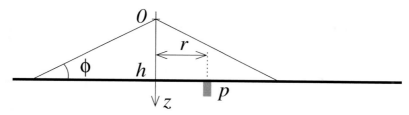

Fig. 14.11 Sketch of the experimental set-up of Vanel *et al.* for the measure of the pressure profile $p(r)$ at the bottom of a sand pile [375]. h and ϕ are the height and the repose angle of the pile. r is the horizontal distance from the centre of the pile to the pressure gauge.

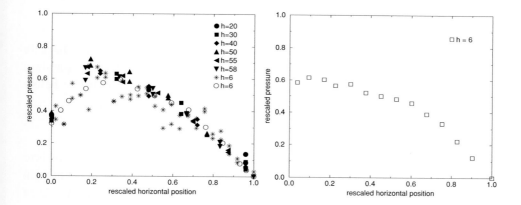

Fig. 14.12 Pressure half profiles at the bottom of sandpiles of various heights h (given here in cm). Besides density normalisation, these data have been rescaled by h and collapse onto a master curve that shows either a 'dip' at the centre ($r = 0$) of the pile (left), or a slight 'hump' (right). In the first case the piles have been built from a hopper, while in the second one the preparation was achieved by successive horizontal layers. The data are from Šmíd and Novosad (filled symbols) [462], Brockbank *et al.* (stars) [388] and Vanel *et al.* (open symbols) [375]. Each data point of Vanel *et al.*'s experiments represents an average over typically ~ 10 different heaps, while the other profiles have been obtained from a single pile and are thus much noisier. In these cases, the granular material is sand with a repose angle of 30–33°.

of the material. More precisely, horizontal lengths have been normalised by the radius of the pile, and stresses have been divided by the total weight of the pile which is the integral of the vertical pressure over r. Each data point of Vanel *et al.*'s experiments represents an average over typically ~ 10 different heaps (there is one single pressure gauge on the rigid bottom plate), while the other profiles have been obtained from a single pile (the whole profile is grabbed at once by a series of captors), and are thus much noisier.

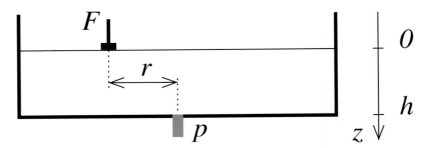

Fig. 14.13 Sketch of the experimental set-up of Reydellet *et al.* for the measure of the pressure profile $p(r)$ at the bottom of a granular layer in response to a localized vertical overload F at its top surface [457, 460]. h is the thickness of the layer. r is the horizontal distance from the overload point to the pressure gauge.

To summarise, two sandpiles that look in appearance indentical may have in fact a rather contrasted distribution of internal stresses due to different preparation histories. In particular, when built from a source point, the pressure profile below the pile shows an interesting minimum right below the apex of the heap.

Response functions

Another way to investigate the effect of the preparation on the stress distribution is to study the response function of a granular layer, i.e. the pressure profile at the bottom of this layer in response to a localised overload at its top surface. The principle of the measurement is shown in Fig. 14.13. We call h the thickness of the layer and r the horizontal distance between the overload point and the pressure gauge position. The applied force F must be small enough to prevent any rearrangement of the initial packing of the grains that we want to probe by this technique.

Experiments have been performed by Reydellet *et al.* [457, 460] with layers of plain sand prepared in two different ways. The packing was either made very dense by successive compressions – Fig. 14.14a – or on the contrary very loose by pulling a sieve through the grains – Fig. 14.14b. It was shown that the pressure profiles present one single broad central peak of width of the order of the layer thickness h. Furthermore, profiles measured on layers of different h reasonably collapse onto the same curve when all lengths are divided by h, and the applied force rescaled to unity. Finally, this master curve is preparation dependent: the graph of Fig. 14.14 clearly shows that the response function of a loose piling is narrower than that of a compressed one.

Complementary experiments have been performed by Geng *et al.* [407, 409] on two-dimensional packings of photoelastic grains. A typical stress response photo is depicted in Fig. 14.15. Averaged over many samples, they showed that the response profile can be very different, depending on whether the grain assembly is ordered

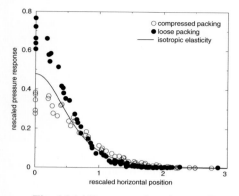

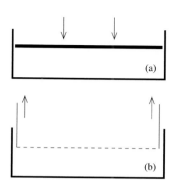

Fig. 14.14 Response function profiles from [460]. The open circle data points have been obtained on dense and compressed layers of grains (a), while the filled ones are from rather loose packings (b). These profiles come from measurements on layers of different heights h. Albeit some (non-systematic) dispersion of the data around $r = 0$, the rescaling is rather correct. The response function of a loose piling is narrower than that of a compressed one. For comparison, the response of a semi-infinite isotropic elastic medium lies in between.

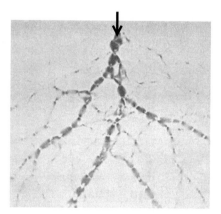

Fig. 14.15 Stress distribution in a two-dimensional packing of photoelastic grains in response to a localised force at the top (the gravitational part has been substracted). Darker zones indicate a larger stress. Chain-like structures are clearly visible. This picture has been obtained by Geng *et al.* [407, 409].

or not. A regular piling of monodisperse beads presents indeed a two lobe response, while that of an amorphous packing of pentagons or polydisperse beads has only one. As a matter of fact, one can continuously change the profile shape – the peaks get closer and closer – by increasing, for instance, the grain size polydispersity. The importance of ordering has been also shown in three dimensions by Mueggenburg *et al.*, who where able to get three peak or ring-like response profiles for respectively FCC and HCP packings [441]. Finally, the skewness of the profile can be affected

  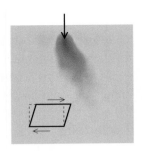

Fig. 14.16 Stress response averaged over ~ 50 pictures like that of Fig. 14.15. The shape of this response shows two lobes for a regular packing of circular monodisperse beads (left), but only one lobe (middle) when the layer is disordered (pentagonal beads). When the layer is sheared beforehand, the response is skewed in the direction of the shearing (right). The typical height of these pictures is $\sim 10 - 15$ grain diameters.

by an initial shearing of the packing: the maximum of the response is then deviated in the direction of the shear stress.

A few two-dimensional numerical simulations have also tackled this stress response problem. Firstly, two extreme situations have been studied: the case of polydisperse and frictionless beads packed under gravity [414] and that of an ordered packing of frictional beads [387]. In both cases, the stress response profile shows a double peaked shape. In the limit of the size of the simulated systems, one can say that the position of these peaks scales like the layer thickness h, while their width grows like $\sqrt{h}$. In the first case the packing is clearly disordered, but, due to the preparation procedure, the contacts between beads are probably distributed in a rather anisotropic way. In the second simulation, the only source of randomness is due to the finite friction between the grains which makes the system hyperstatic. It is observed that the two peaks of the response profile are closer when the friction coefficient is larger. A double peaked response has been also obtained for ordered and frictionless grain layers [449]. Systematic studies of more generic systems – i.e. polydisperse and frictional grains – have shown that the typical stress response shape shows a single broad peak with features similar to those measured in experiments [428, 381], but that sufficiently strong anisotropy can change the response shape from a single to a double peak [413].

In summary, these response experiments and simulations present a very rich phenomenology. The shape of the response function is very sensitive to the frictional properties of the grains, the ordering aspect of the system, as well as the preparation procedure of the packing. It is, for example, possible to relate the dip of pressure under a pile to the skewness of the response curve [382]. The response function is thus a very interesting quantity to study the link between the micro-structure of a granular assembly and its mechanical properties at large scales.

14.2.2 From micro-to macroscopic scales

The aim of this subsection is to show how one can go from the knowledge of contact forces and the geometry at the scale of the grains to a continuum description of a stress field. This is particulary useful when a link should be made between discrete numerical simulations where all microscopic details of the system are known, and large scale experiments or continuous models. Such a coarse-graining procedure is usually called 'homogenisation' and is a full branch of mechanics in itself. Here, we shall dervive in the first part a general expression for the stress tensor and see how this leads to the well-known Born–Huang formula, or to a boundary-like definition. Although elasto-plastic considerations will be discussed in the next subsection, the second part of this one will already concern the basics of the so-called 'effective medium theory' in which microscopic geometrical properties of the grain contacts are related to large-scale elastic coefficients.

From contact forces to stresses

The derivation here will closely follow the work of Goldhirsch *et al.* [295, 410]. For that reason, we shall keep in this subsection to similar notation. Unless denoted otherwise, Greek indices will denote coordinate axis, whereas Latin ones will stand for the number of the particles. For example, $r_{i\alpha}$ is the αth component of the vector refering to the position of the ith particle. This distinction is important as many sums and indices will be involved in the remainder – note that implicit summation on repeated coordinate indices is understood. However, in the next subsection where continuum models are discussed, Latin letters will recover their classical use of coordinate labels.

Let us forget for a little while that we are interested in static grain packings, and let us denote by $\vec{r}_i$ and $\vec{v}_i$ the position and velocity of the ith particle among a system of N other ones. The microscopic mass and momentum density are simply given by

$$\rho_{\text{mic}}(\vec{r}, t) = \sum_{i=1}^{N} m_i\, \delta(\vec{r} - \vec{r}_i(t)), \tag{14.12}$$

$$\vec{p}_{\text{mic}}(\vec{r}, t) = \sum_{i=1}^{N} m_i\, \vec{v}_i(t)\, \delta(\vec{r} - \vec{r}_i(t)). \tag{14.13}$$

In order to go from these δ peaks to a smooth and continuous description, these quantities must be spatially coarse-grained. For that purpose, we introduce a scalar coarse-graining function $\phi(\vec{R})$ which has the property to be positive and normalised so that its integral over space is unity. To make sense, it should also have a 'reasonable' shape with a single maximum in $\vec{R} = \vec{0}$, and a typical example could be a Gaussian $\phi(\vec{R}) = (2\pi\lambda^2)^{-D/2}e^{-R^2/(2\lambda^2)}$, where λ is the coarse-graining length. More anisotropic ϕs will be introduced later on. With this function ϕ, mass and momentum density can be expressed as

$$\rho(\vec{r}, t) = \int d\vec{r}\,'\phi(\vec{r} - \vec{r}\,')\,\rho_{\text{mic}}(\vec{r}\,', t) = \sum_{i=1}^{N} m_i\,\phi(\vec{r} - \vec{r}_i(t)), \qquad (14.14)$$

$$\vec{p}(\vec{r}, t) = \int d\vec{r}\,'\phi(\vec{r} - \vec{r}\,')\,\vec{p}_{\text{mic}}(\vec{r}\,', t) = \sum_{i=1}^{N} m_i\,\vec{v}_i(t)\,\phi(\vec{r} - \vec{r}_i(t)). \quad (14.15)$$

The temporal derivatives of these coarse-grained quantities should yield to the usual conservation equations. As a first example, from

$$\partial_t \phi(\vec{r} - \vec{r}_i(t)) = -\partial_t \vec{r}_i(t) \cdot \vec{\nabla}\phi(\vec{r} - \vec{r}_i(t)) = -v_{i\alpha}\frac{\partial}{\partial r_\alpha}\phi(\vec{r} - \vec{r}_i), \qquad (14.16)$$

it is easy to show that the conservation of mass is recovered as $\partial_t \rho + \vec{\nabla} \cdot \vec{p} = 0$. Similarly, taking the conservation of momentum as the definition of the stress tensor, we have

$$\partial_t p_\alpha + \frac{\partial \sigma_{\alpha\beta}}{\partial r_\beta} = 0. \qquad (14.17)$$

Now, the time derivative of p_α gives

$$\partial_t p_\alpha = \sum_{i=1}^{N} m_i\,\partial_t v_{i\alpha}\,\phi(\vec{r} - \vec{r}_i) - \frac{\partial}{\partial r_\beta}\sum_{i=1}^{N} m_i\,v_{i\alpha}v_{i\beta}\,\phi(\vec{r} - \vec{r}_i). \qquad (14.18)$$

If we come back to the static case $v_{i\alpha} \to 0$, the second sum is of higher order compared to the first one and can be neglected. Furthermore, we can use the fact that

$$m_i\,\partial_t v_{i\alpha} = \sum_{j\neq i} f_{ij\alpha}, \qquad (14.19)$$

where $f_{ij\alpha}$ is αth component of the force exerted by particle j on particle i – we assume pairwise interactions between particles. The goal then is to be able to express this remaining double sum on i and j into a $\partial/\partial r_\beta$ derivative. To do so, several tricks can be used. First, in these sums over particles, i and j are dummy indexes and can be switched. Using the third Newton's law, $f_{ij\alpha} = -f_{ji\alpha}$, we can then write

$$\sum_{i,j,i\neq j} f_{ij\alpha}\,\phi(\vec{r} - \vec{r}_i) = \sum_{i,j,i\neq j} f_{ji\alpha}\,\phi(\vec{r} - \vec{r}_j) = -\sum_{i,j,i\neq j} f_{ij\alpha}\,\phi(\vec{r} - \vec{r}_j). \quad (14.20)$$

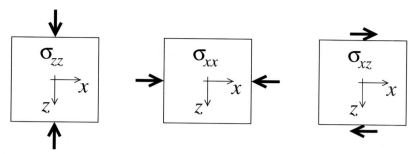

Fig. 14.17 Positive sign convention for the different stress components (2D case).

This is interesting because

$$\phi(\vec{r} - \vec{r}_i) - \phi(\vec{r} - \vec{r}_j) = \int_0^1 ds \, \frac{\partial}{\partial s} \phi(\vec{r} - \vec{r}_i + s \, \vec{r}_{ij}),$$ (14.21)

where we have introduced $\vec{r}_{ij} \equiv \vec{r}_i - \vec{r}_j$, and

$$\frac{\partial}{\partial s} \phi(\vec{r} - \vec{r}_i + s\vec{r}_{ij}) = r_{ij\beta} \frac{\partial}{\partial r_\beta} \phi(\vec{r} - \vec{r}_i + s \, \vec{r}_{ij}),$$ (14.22)

so that the final expression for the stress tensor is

$$\sigma_{\alpha\beta}(\vec{r}) = \frac{1}{2} \sum_{i,j,i\neq j} f_{ij\alpha} \, r_{ij\beta} \int_0^1 ds \, \phi(\vec{r} - \vec{r}_i + s \, \vec{r}_{ij}).$$ (14.23)

Several important remarks should be made at this point. Equation (14.23) is a rigorous expression for the stress tensor *coarse-grained at the scale of the function* ϕ. Changing ϕ and in particular its typical width λ modifies the value of the stress. It is only when λ is much larger than the diameter of the grains (typically several tens of d) that $\sigma_{\alpha\beta}$ becomes independent of ϕ [422–424]. Therefore, in numerical simulations where the number of grains is never that large, the choice of ϕ and especially that of λ is important and must be specified. In particular, it would make no sense to compare stresses coarse-grained over different scales if these are not large enough – note that ensemble average can enlarge the range over which the stress is independent of λ down to scales of a fraction of the grain diameter [414]. Another point is that, with this definition, symmetry of the stress tensor is not automatically ensured – central forces give a symmetric tensor, but this case is not generic. In practice, $\sigma_{\alpha\beta} = \sigma_{\beta\alpha}$ at sufficiently large λ. Finally, the sign convention must be emphasised: normal stresses are counted positive under compression and negative under traction. As for the shear component $\sigma_{\alpha\beta}$, it is positive when the corresponding rotation is clockwise in the direct axis coordinate system (β, α), see Fig. 14.17.

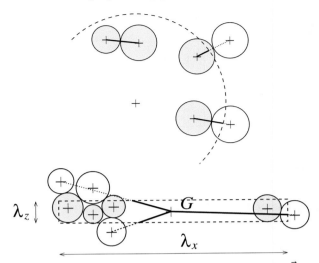

Fig. 14.18 Top: bold lines: visualisation of the branch vectors $\vec{b}$ in the expression (14.24) for a Heaviside coarse-graining function ϕ whose value is finite and constant inside the dashed circle (grey particles), and zero outside (white ones). Bottom: elongated control volume for a boundary-like measure of the stress components σ_{zz} and σ_{xz}.

A particularly common choice for ϕ is to use a Heaviside function: $\phi(\vec{R}) = 1/V$ if $R \leq \lambda = V^{1/3}$, and 0 otherwise. Note that V should be understood as a surface λ^2 in 2D. In that case, the expression (14.23) transforms into the well known Born–Huang formula:

$$\sigma_{\alpha\beta}(\vec{r}) = \frac{1}{2V} \sum_{i,j;i\neq j} f_{ij\alpha}\, b_{ij\beta} = \frac{1}{V} \sum_{\text{contacts } c} f_{c\alpha}\, b_{c\beta}, \qquad (14.24)$$

where $\vec{b}$ is the so-called branch vector. The first sum runs over all pairs of particles for which at least one of them has its centre of mass in this volume V around the point of measure $\vec{r}$, while the second equivalently runs over all contacts between particles of such pairs – this removes the factor of 2 because pairs (i,j) are counted only once and $f_{ij\alpha}\, b_{ij\beta} = f_{ji\alpha}\, b_{ji\beta} = f_{c\alpha}\, b_{c\beta}$. Figure 14.18 illustrates an important subtlety about these branch vectors. When the two particles in contact are inside the control volume V we simply have $\vec{b}_{ij} = \vec{r}_{ij}$, whereas the branch vector is only the fraction of $\vec{r}_{ij}$ which is inside V when the considered contact involves an external grain [383]. Once again, if the control volume V contains a very large number of grains, the contribution to the coarse-grained stress of these 'surface' contacts with respect to the 'bulk' ones is negligible, but it may be of importance for smaller V, such as those usually considered in simulations. Other derivations of this expression for the stress tensor can be found in many other papers, see e.g. [394, 458].

Regarding the way stress is measured in experiments, a natural definition for it involves a boundary of the system. As a matter of fact, at the continuum level, $\sigma_{\alpha\beta} n_\beta \, dS$ is the αth component of the force on the element of surface dS oriented perpendicularly to the vector $\vec{n}$. As an example, in sandpile or response experiments like those of Refs. [375, 457, 460] the vertical normal stress σ_{zz} is measured at the bottom of the layer of grains with a captor of horizontal surface S_{cap}. The vertical component of all vertical forces acting on the captor $\vec{f}_m$ bend its membrane, whose deformation is calibrated with the corresponding variations of its electrical capacitance. In that case, σ_{zz} is thus measured as

$$\sigma_{zz} = \frac{1}{S_{cap}} \sum_{captor} f_{mz} \,. \tag{14.25}$$

A captor sensitive to horizontal forces f_{mx} would equivalently have given the shear stress σ_{xz}, but the measure of the other components σ_{xx} or σ_{zx} would require a vertical membrane. Of course, the number of grains pushing on that captor should be large, but this is typically the case: in experiments cited above, $S_{cap} \sim 1 \text{ cm}^2$ and the grain diameter is of order of 300 μm.

An expression like (14.23) is only valid for positions $\vec{r}$ in the bulk of the packing. This position should be indeed such that the value of the coarse-graining function, centred on $\vec{r}$, is very close to zero at the boundaries of the system – otherwise the normalisation condition of ϕ would not be satisfied. In order to go close to the plate at the bottom of the piling, a possibility is to take an anisotropic ϕ with a vertical extention of the order of one grain diameter, and thus much smaller than the horizontal one. An useful example for two dimensional situations is the following product of Fermi–Dirac functions:

$$\phi(\vec{R}) = B \frac{1}{\left(1 + e^{\frac{|R_x| - \lambda_x/2}{\Delta\lambda}}\right)\left(1 + e^{\frac{|R_z| - \lambda_z/2}{\Delta\lambda}}\right)}, \tag{14.26}$$

where the normalisation factor B is easy to compute. With this choice, it is possible to control independently the coarse-graining lengths λ_x and λ_z in both horizontal and vertical directions as well as the decay length $\Delta\lambda$.

One can go one step further to mimic a captor-like measure of the stress. To do so, let us first go one step backward and consider again the expression for the momentum (14.15) of the system of N particles. Suppose that we divide this system into two parts: particles $j = 1$ to M which are inside a certain volume (that will be identified with the captor), and the others, $J = M + 1$ to N, which are outside. Defining $\vec{G}$ as the centre of mass of the inside particles by $m_{in} G_\alpha = \sum_{j=1}^{M} m_j r_{j\alpha}$, one can compute

$$\tilde{p}_\alpha(\vec{r}) = m_{in}\, \partial_t G_\alpha\, \phi(\vec{r} - \vec{G}) + \sum_{J=M+1}^{N} m_J\, v_{J\alpha}\, \phi(\vec{r} - \vec{r}_J). \tag{14.27}$$

Comparing with p_α we get

$$p_\alpha - \tilde{p}_\alpha = \sum_{j=1}^{M} m_j\, v_{j\alpha}\, \left[\phi(\vec{r} - \vec{r}_j) - \phi(\vec{r} - \vec{G}) \right]. \tag{14.28}$$

The terms between square brackets in the previous sum vanish if ϕ is the Heaviside function which is finite and constant for particles 'inside' and null for those 'outside'. This way, one can therefore gather a group of particles and consider it as if it was a unique object. Now, consider an elongated volume such as that displayed in Fig. 14.18 (bottom). As just discussed, the group of grains whose centre of mass is inside this volume (in grey) can be taken as a single particle of centre G. Using Formula (14.24), the stress $\sigma_{\alpha z}$ coarse-grained over this volume is

$$\sigma_{\alpha z} = \frac{1}{\lambda_x \lambda_z} \sum_{\text{contacts } c} f_{c\alpha}\, b_{cz}. \tag{14.29}$$

If this volume is large enough, the centre of mass G is reasonably located in its geometrical centre so that, as drawn in Fig. 14.18, the z-component of the branch vectors is $\lambda_z/2$ for up-contacts and $-\lambda_z/2$ for down ones. Besides, if $\lambda_z \ll \lambda_x$, the contribution of lateral contacts is negligible. Since this volume is supposed at rest, the total force must balance, i.e. $\sum_c f_{c\alpha}$ on up-contacts equals $-\sum_c f_{c\alpha}$ on down ones, so that we end up with

$$\sigma_{\alpha z} = \frac{1}{\lambda_x} \sum_{\text{up-contacts } c} f_{c\alpha}, \tag{14.30}$$

which is very similar to the boundary stress, Expression (14.25). Note that such a quasi-horizontal control volume is not very well adapted to the computation of the two other components of the stress $\sigma_{x\alpha}$, for which a vertically elongated shape is required.

In summary, we have explicited several formulae to compute the stress tensor from the knowledge of grain force contacts. These expressions involve a coarse-function ϕ which makes a spatial average of the forces over a length scale λ in a given packing – ensemble average over samples can be computed in addition. When λ is large in comparison to the grain diameter the stress values are independent of the choice for ϕ and λ. For smaller coarse-graining length, details of the averaging procedure become more important and comparisons must be done with care. Finally, note that a similar analysis can be done for the computation of the strain tensor [410, 411].

The effective medium theory

We would like to illustrate here another classical coarse-graining procedure, which consists of computing the effective elastic coefficients that describe the average mechanical behaviour of a packing of grains. This requires us to introduce the displacement field of the grains when they are submitted to some external load. Although this seems to be one step beyond the description of packings at rest, we shall see in the next section that elasto-plasticity is in fact related to the statics.

The following calculations will be very basic, and give only a flavour of what can be done in this domain, see e.g. [477, 385, 390, 391]. More precisely, we will show how these elastic coefficients can be related to a fourth-order texture tensor, and discuss afterwards the limitations of this approach. Lastly, it should be emphasised that the index conventions – Greek versus Latin letters – will be kept the same as in the previous part on coarse-grained stresses.

Let us start with some notation, and call $\vec{r}_i{}^0$ the reference position of particle i, and $\vec{r}_i(t)$ that at a later time t. We define the displacement vector as $\vec{u}_i = \vec{r}_i(t) - \vec{r}_i{}^0$. The relative displacement between two grains i and j is thus $\vec{u}_{ij} = \vec{u}_i - \vec{u}_j$, and was $\vec{r}_{ij}^{\,0} = \vec{r}_i{}^0 - \vec{r}_j{}^0$ at the initial time. Later on, we will also need the unit vector $\vec{n}_{ij}^{\,0} = \vec{r}_{ij}^{\,0}/\|\vec{r}_{ij}^{\,0}\|$ along this last direction. In all the following, we will implicitly restrict the analysis in the limit of small displacement for which $\vec{r}_i{}^0$ and $\vec{r}_i$ are very close vectors.

The main so-called 'affinity' assumption of this approach is that the relative displacements between the individual particles follow the macroscopic strain of the global system, i.e. that we can write

$$u_{ij\alpha} = \epsilon_{\alpha\beta} r_{ij\beta}^{0}, \tag{14.31}$$

where $\epsilon_{\alpha\beta}$ is the macroscopic strain tensor. Now, we suppose that no contact is either created or removed between two particles, so that the interaction between the grains i and j can be linearised and modelled as a spring of stiffness K_{ij}. The force $\vec{f}_{ij}$ is then

$$f_{ij\alpha} \sim K_{ij} \left(\vec{n}_{ij}^{\,0} \cdot \vec{u}_{ij} \right) n_{ij\alpha}^{0}, \tag{14.32}$$

where we have projected the displacement $\vec{u}_{ij}$ on $\vec{n}_{ij}^{\,0}$. As we have seen in the previous part, the computation of the stress tensor from the contact forces requires a fine analysis, but for the sake of simplicity let use the formula $\sigma_{\alpha\beta} = \dfrac{1}{2V} \sum_{i \neq j} f_{ij\alpha}\, r_{ij\beta}^{0}$.

With the previous expression for $\vec{f}_{ij}$, and using the mean field assumption for the

Static properties of granular materials

displacement vectors, each term of the sum can be written as

$$f_{ij\alpha}\, r_{ij\beta}^0 = K_{ij}\, n_{ijy}^0\, u_{ijy}\, n_{ij\alpha}^0\, r_{ij\beta}^0 = K_{ij}\, n_{ijy}^0\, \epsilon_{\gamma\delta}\, r_{ij\delta}^0\, n_{ij\alpha}^0\, r_{ij\beta}^0. \tag{14.33}$$

In the simplest case where $K_{ij} = K$ and $||\vec{r}_{ij}^{\,0}|| = d$, we then end up with a standard linear stress–strain relation:

$$\sigma_{\alpha\beta} = C_{\alpha\beta\gamma\delta}\, \epsilon_{\gamma\delta} \quad \text{with} \quad C_{\alpha\beta\gamma\delta} = \frac{K d^2}{2V} \sum_{i,j,i\neq j} n_{ij\alpha}^0\, n_{ij\beta}^0\, n_{ijy}^0\, n_{ij\delta}^0. \tag{14.34}$$

$C_{\alpha\beta\gamma\delta}$ is a fourth-order texture tensor which contains all the effective elastic coefficients of the considered system, as a function of the geometry of the packing arrangements.

In general, only bounds for these coefficients can be found, but the computation of the values of $C_{\alpha\beta\gamma\delta}$ can be done analytically in simple arrangements such as regular pilings or by the use of an additional ensemble averaging in the case of disordered packings. One can illustrate this last point with a basic example. An ensemble average indeed allows us to replace the expression for $C_{\alpha\beta\gamma\delta}$ in (14.34) by

$$C_{\alpha\beta\gamma\delta} = K d^2 \rho_c \int d\Omega\, P(\Omega)\, n_\alpha\, n_\beta\, n_\gamma\, n_\delta, \tag{14.35}$$

where ρ_c is the number of contacts per unit volume. The integration runs over all possible orientations of the unit vector $\vec{n}$. For a 2D (x, z) arrangement, the running parameter is simply the contact angle θ already introduced in Subsection 14.1.3 with its probability distribution $Q(\theta)$, which describes the anisotropy of the packing texture. Suppose we handle an isotropic system; $Q(\theta)$ is then constant and we have to compute

$$C_{\alpha\beta\gamma\delta} = K d^2 \rho_c \int_0^{2\pi} \frac{d\theta}{2\pi}\, n_\alpha\, n_\beta\, n_\gamma\, n_\delta, \tag{14.36}$$

where $n_x = \cos\theta$ or $n_z = \sin\theta$, which finally gives

$$C_{xzzz} = C_{zxzz} = C_{zzxz} = C_{zzzx} = 0, \tag{14.37}$$

$$C_{zxxx} = C_{xzxx} = C_{xxzx} = C_{xxxz} = 0, \tag{14.38}$$

$$C_{xxxx} = C_{zzzz} = \frac{3\pi}{4}\, C_0, \tag{14.39}$$

$$C_{xxzz} = C_{zzxx} = \frac{\pi}{4}\, C_0, \tag{14.40}$$

where $C_0 = K d^2 \rho_c/(2\pi)$. Comparing with elastic stress–strain relations expressed in terms of the Young E and shear G moduli, as well as the Poisson ratio ν, we have

$$C_{xxxx} = C_{zzzz} = \frac{E}{1 - \nu^2} , \qquad (14.41)$$

$$C_{xxzz} = C_{zzxx} = \frac{\nu E}{1 - \nu^2} , \qquad (14.42)$$

$$C_{xzxz} + C_{xzzx} = 2G , \qquad (14.43)$$

which gives $\nu = 1/3$, $E = 2\pi \, C_0/3$ and $G = \pi \, C_0/4$. Note that, as expected for isotropic elasticity, the relation $G = E/(2(1 + \nu))$ is recovered. More complex calculations have been performed in many other situations. For example, the above formulae can be generalised to the case where the contact force is described by a tangential as well as a normal spring [385, 393].

How well does this type of approach work for real granular media? It has been tested by the use of numerical simulations, see e.g. [433, 420, 434], and although a fair agreement can be found for the Young modulus, the match with G is in gereral not very good. The main reason is that the affinity assumption is usually wrong, as, even for small external perturbations, some local rearrangments of the grain packing do not follow the global strain field. In fact, this observation of local so-called 'shear transformation zones' [404] is precisely at the source of recent theories to describe the plasticity of granular media [426]. Besides, even with a more sophisticated definition of the coarse-grained strain tensor [411], which would be the equivalent of the expression we have for the stress (14.23), a linear relationship between $\sigma_{\alpha\beta}$ and $u_{\alpha\beta}$ is in general *not* verified when the coarse graining length is smaller than typically several tens of particle diameters. Lastly, note that this is not specific to granular materials as a similar small size effect can be seen in Lenard-Jones glasses [481, 468].

14.2.3 Theoretical descriptions

We shall present hereafter the theoretical approaches which have been used in the field of granular media for a very long time. Precisely for that reason, they are well documented and detailed in many books. It is not the ambition of the author to produce an additional complete review of these approaches. Rather, we would like to present a concise and practical summary in relation to the other parts of this chapter, and the reader interested in more details is invited to consult the cited references. More precisely, we shall first describe very qualitatively the general framework of elasto-plasticity of granular assemblies. We continue with a more technical presentation of the linear elasticity and the Mohr–Coulomb yield criterion. Then, the Janssen model for silos is briefly described. Finally, the last short part is devoted to a more recent phenomenological approach for stress distribution.

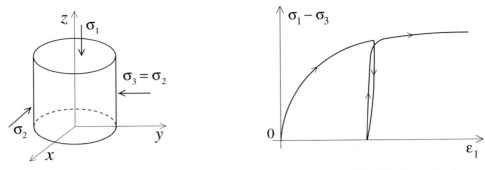

Fig. 14.19 Left: typical triaxial test in soil mechanics: the cylindrical sample is compressed at an increasing controlled stress σ_1 or strain ϵ_1 along the z-axis, while the pressure $\sigma_2 = \sigma_3$ is kept constant in the x, y-plane. Right: deviatoric stress $\sigma_1 - \sigma_3$ as a function of the strain along the compression direction ϵ_1. For a granular material, this curve is nonlinear and saturates at large strain ('plastic' behaviour). As evidenced on the graph, it is also irreversible and hysteretic when the sample is unloaded and reloaded after a finite deformation.

Elasto-plasticity

Elasto-plasticity is the very general term which designates the standard framework for the modelling of the mechanical behaviour of solids, soils and granular materials in particular. Very briefly speaking, research in this field consists of finding the constitutive law of the studied material, i.e. the mathematical expression which tells how much this material deforms when submitted to a given increment of stress. In order to go beyond the simple qualitative description of the elasto-plastic behaviour of granular assemblies that will follow, one can refer, for example, to the books [405, 418, 482].

The famous experiment associated with the search for constitutive laws of materials is the so-called 'triaxial test'. A schematic description of this test in the usual axi-symmetrical geometry is depicted in Fig. 14.19 (left): a cylindrical sample, initialy loaded under an isotropic pressure, is compressed along its symmetry direction 1, while the stresses in the perpendicular directions 2 and 3 are kept constant. This compression can be realised by applying a stress increment $\Delta\sigma_1$, and the corresponding strains $\Delta\epsilon_1$ and $\Delta\epsilon_2 = \Delta\epsilon_3$ are then measured. Alternatively, a strain increment $\Delta\epsilon_1$ can be imposed on the sample, leading to some $\Delta\sigma_1$ and $\Delta\epsilon_2 = \Delta\epsilon_3$. Suppose that the tested sample is an ideal isotropic linear elastic material. In that case, simple proportionality relations are observed: $\Delta\sigma_1 = E\,\Delta\epsilon_1$ and $\Delta\epsilon_2 = \nu\,\Delta\epsilon_1$, where E and ν denote respectively the Young modulus and the Poisson ratio of this material. In elasticity, such a compression is of course reversible: a decrease of σ_1 to its initial value σ_3 removes all strains.

Table 14.1 *Typical values of E and v for loose and dense sand. Note that the Young modulus of the individual quartz grains is much higher than that of the grain packing.*

loose sand	$E \sim 10$ MPa	$v \sim 0.3$
dense sand	$E \sim 100$ MPa	$v \sim 0.3$
silice/quartz	$E \sim 10^4 - 10^5$ MPa	

In comparison, a typical stress–strain curve for a granular assembly obtained by a quasistatic addition of successive small stress increments $\Delta\sigma_1$ is shown in Fig. 14.19 (right). Several observations should be mentioned: first of all, the curve is not linear and even saturates at large ϵ_1. Second, irreversible grain rearrangements occur in the packing. This can be evidenced by the unloading of σ_1 to σ_3 at a finite value of ϵ_1, which does not remove all the deformation of the sample. Besides, loading–unloading cycles show some hysteresis effects. Finally, the shape of this stress–strain curve also depends on the preparation of the sample. In fact, the curve of Fig. 14.19 is what would be measured with a loose packing of grains. A dense one would rather produce a curve with an 'overshoot' of the saturation value.

The slope of this curve close to the origin $\epsilon_1 = 0$ would give the effective Young modulus of the sample at the beginning of the test – see typical values of E and v for sand in Table 14.1. However, as the axial compression goes on, due to grain rearrangements and packing restructuration under loading, this slope decreases, leading to an effective (anisotropic) elasticity with new coefficients. The description of the evolution of these new coefficients – and the corresponding microscopic texture – with the variation of ϵ_1 or σ_1 is beyond the scope of this chapter, and is in fact still a matter of current research, see e.g. [432]. After a large deformation, the system is described as 'plastic', as a flat stress–strain curve means that a very small stress increment would deform the sample a lot.

So far this discussion looks focused on the quasistatic rheology of a granular system, so why is all of this related to the description of its statics? As a matter of fact, the static properties of the studied sample are responsible for its mechanical behaviour at the very starting point of the curve, i.e. close to $\epsilon_1 = 0$. However, any stress state could be, at least in principle, reached by an appropriate stress–strain path starting from some well-defined and controlled point – say, e.g., a homogeneous and isotropic packing of known E and v values. Therefore, although stress–strain tests on granular assemblies show rather complex features, the simplest idea is to describe their static properties as either an effective anisotropic elastic medium or a plastic material. Triaxial experiments then test the incremental and global mechanical response of the considered sample and allow one to get a measure of its constitutive effective coefficients. In the following, we shall crudely make explicit the equations for the computation of stress distribution (i) in the framework of anisotropic elasticity, and (ii) using the Mohr–Coulomb criterion for the plasticity of granular materials.

Before turning to these mathematical descriptions, we would like to make several important remarks. First of all, the true elastic regime of an assembly of grains is probably restricted to extremely small strains (typically 10^{-5} [469, 415]): this means that if none of the grains changes contact with its neighbours or slides, so that the global strain is only due to the deformation of the grains themselves. In fact, even a very small increment of σ_1 almost always produces some irreversible packing rearrangement – i.e. a plastic event – and the corresponding $\Delta\epsilon_1$ reflects the corresponding motion of these grains. In that respect, the study of the biaxial compression of a two-dimensional system of polydisperse frictionless discs by Combe and Roux [397] is particulary enlightening, as they study the statistics of the strain jumps in response to a stress loading. Another point is that, in these triaxial tests, it is very crucial and difficult to have a good control of the homogeneity of the deformation induced by the compression. The classical example is the progressive appearance of shear bands, where the deformation is localised [399]. In that case, the different directions of the sample can behave in very different ways – think of the direction perpendicular to those bands compared to the others. Finally, full three-dimensional tests, i.e. without the axi-symmetry, are even more complicated to analyse, as to reach a state $(\sigma_1, \sigma_2, \sigma_3)$ many 'paths' are possible, which may induce different strain variations.

Elasticity formalism

Suppose that one finds it appropriate to describe a granular packing as a global effective elastic medium. This may be particulary valid for a system submitted to very small external perturbations such as those considered in the response function experiment evoked in one of the previous subsections – note that, as already mentioned in the part devoted to the 'effective medium theory', a proper justification of such an assumption is a subject of on-going research. This part will provide the standard formalism of the isotropic and anisotropic elasticity in a brief and pragmatic form. Much more about this very wide subject can be found in books such as [424, 405, 418].

The following equations will involve two main quantities: the stress σ_{ij} and the strain u_{ij} tensors. We deliberately use the letter u for the strain field in order to match Landau's notations $u_{ij} = (\partial_i u_j + \partial_j u_i)/2$, where u_i is the displacement vector. The very same quantity was, however, denoted ϵ in the previous subsection in accordance with soil mechanics use. At mechanical equilibrium, the stresses must verify the force balance equations

$$\partial_i \sigma_{ij} = f_j^b, \tag{14.44}$$

where f_j^b is an external body force per unit volume applied to the system (e.g. gravity ρg_j). Torque balance is ensured by the symmetry of the stress tensor $\sigma_{ij} = \sigma_{ji}$. A reference state for the displacements for which both σ_{ij} and u_{ij} are set to zero must be defined. For an isotropic and linear elastic material, stress and strain tensors are linked by the relation

$$\sigma_{ij} = \frac{E}{1+\nu} \left(u_{ij} + \frac{\nu}{1-(D-1)\nu} u_{kk} \delta_{ij} \right), \tag{14.45}$$

or conversely

$$u_{ij} = \frac{1}{E}[(1+\nu)\sigma_{ij} - \nu \sigma_{kk} \delta_{ij}], \tag{14.46}$$

where D is the space dimension – note that it appears only in (14.45). Only two phenomenological parameters enter these relations: E and ν, the Young modulus and the Poisson ratio of the material. Because the strain *tensor* u_{ij} has been constructed from the *vector* u_i, its components verify the so-called 'compatibility relation':

$$\frac{\partial^2 u_{ik}}{\partial x_l \partial x_m} + \frac{\partial^2 u_{lm}}{\partial x_i \partial x_k} = \frac{\partial^2 u_{il}}{\partial x_k \partial x_m} + \frac{\partial^2 u_{km}}{\partial x_i \partial x_l}. \tag{14.47}$$

We emphasise the fact that it is a pure mathematical identity only. Using the stress–strain relation and (derivatives of) the force balance equation we can eliminate the u_{ij}s and find

$$(1+\nu)\Delta\sigma_{ij} + [1+(3-D)]\frac{\partial^2 \sigma_{kk}}{\partial x_i \partial x_j} = 0. \tag{14.48}$$

Notice that this relation (14.48) is not valid in the case of a non-uniform external body force, see [424]. Contracting indices i and j, we see that the trace of the stress tensor is a harmonic function, i.e. $\Delta\sigma_{kk} = 0$. Taking the Laplacian of (14.48), we also see that all the stress components are biharmonic: $\Delta\Delta\sigma_{ij} = 0$. These relations then provide a set of closed equations for all the stresses – Eqs. (14.44) alone are indeterminate – that can be solved for given boundary conditions. As for terminology, the differential equations are of elliptic type.

As a simple example, let us focus on the case of a two-dimensional slab of finite depth h along the axis z, but of infinite horizontal extension along x. For 2D elasticity, there exist sophisticated solving techniques involving holomorphic complex functions that won't be described here. We will rather look here for Fourier modes solutions which are well adapted to this case – more convenient biharmonic function bases could be chosen for other geometries. Suppose we are interested in the stress distribution in this slab due to, say, some overloading at its top surface ($z = 0$) but not in the effect of uniform gravity – we take $f_i^b = 0$ in (14.44). A general solution of the problem is given by

$$\sigma_{zz} = \int_{-\infty}^{+\infty} dq\, e^{iqx} \left[(a_1 + qza_2)e^{qz} + (a_3 + qza_4)e^{-qz} \right],$$ (14.49)

$$\sigma_{xz} = i \int_{-\infty}^{+\infty} dq\, e^{iqx} \left[(a_1 + a_2 + qza_2)e^{qz} + (-a_3 + a_4 - qza_4)e^{-qz} \right],$$ (14.50)

$$\sigma_{xx} = - \int_{-\infty}^{+\infty} dq\, e^{iqx} \left[(a_1 + 2a_2 + qza_2)e^{qz} + (a_3 - 2a_4 + qza_4)e^{-qz} \right].$$ (14.51)

The four functions $a_k(q)$ have to be determined by the boundary conditions. A condition on the stresses, such as an overload at the top of the layer, is very simple to express as soon as its Fourier transform is known. A constraint on the displacements such as $u_i = 0$ on a rigid and rough bottom plate can be transformed into a condition on the σ_{ij}s or their derivatives by the use of the stress–strain relation.

A classical application of the previous example is that of the stress distribution in a semi-infinite medium in response to a point force load at the surface. If that force $\vec{F_0}$ makes an angle θ_0 with the vertical direction, the top conditions are $\sigma_{zz} = F_0 \cos \theta_0\, \delta(x)$ and $\sigma_{xz} = F_0 \sin \theta_0\, \delta(x)$, and vanishing stresses are required for $z \to \infty$. In that case the a_ks are very simple and the integrals in (14.49)–(14.51) can be computed explicitly so that the vertical normal stress component reads

$$\sigma_{zz} = \frac{2F_0}{\pi} \frac{z^3}{(x^2 + z^2)^2}.$$ (14.52)

The result for the more general case of a slab of finite thickness can be found for instance in [460].

Full three-dimensional elastic systems are of course more difficult to handle. However, the case of a vertical axi-symmetric situation is tractable, and solutions similar to those given by (14.49)–(14.51) can be found where, broadly speaking, the e^{iqx} must be replaced by Bessel functions, see e.g. [460] for more details. The point force load on a semi-infinite medium leads in this case to a stress distribution given by the Boussinesq and Cerruti's formulae [418] – this solution is what is labelled 'elasticity' in Fig. 14.14. For instance, the σ_{zz} component reads

$$\sigma_{zz} = \frac{3F_0}{2\pi} \frac{z^3}{(r^2 + z^2)^{5/2}}.$$ (14.53)

For the case of granular pilings which can be textured with preferred orientations, it is useful to generalise these calculations to the situation where the elastic material is anisotropic. Let us consider for simplicity the case of a two-dimensional system with a uniaxial symmetry, where the vertical z and the horizontal x directions are along the principal axes of the anisotropy. This kind of anisotropy has already five phenomenological parameters, and the equivalent of the relation (14.45) can be represented by a matrix Λ that relates the 'vectors' $\Sigma = (\sigma_{xx}, \sigma_{zz}, \sigma_{xz})$ and $U = (u_{xx}, u_{zz}, u_{xz})$ by $\Sigma = \Lambda U$, with

$$\Lambda = \frac{1}{1 - \nu_{xz}\nu_{zx}} \begin{pmatrix} E_x & \nu_{zx}E_x & 0 \\ \nu_{xz}E_z & E_z & 0 \\ 0 & 0 & (1 - \nu_{xz}\nu_{zx})2G_{xz} \end{pmatrix}. \tag{14.54}$$

E_x, E_z and G_{xz} are the Young and shear moduli, and ν_{xz}, ν_{zx} the Poisson ratios. The first three coefficients encode the stiffness of the material under x or z uniaxial loading, or shearing. ν_{ij} quantifies the transverse extension $-u_{jj}$ with respect to the compression u_{ii} in the direction of the loading. Isotropy gives $E_x = E_z = E$, $\nu_{xz} = \nu_{zx} = \nu$ and $G = E/(2(1 + \nu))$. Note that the matrix Λ is symmetrical and these five parameters are not independant. They satisfy the extra relation

$$\frac{\nu_{zx}}{E_z} = \frac{\nu_{xz}}{E_x}. \tag{14.55}$$

Besides, the elastic energy is well defined – i.e. is a quadratic and positive function of the strain variables so that the material is stable – if all moduli are positive and if the Poisson ratios verify $\nu_{zx}\nu_{xz} < 1$. With this anisotropic stress–strain relation, the equivalent of the bi-Laplacian equation for the σ_{ij} is now

$$\left(\partial_z^4 + 2r\partial_x^2\partial_z^2 + s\partial_x^4\right)\sigma_{ij} = 0, \tag{14.56}$$

where r and s are given by $r = E_x \left(1/G_{xz} - \nu_{xz}/E_x - \nu_{zx}/E_z\right)/2$ and $s = E_x/E_z$, and whose solutions can be expressed in Fourier modes in a similar way to the isotropic case. The corresponding analytic solutions in the case of the point force load on a semi-infinite slab can be found in [450], even including the situation where the anisotropy axis makes an angle with respect to the vertical and horizontal directions. Depending on the values of r and s, the shape of the stress profiles can be quite different, and show for instance either one broad peak or two distinct maxima – hyperbolic behaviour can be obtained as a limit of anisotropic elasticity, see also [413].

Much more could be said about elasticity in general, and anisotropic elasticity in particular. In 3D, for example, the equivalent of the matrix (14.54) has nine independent parameters: three Young moduli, three shear moduli, and six Poisson ratios but with three symmetry extra relations like (14.55). Finally, situations where the rotations of individual grains are important lead to so-called Cosserat elasticity. The large-scale behaviour of Cosserat-type or micro-polar granular assemblies have been recently studied in, e.g. [421, 406]. The stress response function of such a material is analysed in [476].

Mohr–Coulomb yield criterion

An important issue about static granular pilings is that of their mechanical stability: when a given assembly is submitted to an increasing shear stress, until when will it support the load without (major) rearrangements? A situation where small stress increments produce large strain changes is the so-called 'plastic zone' of the stress–strain curve in Fig. 14.19. The plasticity of solids and soils is a vast field which aims to describe when and how these systems yield and flow [405, 482]. In this subsection, we present the Mohr–Coulomb yield criterion (here for two-dimensional situations, i.e. plane stress), which is based on a solid friction-like criterion. Note that it does not say anything about the flow and in particular about its orientation beyond the yielding point.

The Mohr–Coulmb assumption is that an assembly of grains is stable if, for any coordinate axis orientation (n, t), the stress components satisfy

$$|\sigma_{nt}| \leq \tan \phi \, \sigma_{nn} \tag{14.57}$$

at any point. Considering a unit length of the line along the t-axis, it means that the ratio of the tangential force to the normal one should not exceed a given maximum $\mu = \tan \phi$. μ is called the internal friction coefficient, and ϕ the internal friction angle. This criterion can be expressed by a relation on the stress components valid in any coordinate system. An elegant way to do so is to represent the σ_{ij} in a geometrical way. This is called the Mohr circle. This circle is a tool used to determine transformations of a symmetrical tensor of rank 2 (such as the stress in 2D) under rotation. Here briefly follows the derivation leading to this circle.

Suppose the eigendirections of the stress tensor are the 1 and 2 axis. We call σ_1 and σ_2 the major and minor eigenvalues. The force acting on a length element $\delta\ell$ whose normal vector is n_i can be expressed in terms of the stress tensor as $f_i = \delta\ell \, \sigma_{ij} n_j$. Take the case of this element making an angle θ with direction 2, we can choose $n_1 = -\cos\theta$, $n_2 = \sin\theta$, $t_1 = \sin\theta$ and $t_2 = \cos\theta$ for the components of the normal and tangential vectors – see Fig. 14.20. On this θ oriented line, the normal and shear stresses are then respectively given by $\sigma = \sigma_{nn} = f_k n_k / \delta\ell$ and $\tau = \sigma_{nt} = f_k t_k / \delta\ell$, i.e. after some elementary trigonometric calculations,

$$\sigma = \frac{\sigma_1 + \sigma_2}{2} + \frac{\sigma_1 - \sigma_2}{2} \cos 2\theta, \tag{14.58}$$

$$\tau = -\frac{\sigma_1 - \sigma_2}{2} \sin 2\theta. \tag{14.59}$$

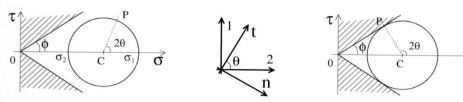

Fig. 14.20 Graphical representation of the stress tensor on the Mohr circle. σ_1 and σ_2 are the major and minor principal stress values. The coordinates of the point P give the stress components σ_{nn} and σ_{nt}, where n and t are the normal and tangential directions of the line making an angle θ with the minor principal direction. Left: for any orientation θ the shear is smaller in absolute value than $\tan\phi$. Right: the circle and the lines $\tau = \pm\tan\phi$ are tangent to the two points which are called the active and passive states of yield.

This is precisely the parametric representation of a circle in the plane (σ, τ) of centre C $(\sigma_C = (\sigma_1 + \sigma_2)/2, \tau_C = 0)$ and radius $R = (\sigma_1 - \sigma_2)/2$, see Fig. 14.20. In this graph, the criterion (14.57) requires that this circle is everywhere in between the two symmetric lines $\tau \pm \tan\phi\,\sigma$. When the circle and the lines are tangent, we have $R = \sin\phi\,\sigma_C$, so that the stability condition is $R \leq \sin\phi\,\sigma_C$. Now, for a representation of the stress tensor in an arbitrary coordinate system, the principal values can be computed from the three components $\sigma_{aa}, \sigma_{bb}, \sigma_{ab}$, and we get

$$\sigma_{1,2} = \frac{\sigma_{aa} + \sigma_{bb}}{2} \pm \sqrt{\left(\frac{\sigma_{aa} - \sigma_{bb}}{2}\right)^2 + \sigma_{ab}^2}, \qquad (14.60)$$

so that the Mohr–Coulomb criterion for stability can be finally written as

$$(\sigma_{aa} - \sigma_{bb})^2 + 4\sigma_{ab}^2 \leq \sin^2\phi\,(\sigma_{aa} + \sigma_{bb})^2. \qquad (14.61)$$

The two points where the circle and the lines are tangent are called the active and passive states. The first one corresponds to the positive value for the shear, i.e. $\theta = \pi/4 + \phi/2$, so that the slip direction (that of vector t) makes, with the major principal axis, the angle $-\pi/4 + \phi/2$. For the passive case, the shear stress is negative and the orientation of the slip plane with respect to direction 1 is opposite, i.e. $\pi/4 - \phi/2$.

The Mohr–Coulomb criterion can be used with equality – i.e. assume a constant ratio $(1 + \sin\phi)/(1 - \sin\phi)$ between the principal stresses – as a closure relation to the mechanical equilibrium equations. This means that this 'Mohr–Coulomb material' is just about to yield everywhere. Such an assumption is useful to give an estimate of the stresses and/or give a bound below which failure does not occur. With this relation, one does not need to consider strain variables, and boundary conditions must be specified in terms of the σ_{ij}. The differential equations, which are hyperbolic, can be solved by the method of characteristics, see e.g. [443]. As a final remark, note that it is possible to mix both elliptic equations of elasticity and hyperbolic ones of plasticity, as in the example treated in [400, 442] where a sandpile is composed of an outer plastic region which matches an inner elastic one.

Janssen's approach

Janssen's model is a classical approach to describe the screening of stresses in silos. The original papers date from 1895 and became very popular, certainly because the model could reproduce the correct phenomenology with an extremely simple mathematical framework. As a reference book for this model, one can, e.g., look at that of Nedderman [443].

As depicted in Fig. 14.10, we consider a column of height h and radius R. For the sake of simplicity, the model neglects horizontal stress dependency and considers slices at a given height. This means that σ_{zz} depends on the vertical variable z only and represents the average vertical pressure of the slice. We can write the force balance on a slice of thickness dz, it reads:

$$\sigma_{zz}(z + dz)S = \sigma_{zz}(z)S + \rho g dz S - \tau L dz, \qquad (14.62)$$

where $S = \pi R^2$ is the section area of the column and $L = 2\pi R$ its perimeter. ρ is the density of the granular material and g is the gravity acceleration. Finally, τ is stress due to the friction between the grains and the wall of the silo.

The central trick of Janssen's model is to be able to close equation (14.62), i.e. to express τ as a function of σ_{zz}, thanks to several crucial assumptions. In a full tensorial description of the stresses into such a silo, τ would simply be the shear stress σ_{rz} at the wall, i.e. taken at $r = R$, where r is the horizontal position variable. If this friction is 'fully mobilised', i.e. if the grains at the wall are just about to slip, a Coulomb-like description of the solid friction gives $\sigma_{rz} = \mu_w \sigma_{rr}$ (again at $r = R$), where μ_w is the corresponding friction coefficient. Now we assume on top of this that the overall horizontal pressure of the slice is simply proportional to the vertical one:

$$\sigma_{rr} = K\sigma_{zz}. \qquad (14.63)$$

K is the so-called Janssen's constant. Because r-dependency of σ_{zz} is neglected, we have then closed Eq. (14.62), which can be rewritten as

$$\partial_z \sigma_{zz} + \frac{2\mu_w K}{R} \sigma_{zz} = \rho g. \qquad (14.64)$$

It is easy to integrate this first-order differential equation with the condition that $\sigma_{zz} = Q$ at the top surface $z = 0$, we get

$$M_{\text{app}} = M_{\text{sat}} \left(1 - e^{-M_{\text{fill}}/M_{\text{sat}}}\right) + Q e^{-M_{\text{fill}}/M_{\text{sat}}}, \qquad (14.65)$$

where, according to the previous experimental section, we defined the 'apparent mass' at the depth h as $M_{\text{app}} = \sigma_{zz}(h)S/g$, and the corresponding 'filling mass' as $M_{\text{fill}} = \rho Sh$. For a tall column (large h), M_{app} saturates to the value $M_{\text{sat}} = \rho SR/(2\mu_w K)$.

This saturation is, in comparison to experimental measurements (see Fig. 14.10), the main success of Janssen's model. In particular, it gives the correct scaling for $M_{\text{sat}} \propto R^3$ – note that for two dimensions columns $S = 2R$, $L = 2$ and $M_{\text{sat}} = \rho 2R^2/(\mu_w K)$. More quantitatively, the unoverloaded ($Q = 0$) data of Ovarlez *et al.* [452] are very well fitted by a relation like Eq. (14.65). Of course, the quality of such a comparison is crucially dependent on the experimental control of the packing density and the preparation procedure (which both govern the redistribution effect, i.e. the value of K), as well as the mobilisation of the friction at the wall (i.e. the value of μ_w).

In contrast, the presence of a finite overload Q is badly reproduced by the model. In particular, it predicts that M_{app} becomes independent of depth if this overload is precisely chosen such that $Q = M_{\text{sat}}$. This is not what is measured experimentaly where an 'overshoot' is observed, see Fig. 14.10.

Finally, it must be noted that no real 'granular features' are included in this approach. It is rather a model of screening effect. As a matter of fact, an elastic material confined into a rough rigid column would also show a saturation curve due to the Poisson effect which couples vertical and horizontal normal stresses. One can in particular compute the large-scale effective Janssen coefficient K in the framework of the linear isotropic elasticity. One gets $K = \nu$ and $K = \nu/(1 - \nu)$ in two and three dimensions respectively (ν is as usual the Poisson ratio) – see [451] for an elastic analysis of the Janssen experiment.

OSL model

In the last subsection of this chapter, we would like to present the phenomenological so-called 'OSL' model which was introduced a few years ago in the context of the sandpile dip problem – see the corresponding part in Subsection 14.2.1 and references [370, 480, 479]. We shall briefly start with the description of the basic assumptions of this approach, we then show its main results and finally discuss its relevence to experiments and simulations.

The simplest version of this model assumes a local Janssen-like relation between the stress components, e.g. a proportionality between horizontal and vertical normal stresses: $\sigma_{xx} = K\sigma_{zz}$. This must be not confused with a Mohr–Coulomb type of assumption for which the ratio of the two principle stresses is taken constant (see above), as the x and z axis may not be the eigendirections of the stress tensor. In fact, this model will give very similar mathematical features to those of a Mohr–Coulomb material – hyperbolic equations with characteristic lines – but here the yield criterion is used as an additional and independent constraint on the stresses. In the context of the sandpile construction, the argument for such a linear relationship between the σ_{ij} was that some 'stress state' of the static granular material, just after the surface avalanche has jammed, remains 'frozen' when the grains are buried by the next successive avalanche. As the avalanche surface must be a Coulomb yield plane, the principal directions of the stress tensor can be computed as well as the Mohr–Coulomb stress ratio. Assuming that (i) the internal angle of friction ϕ is equal to the angle of repose of the pile and that (ii) the packing keeps the memory of these eigendirections all through the pile, we end up with a relation of the form $\sigma_{xx} = \sigma_{zz} - 2 \tan \phi \, \sigma_{xz}$. Note that here we implicitly present the model in two dimensions (x, z) and that $x = 0$ is the horizontal position of the apex of the pile – also, changing the sign of x changes that of the shear σ_{xz}. More generally, the stress state of the jammed grain packing is assumed to be of the form

$$\sigma_{xx} = \eta \sigma_{zz} + \mu \sigma_{xz}, \qquad (14.66)$$

where η and μ are two phenomenological parameters whose values depend on the way the considered system (pile, silo) has been prepared. Because this expression can be seen as a Janssen-like relation in a given and fixed coordinate axis (n, t), this approach has been called the 'oriented stress linearity' (OSL) model. Together with the force balance condition under gravity, $\partial_j \sigma_{ij} = \rho g_i$, the stress components satisfy a wave-like equation

$$(\partial_z + c_+ \partial_x)(\partial_z + c_- \partial_x) \, \sigma_{ij} = 0, \qquad (14.67)$$

where $c_\pm = (\mu \pm \sqrt{\mu^2 + 4\eta})/2$. As for a Mohr–Coulomb material, no specification of the strain variables is needed and these hyperbolic equations can be solved by the method of characteristics – which are simply straight lines in this linear model. Note that, although no explicit link has been established, these characteristics were intuitively thought to be related to the mesoscopic 'force chain' network whose structure and orientation are shaped by the previous history of the granular assembly.

Regarding to stress measurements below a sandpile [462, 388, 375], the 3D version of the OSL model for which $\eta \sim 1$ and $\mu \sim -2 \tan \phi$ (fixed principal axis) gives a remarkable fit to the experimental data. For the silo geometry, the Janssen curve is also well reproduced [475], but the predicted quasi-oscillations of the apparent mass on the bottom of the column as a function of the filling mass when a top overload is applied is not observed in careful experiments such as [452]. Lastly, the most striking feature of hyperbolic models is certainly the prediction of two peaks (a ring in 3D) for the shape of the vertical normal stress profile at the bottom of a grain layer in response to a localised overload at the top. But, as discussed in the above corresponding subsection, the typical stress response profile of a disordered layer of grains shows a single broad peak whose shape and scalings are rather in accordance with an elastic description, i.e. elliptic equations, see e.g. [457, 460]. However, the stress response of ordered packings [407, 409, 441, 387] as well as disordered anisotropic isostatic systems and models [472, 473, 414, 439, 368, 402, 363] show some hyperbolic-like behaviours.

Another interesting issue that has been tackled by this approach is the question of mechanical compatibility between the external load applied to the system and the internal structure of the packing. Although the term may be a bit misleading, a 'fragile' character of granular materials was put forward in [392]. As a matter of fact, because, by contrast to elliptic equations, hyperbolic ones require the specification of the stress values on half of the boundary conditions only, the prediction is that incompatible loads on both halves must lead to major rearrangements. Interestingly, this is in a sense what is observed in the biaxial test realised by Combe and Roux on a polydisperse system of frictionless and rigid particles [397].

14.3 Conclusion

In this chapter we have presented the static properties of granular materials at different scales. At the level of the grains, we discussed the contact forces and their statistical distribution. At the large scale, we have reviewed macroscopic experiments where the stresses were measures in different geometries (pile, silo, layer). We have shown how to go from these discrete forces to a continuum stress field, and introduced the theoretical framework of elasto-plasticity in which these systems are usually described.

As has been seen, this subject consists of a peculiar mixture of some rather old but still up-to-date parts (e.g. Coulomb criterion, Janssen's approach or elasticity), with new features (e.g. force probability distribution). It is very likely that it will further evolve on a short timescale. In particular, progress will probably be made thanks to the understanding of the jamming/unjamming processes which control how grains come to rest or on the contrary get out of mechanical equilibrium. In that respect, the study of grain rearrangements in response to perturbations [423, 440] as well as slow dense granular flows will be of great interest.

References

[1] Reynolds, O. 1885. On the dilatancy of media composed of rigid particles in contact, with experimental illustrations. *Phil. Mag.* **20** 469–481.

[2] De Gennes, P. G. 1966. *Superconductivity of Metals and Alloys* (New York: Benjamin) 83.

[3] Briscoe, B. J. and Adams, M. J., eds. 1987. *Tribology in Particulate Technology* (Bristol: Adam Hilger).

[4] Savage, S. B. 1984. The mechanics of rapid granular flows. *Adv. Appl. Mech.* **24** 289–366; 1988. Streaming motions in a bed of vibrationally fluidized dry granular material *J. Fluid Mech.* **194** 457–478.

[5] Bagnold, R. A. 1954. Experiments on a gravity free dispersion of large solid particles in a Newtonian fluid under shear. *Proc. Roy. Soc. London* **A225** 49–63.

[6] Bagnold, R. A. 1966. The shearing and dilatation of dry sand and the 'singing' mechanism. *Proc. Roy. Soc. London* **A295** 219–232.

[7] Jackson, R. 1983. Some mathematical and physical aspects of continuum models for the motion of granular materials. In *Theories of Dispersed Multiphase Flow*, ed. R. E. Meyer (New York: Academic) 291–337.

[8] Haff, P. K. 1983. Grain flow as a fluid-mechanical phenomenon. *J. Fluid Mech.* **134** 401–430.

[9] Edwards, S. F. 1990. The theory of powders. In *Proceedings of the International School of Physics: Enrico Fermi*, eds. G. E. Chiarotti, E. Fumi and M. P. Tosi **106** (New York: North Holland) 849.

[10] Bernal, J. D. 1964. The structure of liquids. *Proc. Roy. Soc. London* **A280** 299–320.

[11] Onoda, G. Y. and Liniger, E. G. 1990. Random loose packings of uniform spheres and the dilatancy onset. *Phys. Rev. Lett.* **64** 2727–2730.

[12] Shapiro, A. P. and Probstein, R. F. 1992. Random packings of spheres and fluidity limits of monodisperse and bidisperse suspensions. *Phys. Rev. Lett.* **68** 1422–1425.

[13] Edwards, S. F. 1990. The rheology of powders. *Rheol. Acta.* **29** 493–499.

[14] Edwards, S. F. 1990. The flow of powders and of liquids of high viscosity. *J. Phys.: Condens. Matter* **2** SA63–SA68.

[15] Edwards, S. F. 1993. The role of entropy in the specification of a powder. In *Granular Matter: An Interdisciplinary Approach*, ed. Anita Mehta (New York: Springer-Verlag), 121–140.

[16] Mehta, Anita and Edwards, S. F. 1990. A new statistical approach to granular mixtures. In *Disorder in Condensed Matter Physics*, ed. J. Blackman and J. Tagüeña (Oxford: Oxford University Press) 155–170.

[17] Mehta, Anita and Edwards, S. F. 1989. Statistical mechanics of powder mixtures. *Physica* **A 157** 1091–1100.

[18] Baxter, R. J. 1971. Eight-vertex model in lattice statistics. *Phys. Rev. Lett.* **26** 832–833.

[19] Bartlett, P. and van Megen, W. 1993. Physics of hard-sphere colloidal suspensions. In *Granular Matter: an Interdisciplinary Approach*, ed. Anita Mehta (New York: Springer-Verlag), 195–258.

[20] Coulomb, C. A. 1773. Essai sur une application des règles de maximis et minimis a quelques problèmes de statique relatifs a l'architecture. In *Memoires de Mathématiques et de Physique Presentés l'Academie Royale des Sciences par divers Savants et Lus dans les Assemblées* (Paris: L'Imprimerie Royale), 343–382.

[21] Brown, R. L. and Richards, J. C. 1966. *Principles of Powder Mechanics* (Oxford: Pergamon Press).

[22] Barker, G. C. and Mehta, Anita 1996. Rotated sandpiles: the role of grain reorganization and inertia. *Phys. Rev. E* **53**, 5704–5713.

[23] Luck, J. M. and Mehta, Anita 2004. Dynamics at the angle of repose: jamming, bistability, and collapse. *JSTAT*, P10015.

[24] Hagen, G. H. L. 1852. *Berichtüber die zur Bekanntmachung geeigneten Verhandlungen der K oniglich Preussischen*. Berlin: Akademie der Wissenschaften zu Berlin, 35–42.

[25] Baxter, G. W., Behringer, R. P., Fagert, T., and Johnson, G. A. 1989. Pattern formation in flowing sand. *Phys. Rev. Lett.* **62** 2825–2828.

[26] Evesque, P. 1992. Shaking dry powders and grains. *Contemporary Phys.* **33** 245–262.

[27] Mehta, Anita and Halsey, T. C. 2003. eds. *Challenges in Granular Physics* (Singapore: World Scientific).

[28] Jaeger, H. M., Nagel, S. R., and Behringer, R. P. 1996. Granular solids, liquids, and gases. *Rev. Mod. Phys.* **68** 1259–1273.

[29] Hui, K. and Haff, P. K. 1986. Kinetic grain flow in a vertical channel. *Int. J. Multiphase Flow* **12** (2) 289–298.

[30] Caram, H. and Hong, D. C. 1991. Random-walk approach to granular flows. *Phys. Rev. Lett.* **67** 828–831.

[31] Baxter, G. W. and Behringer, R. P. 1990. Cellular automata models of granular flow. *Phys. Rev.* **A 42** 1017–1020.

[32] Behringer, R. P. and Baxter, G. W. 1993. Pattern formation and complexity in granular flows. In *Granular Matter: an Interdisciplinary Approach*, ed. Anita Mehta (New York: Springer-Verlag) 85–120.

[33] Mehta, A., Barker, G. C., and Luck, J. M. 2004. Cooperativity in sandpiles: statistics of bridge geometries. *JSTAT*, P10014.

[34] Mehta, A. and Luck, J. M. 2003. Why shape matters in granular compaction. *J. Phys.* **A 36** L365–L372.

[35] Luck, J. M. and Mehta, A. 2003. A column of grains in the jamming limit: glassy dynamics in the compaction process. *Eur. Phys. J.* **B 35** 399–411.

[36] Walker, J. 1982. What happens when water boils is a lot more complicated than vou might think. *Sci. Am.* **247** 6 162–171.

[37] Evesque, P. and Rajchenbach, J. 1989. Instability in a sand heap. *Phys. Rev. Lett.* **62** 44–46.

[38] Laroche, C., Douady, S., and Fauve, S. 1989. Convective flow of granular masses under vertical vibrations. *J. Physique* **50** 699–706.

[39] Faraday, M. 1831. On the forms and states of fluids on vibrating elastic surfaces. *Philos. Trans. R. Soc. London* **52** 319–340.

[40] Batchelor, G. K. 1988. A new theory of the instability of a uniformly fluidized bed. *J. Fluid Mech.* **193** 75–110.

[41] Evesque, P., Szmatula, E., and Denis, J. P. 1990. Surface fluidization of a sand pile. *Europhys. Lett.* **12** 623–627.

[42] Mehta, Anita, Needs, R. J., and Dattagupta, S. 1992. The Langevin dynamics of vibrated powders. *J. Stat. Phys.* **68** (5/6) 1131–1141.

[43] Zik, O. and Stavans, J. 1991. Self-diffusion in granular flows. *Europhys. Lett.* **16** 255–258.

[44] Taguchi, Y. H. 1992. New origin of a convective motion: elastically induced convection in granular materials. *Phys. Rev. Lett.* **69** 1367–1370.

[45] Gallas, J. A. C., Herrmann, H. J., and Sokolowski, S. 1992. Convection cells in vibrating granular media. *Phys. Rev. Lett.* **69** 1371–1374.

[46] Pak, H. K. and Behringer, R. P. 1993. Surface waves in vertically vibrated granular materials. *Phys. Rev. Lett.* **71** 1832–1835

[47] Melo, F., Umbanhowar, P. B., and Swinney, H. L. 1995. Hexagons, kinks, and disorder in oscillated granular layers. *Phys. Rev. Lett.* **75** 3838–3841.

[48] Umbanhowar, P. B. and Swinney, H. L. 2000. Wavelength scaling and square/stripe and grain mobility transitions in vertically oscillated granular layers. *Physica* **A 288** 345.

[49] Umbanhowar, P. B., Melo, F., and Swinney, H. L. 1996. Localized excitations in a vertically vibrated layer. *Nature*, **382** 793–796.

[50] Barker, G. C. and Mehta, Anita 1993. Size segregation in powders. *Nature* **361** 308.

[51] Mehta, Anita and Luck, J. M. 1990. Novel temporal behavior of a nonlinear dynamical system: the completely inelastic bouncing ball. *Phys. Rev. Lett.* **65** 393–396.

[52] Luck, J. M. and Mehta, Anita 1993. Bouncing ball with a finite restitution: chattering, locking, and chaos. *Phys. Rev. E* **48** 3988–3997.

[53] Matyas, L. and Klages, M. 2004. Irregular diffusion in the bouncing ball billiard. *Physica D* **187** 165–183.

[54] Knight, J. B., Jaeger, H. M., and Nagel, S. R. 1993. Vibration-induced size separation in granular media: the convection connection. *Phys. Rev. Lett.* **70** 3728–3731.

[55] Kudrolli, A. 2004. Size separation in vibrated granular matter. *Rep. Prog. Phys.* **67** 209–247.

[56] Williams, J. C. 1976. Segregation of particulate materials – a review. *Powder Tech.* **15** 245–251.

[57] Bridgwater, J. 1976. Fundamental powder mixing mechanisms. *Powder Tech.* **15** 215–236.

[58] Rosato, A., Strandburg, K. J., Prinz, F., and Swendsen, R. H. 1987. Why the brazil nuts are on top: size segregation of particulate matter by shaking. *Phys. Rev. Lett.* **58** 1038–1042.

[59] Barker, G. C. and Grimson, M. J. 1990. Theory of sedimentation in colloidal suspensions. *Colloids and Surfaces* **43** 55–66.

[60] Cichocki, B. and Hinsen, K. 1990. Dynamic computer simulation of concentrated hard sphere suspensions I. Simulation technique and mean square displacement data. *Physica* **A166** 473–491.

[61] Mehta, Anita and Barker, G. C. 1991. Vibrated powders: a microscopic approach. *Phys. Rev. Lett.* **67** 394–397.

[62] Barker, G. C. and Mehta, Anita 1992. Vibrated powders: structure, correlations, and dynamics. *Phys. Rev.* **A 45** 3435– 3446.

[63] Barker, G. C., Mehta, Anita, and Grimson, M. J. 1993. Comment on Three-dimensional model for particle size segregation by shaking. *Phys. Rev. Lett.* **70** 2194.

[64] Barker, G. C., Grimson, M. J., and Mehta, Anita. 1993. Segregation phenomena in vibrated powders. *Powders and Grains '93*, ed. C. Thornton (Rotterdam: Balkema), 253–257.

[65] Bak, P., Tang, C., and Wiesenfeld, K. 1987. Self-organized criticality: an explanation of the 1/f noise. *Phys. Rev. Lett.* **59** 381–384; 1988. Self-organized criticality. *Phys. Rev.* **A38** 364–374.

[66] Mehta, Anita 1992. Real sandpiles: dilatancy, hysteresis and cooperative dynamics. *Physica* **A 186** 121–153.

[67] Mehta, Anita and Barker, G. C. 1991. The self-organising sand pile. *New Scientist* **1773** 40–43.

[68] Barker, G. C. and Mehta, Anita 1993. Avalanches in real sandpiles – the role of disorder. In *Powders and Grains*, ed. Thornton, C. (Rotterdam: Balkema), 315–320.

[69] Mehta, Anita, ed. 1994. *Granular Matter: an Interdisciplinary Approach* (New York: Springer-Verlag).

[70] Nagel, S. R. 1992. Instabilities in a sandpile. *Rev. Mod. Phys.* **64** 321–325.

[71] Jaeger, H. M. and Nagel, S. R. 1992. Physics of the granular state. *Science* **255** 1523–1531.

[72] Jaeger. H. M., Liu, C., and Nagel, S. R. 1989. Relaxation at the angle of repose. *Phys. Rev. Lett.* **62** 40–43.

[73] Bak, P. and Chen, K. 1989. The physics of fractals. *Physica* **D 38** 5–12.

[74] Held, G. A., Solina, D. H., Keane, D. T., Haag, W. J., Horn, P. M., and Grinstein, G. 1990. Experimental study of critical-mass fluctuations in an evolving sandpile. *Phys. Rev. Lett* **65** 1120–1123.

[75] Mehta, Anita and Barker, G. C. 1994. Disorder, memory and avalanches in sandpiles. *Europhys. Lett.* **27** 501–506.

[76] Liu, C., Jaeger, H. M., and Nagel, S. R. 1991. Finite-size effects in a sandpile. *Phys. Rev.* **A 43** 7091–7092.

[77] Kadanoff, L. P., Nagel, S. R., Wu, L., and Zhou, S. 1989. Scaling and universality in avalanches. *Phys. Rev.* **A39** 6524–6537.

[78] Frette, V., Christensen, K., Malthe-Sprenssen, A., Feder, J., Jossang, T., Meakin, P., 1996. Avalanche dynamics in a pile of rice. *Nature* **379** 49–52.

[79] Prado, C. P. C. and Olami, Z. 1992. Inertia and break of self-organized criticality in sandpile cellular-automata models. *Phys. Rev.* **A45** 665–669.

[80] Ding, E. J., Lu, Y. N., and Ouyang, H. F. 1992. Theoretical sandpile with stochastic slide. *Phys. Rev.* **A46** R6136–R6139.

[81] Christensen, K., Olami, Z., and Bak, P. 1992. Deterministic 1/f noise in nonconservative models of self-organized criticality. *Phys. Rev. Lett.* **68** 2417–2420.

[82] Socolar, J. E. S., Grinstein, G., and Jayaprakash, C. 1993. On self-organized criticality in nonconserving systems. *Phys. Rev.* **E47** 2366–2376.

[83] Barker, G. C. and Mehta, Anita 2000. Avalanches at rough surfaces. *Phys. Rev.* **E 61** 6765–6772.

[84] Langer, J. S. 1980. Instabilities and pattern formation in crystal growth. *Rev. Mod. Phys.* **52** 1–28.

[85] Edwards, S. F. and Wilkinson, D. R. 1982. The surface statistics of a granular aggregate. *Proc. Roy. Soc.* **A 381** 17–31.

[86] Kardar, M., Parisi, G., and Zhang, Y. 1986. Dynamic scaling of growing interfaces. *Phys. Rev. Lett.* **56** 889–892.

[87] Burgers, J. M. 1974. *The Nonlinear Diffusion Equation* (Boston: Reidel).

[88] Forster, D., Nelson, D. R., and Stephen, M. J. 1977. Large-distance and long-time properties of a randomly stirred fluid. *Phys. Rev.* **A 16** 732–749.

[89] Maritan, A., Toigo, F., Koplik, J., and Banavar, J. R. 1992. Dynamics of growing interfaces. *Phys. Rev. Lett.* **69** 3193–95.

[90] Hwa, T. and Kardar, M. 1989. Dissipative transport in open systems: an investigation of self-organized criticality *Phys. Rev. Lett.* **62** 1813–1816.

[91] Grinstein, G. and Lee, D. H. 1991. Generic scale invariance and roughening in noisy model sandpiles and other driven interfaces. *Phys. Rev. Lett.* **66** 177–180.

[92] Grinstein, G., Lee, D. H., and Sachdev, S. 1990. Conservation laws, anisotropy, and 'self-organized criticality¬í in noisy nonequilibrium systems. *Phys. Rev. Lett.* **64** 1927–1930.

[93] Kirkpatrick, T. R., Cohen, E. G. D., and Dorfman, J. 1982. Fluctuations in a nonequilibrium steady state: basic equations. *Phys. Rev.* **A 26** 950–971.

[94] Toner, John 1991. Dirt roughens real sandpiles. *Phys. Rev. Lett.* **66** 679–682.

[95] Mehta, Anita, Luck, J. M., and Needs, R. J. 1996. Dynamics of sandpiles: physical mechanisms, coupled stochastic equations, and alternative universality classes. *Phys. Rev.* **E 53** 92–102.

[96] Biswas, P., Majumdar, A., Mehta, Anita, and Bhattacharjee, J. K. 1998. Smoothing of sandpile surfaces after intermittent and continuous avalanches: three models in search of an experiment. *Phys. Rev.* **E58** 1266–1285.

[97] Malthe-Sorensen, A., Feder, J., Christensen, K., Frette, V., and Jossang, T. 1999. Surface fluctuations and correlations in a pile of rice. *Phys. Rev. Lett.* **83** 764–767.

[98] Finney, J. L. 1977. Modelling the structures of amorphous metals and alloys. *Nature* **266** 309–314.

[99] Berryman, J. G. 1983. Random close packing of hard spheres and disks. *Phys. Rev.* **A27** 1053–1061.

[100] Torquato, S., Truskett, T. M., and Debenedetti, P. G. 2000. Is random close packing of spheres well defined? *Phys. Rev. Lett.* **84** 2064–2067.

[101] Barker, G. C. and Grimson, M. J. 1989. Sequential random close packing of binary disc mixtures. *J. Phys. Condens. Matter* **1** 2779–2789.

[102] Yen, K. Z. Y. and Chaki, T. K. 1992. A dynamic simulation of particle rearrangement in powder packings with realistic interactions. *J. Appl. Phys.* **71** 3164–3173.

[103] Meakin, P. and Jullien, R. 1987. Restructuring effects in the rain model for random deposition. *Journal de Physique* **48** 1651–1662 .

[104] Bennett, C. H. 1972. Serially deposited amorphous aggregates of hard spheres. *J. Appl. Phys.* **43** 2727–2734.

[105] Visscher, W. M. and Bolsterli, M. 1972. Random packing of equal and unequal spheres in two and three dimensions. *Nature* **239** 504–507.

[106] Mehta, Anita and Barker, G. C. 1994. The dynamics of sand. *Rep. Prog. Phys.* **57** 383–416.

[107] Jodrey, W. S. and Tory, E. M. 1985. Computer simulation of close random packing of equal spheres. *Phys. Rev. A* **32** 2347–2351.

[108] Nolan, G. T. and Kavanagh, P. E. 1992. Computer simulation of random packing of hard spheres. *Powder Technology* **72** 149–155.

[109] Soppe, W. 1990. Computer simulations of random packings of hard spheres. *Powder Technology* **62** 189–197.

[110] Allen, M. P. and Tildesley, D. J. 1987. *Computer Simulation of Liquids* (Oxford: Oxford University Press).

[111] Bizon, C., Shattuck, M. D., Swift, J. B., and Swinney, H. L. 1999. Transport coefficients for granular media from molecular dynamics simulations. *Phys. Rev.* **E60** 4340–4351

[112] Cafiero, R., Luding, S., and Herrmann, H. J. 2002. Rotationally driven gas of inelastic rough spheres. *Europhys. Lett.* **60** 854–860.

[113] Walton, O. R. 1984. Application of molecular dynamics to macroscopic particles. *Int. J. Eng. Sci.* **22** 1097–1107.

[114] Walton, O. R. and Braun, R. L. 1986. Viscosity, granular-temperature, and stress calculations for shearing assemblies of inelastic, frictional disks. *J. Rheol.* **30** 949–980.

[115] Campbell, C. S. and Brennen, C. E. 1985. Chute flows of granular material: some computer simulations. *J. Appl. Mech.* **52** 172–178; 1985. Computer simulations of granular shear flows. *J. Fluid Mech.* **151** 167–188.

[116] Rothman, D. H. and Keller, J. M. 1988. Immiscible cellular-automaton fluids. *J. Stat. Phys.* **52** 1119–1127.

[117] Gutt, G. M. and Haff, P. K. 1990. An automata model of granular materials. In *Proceedings of the Fifth Distributed Memory Computing Conference* (IEEE Computer Society Press), 522–529.

[118] Fitt, A. D. and Wilmott, P. 1992. Cellular-automaton model for segregation of a two-species granular flow. *Phys. Rev.* **A45** 2383–2388.

[119] Cundall, P. A. and Strack, O. D. L. 1979. A discrete numerical model for granular assemblies. *Geotechnique* **29** 47–65.

[120] Silbert, L. E., Ertas, E., Grest, G. S., Halsey, T. C., and Levine, D. 2002. Geometry of frictional and frictionless sphere packings. *Phys. Rev.* **E 65** 031304.

[121] Duke, T. A. J., Barker, G. C., and Mehta, Anita 1990. A Monte Carlo study of granular relaxation. *Europhys. Lett.* **13** 19–24.

[122] Mehta, Anita 1990. The physics of powders. In *Correlations and Connectivity; Geometrical Aspects of Physics, Chemistry and Biology*, eds. H. E. Stanley and N. Ostrowsky (Dordrecht: Kluwer Academic Press), 88–108.

[123] Kob, W., Donati, C., Plimpton, S. J., Poole, P. H., and Glotzer, S. C. 1997 Dynamical heterogeneity in a supercooled Lennard-Jones mixture. *Phys. Rev. Lett.* **79** 2827–2830.

[124] Weeks, E. R., Crocker, D. E., Levitt, A. C., Schofield, A., and Weitz, D. A., 2002. Three-dimensional direct imaging of structural relaxation near the colloidal glass transition. *Science* **287** 627–631.

[125] Tassopoulos, M. and Rosner, D. E. 1992. Microstructural descriptors characterizing granular deposits. *A. I. Ch. E.* **38** 15–25.

[126] Mehta, Anita and Barker, G. C. 2001. Bistability and hysteresis in tilted sandpiles. *Europhys. Lett.* **56** 626–632.

[127] Lawlor, A., De Gregorio, P., Dawson, K. A. (2004) The geometry of empty space is the key to arresting dynamics. *J. Phys. - Cond. Mat.* **16** S4841–S4848.

[128] Philippe, P. and Bideau, D. 2001. Numerical model for granular compaction under vertical tapping. *Phys. Rev.* **E 63** 051304–051313.

[129] Harnby, N., Hawkins, A. E., and Vandamme, D. E. 1987. The use of bulk density determination as a means of typifying the flow characteristics of loosely compacted powders under conditions of variable relative humidity. *Chem. Eng. Sci.* **42** 879–888.

[130] Barker, G. C. and Mehta, Anita 1993. Transient phenomena, self-diffusion, and orientational effects in vibrated powders. *Phys. Rev.* E **47** 184–188.

[131] Barker, G. C. and Mehta, Anita 2002. Inhomogeneous relaxation in vibrated granular media:consolidation waves, *Phase Transitions* **75** 519–528.

[132] Mehta, Anita and Barker, G. C. 2000. Glassy dynamics of granular compaction. *J. Phys. - Cond. Mat.* **12** 6619–6628.

[133] Villarruel, F. X., Lauderdale, B. E., Mueth, D. E., and Jaeger, H. E. 2000. Compaction of rods: relaxation and ordering in vibrated, anisotropic granular material. *Phys. Rev.* E **61** 6914–6921.

[134] Luck, J. M. and Mehta, Anita 2007. Jamming and metastability in a column of grains. Submitted to *European Journal of Physics B*.

[135] Jullien, R. and Meakin, P. 1990. A mechanism for particle size segregation in three dimensions. *Nature* 425–427.

[136] Barker, G. C. and Grimson, M. J. 1990. The physics of muesli. *New Scientist* **1718** 37–40.

[137] Devillard, P. 1990. Scaling behaviour in size segregation (Brazil nuts). *Journal de Physique* **51** 369–373.

[138] Ehrichs, E. E., Jaeger, H. M., Karczmar, G. S., Knight, J. B., Kuperman, V. Y., and Nagel, S. R. 1995. Granular convection observed by magnetic resonance imaging. *Science* **267** 1632–1634.

[139] Mobius, M. E., Lauderdale, B. E., Nagel, S. R., and Jaeger, H. M. 2001. Size separation of granular particles. *Nature* **414** 270.

[140] Hong, D. C., Quinn, P. V., and Luding, S. 2001. Reverse Brazil nut problem: competition between percolation and condensation. *Phys. Rev. Lett.* **86** 3423–3426.

[141] Shinbrot, T. 2004. The Brazil nut effect – in reverse. *Nature* **429** 352–353.

[142] Baxter, J., Tuzun, U., Heyes, D., Hayati, I., and Fredlund, P. 1998. Stratification in poured granular heaps. *Nature* **391** 136.

[143] Makse, H. A., Havlin, S., King, P. R., and Stanley, H. E. 1997. Spontaneous stratification in granular mixtures. *Nature* **386** 379–382.

[144] Li, H. and McCarthy, J. J. 2003. Controlling cohesive particle mixing and segregation. *Phys. Rev. Lett.* **90** 184301.

[145] Shinbrot, T. and Muzzio, F. J. 2000. Nonequilibrium patterns in granular mixing and segregation. *Physics Today* 25–30.

[146] Seymour, J. D., Caprihan, A., Altobelli, S. A., and Fukushima, E. 2000. Pulsed gradient spin echo nuclear magnetic resonance imaging of diffusion in granular flow. *Phys. Rev. Lett.* **84** 266–269.

[147] Conway, S. L., Shinbrot, T., and Glasser, B. J. 2004. A Taylor vortex analogy in granular flows. *Nature* **431** 433–437.

[148] Mézard, M., Parisi, G., and Virasoro, M. A. 1987. *Spin Glass Theory and Beyond* (Singapore: World Scientific).

[149] Marinari, E., Parisi, G., Ricci-Tersenghi, F., and Zuliani, F. 2001. The use of optimized Monte Carlo methods for studying spin glasses. *J. Phys.* A **34** 383–390.

[150] Biroli, G. and Mézard, M. 2001. Lattice glass models. *Phys. Rev. Lett.* **88** 025501.

[151] De Gennes, P. G. 1999. Granular matter: a tentative view. *Rev. Mod. Phys.* **71** S374–S382.

[152] Berg, J. and Mehta, Anita 2001. On random graphs and the statistical mechanics of granular matter. *Europhys. Lett.* **56** 784–790.

[153] Berg, J. and Mehta, Anita 2002. Glassy dynamics in granular compaction: sand on random graphs. *Phys. Rev.* E **65** 031305.

[154] Edwards, S. F. 1998. The equations of stress in a granular material. *Physica* **A 249** 226–231.

[155] Donev, A., Cisse, I., Sachs, D., *et al.* 2004. Improving the density of jammed disordered packings using ellipsoids. *Science* **303** 990–993.

[156] Doi, M. and Edwards, S. F. 1986. *The Theory of Polymer Dynamics* (Oxford: Clarendon).

[157] To, K., Lai, P. Y., and Pak, H. K. 2001. Jamming of granular flow in a two-dimensional hopper. *Phys. Rev. Lett.* **86** 71–74.

[158] Liu, C.-H., Nagel, S. R., Schecter, D. A., Coppersmith, S. N., Majumdar, S., Narayan, O., and Witten T. A. 1995. Force fluctuations in bead packs. *Science* **269** 513–515.

[159] Mueth, D. M., Jaeger, H. M., and Nagel, S. R. 1998. Force distribution in a granular medium. *Phys. Rev.* **E 57** 3164–3169.

[160] Majmudar, T. S. and Behringer, R. P. 2005. Contact force measurements and stress-induced anisotropy in granular materials. *Nature* **435** 1079–1082.

[161] Erikson, J. M., Mueggenburg, N. W., Jaeger, H. M., and Nagel, S. R. 2002. Force distributions in three compressible granular packs. *Phys. Rev.* **E 66** 040301(R).

[162] O'Hern, C. S., Langer, S. A., Liu, A. J., and Nagel, S. R. 2002. Random packings of frictionless particles. *Phys. Rev. Lett.* **88** 075507.

[163] See chapters by Fukushima, E. Seidler, G. T., *et al.* 2003. In *Challenges in Granular Physics*, ed. Mehta, Anita and Halsey, T. C. (Singapore: World Scientific).

[164] Uhlenbeck, G. E. and Ornstein, L. S. 1930. On the theory of the Brownian motion. *Phys. Rev.* **36** 823–841; Wang, M. C. and Uhlenbeck, G. E. 1945. On the theory of the Brownian motion II. *Rev. Mod. Phys.* **17** 323–343.

[165] Barker, G. C. and Mehta, Anita 2000. Two types of avalanche behaviour in model granular media. *Physica* **A 283** 328–336.

[166] Daerr, A. and Douady, S. 1999. Two types of avalanche behaviour in granular media. *Nature* **399** 241–243.

[167] Smoluchowski, M. V. 1916. Three lectures on diffusion, Brownian molecular movement and coagulation of colloidal particles *Z. Phys.* **17** 557–571.

[168] Mehta, Anita, Barker, G. C., Luck, J. M., and Needs, R. J. 1996. The dynamics of sandpiles: the competing roles of grains and clusters. *Physica* **A224** 48–67.

[169] Krug, J. 1997. Origins of scale invariance in growth processes. *Advances in Physics* **46** 139–282.

[170] Stauffer, D. and Aharony, A. 1994. *Introduction to Percolation Theory* (London: Taylor and Francis).

[171] Barma, M. and Ramaswamy, R. 1986. On backbends on percolation backbones. *J. Phys.* **A 19** L605–611.

[172] Nowak, E. R., Knight, J. B., Povinelli, M., Jaeger, H. M., and Nagel, S. R. 1997. Reversibility and irreversibility in the packing of vibrated granular material. *Powder Technology* **94** 79–83.

[173] Nowak, E. R., Knight, J. B., Ben-Naim, E., Jaeger, H. M., and Nagel, S. R.. 1998. Density fluctuations in vibrated granular materials. *Phys. Rev.* **E 57** 1971–1982.

[174] Stadler, P. F., Luck, J. M., and Mehta, Anita. 2002. Shaking a box of sand. *Europhys. Lett.* **57** 46–52.

[175] Caglioti, E., Loreto, V., Herrmann, H. J., and Nicodemi, M. 1997. A 'Tetris- like' model for the compaction of dry granular media. *Phys. Rev. Lett.* **79** 1575–1578.

[176] Kurchan, J. 2000. Emergence of macroscopic temperatures in systems that are not thermodynamical microscopically: towards a thermodynamical description of slow granular rheology. *J. Phys. Cond. Mat.* **12** 6611–6617.

[177] Bollobas, B. 1985. *Random Graphs* (London: Academic Press).

[178] Kob, W. and Andersen, H. C. 1993. Kinetic lattice-gas model of cage effects in high-density liquids and a test of mode-coupling theory of the ideal-glass transition. *Phys. Rev.* **E 48** 4364–4377.

[179] Viana, L. and Bray, A. J. 1985. Phase diagrams for dilute spin glasses. *J. Phys.* **C 18** 3037–3051.

[180] Monasson, R., Kirkpatrick, S., Selman, B., Troyansky, L., and Zecchina, R. 1999. Determining computational complexity from characteristic 'phase transitions'. *Nature* **400** 133–137.

[181] Ricci-Tersenghi, F., Weigt, M., and Zecchina, R. 2001. Simplest random K-satisfiability problem. *Phys. Rev.* **E 63** 026702.

[182] Newman, M. and Moore, C. 1999. Glassy dynamics in an exactly solvable spin model. *Phys Rev.* **E 60** 5068–5072.

[183] Dean, A. S. and Lefèvre, A. 2001. Tapping spin glasses and ferromagnets on random graphs. *Phys. Rev. Lett.* **86** 5639–5642.

[184] Nowak, E. R., Grushin, A., Barnum, A. C. B., and Weissman, M. B. 2001. Density-noise power fluctuations in vibrated granular media. *Phys. Rev.* **E 63** 020301.

[185] Barrat, A. and Zecchina, R. 1999. Time scale separation and heterogeneous off-equilibrium dynamics in spin models over random graphs. *Phys. Rev.* **E 59** R1299–R1302.

[186] Monasson, R. 1995. Structural glass transition and the entropy of the metastable states. *Phys. Rev. Lett.* **75**, 2847–2850.

[187] Franz, S. and Parisi, G. 1995. Recipes for metastable states in spin glasses. *J. Physique* **5** 1401–1415.

[188] Bouchaud, J.-P., Cugliandolo, L. F., Kurchan, J., and Mezard, M. 1997. Out of equilibrium dynamics in spin-glasses and other glassy systems. In *Spin Glasses and Random fields*, ed. Young, A. P. (Singapore: World Scientific).

[189] Kolan, A. J., Nowak, E. R., and Tchakenko, A. J. 1999. Glassy behavior of the parking lot model. *Phys. Rev.* **E 59** 3094–3099.

[190] Coniglio, A. and Nicodemi, M. 2000. The jamming transition of granular media. *J. Phys.* **C 12** 6601–6610.

[191] Liu, A. J. and Nagel, S. R. 1998. Jamming is not just cool anymore. *Nature* **396** 21–22.

[192] Berg, J. and Mehta, Anita 2003. Spin-models of granular compaction: from one-dimensional models to random graphs. In *Challenges in Granular Physics*, ed. Mehta, Anita and Halsey, T. C. (Singapore: World Scientific).

[193] Van Noije, T. P. C., Ernst, M. H., and Brito, R. 1998. Spatial correlations in compressible granular flows. *Phys. Rev.* **E 57** R4891–R4894.

[194] Campbell, C. S. 1990. Rapid granular flows. *Ann. Rev. Fluid Mech.* **22** 57–92.

[195] Tan, M.-L. and Goldhirsch, I. 1998. Rapid granular flows as mesoscopic systems. *Phys. Rev. Lett.* **81(14)** 3022–3025.

[196] McNamara, S. and Young, W. R. 1994. Inelastic collapse in two dimensions. *Phys. Rev.* **E50(1)** 28–31.

[197] Brey, J. J., Moreno, F., and Dufty, J. W. 1996. Model kinetic equation for low density granular flow. *Phys. Rev.* **E54(1)** 445–456.

[198] Monasson, R. and Pouliquen, O. 1997. Entropy of particle packings: an illustration on a toy model. *Physica* **A 236** 395–410.

[199] Majumdar, A. S., Mehta, Anita, and Luck, J. M. 2005. Interacting black holes on the brane: the seeding of binaries. *Phys. Lett.* **B 607** 219–224.

[200] Luck, J. M. and Mehta, Anita 2005. A deterministic model of competitive cluster growth: glassy dynamics, metastability and pattern formation. *European Phys. J.* **B 44** 79–92.

[201] Caballero, G., Lindner, A., Ovarlez, G., Reydellet, G., Lanuza, J., and Clement, E. 2004. Experiments in randomly agitated granular assemblies close to the jamming transition. In *Unifying Concepts in Granular Media and Glasses*, ed. Coniglio, A., Fierro, A., Herrmann, H. J., and Nicodemi, M. (Amsterdam: Elsevier) 77–87.

[202] Frisch, U., d'Humieres, D., Hasslacher, B., Lallemand, P., Pomeau, Y., and Rivet, J. P. 1987. Lattice gas hydrodynamics in two and three dimensions. *Complex Systems* **1** 649–707.

[203] Wolfram, S. 1986. Cellular automaton fluids: basic theory. *J. Stat. Phys.* **45** 471–526.

[204] Snyder, R. E. and Ball, R. C. 1994. Self-organized criticality in computer models of settling powders. *Phys. Rev.* **E49** 104–109.

[205] Mehta, Anita 1999. Smoothing of sandpiles after avalanche propagation. In *Structure and Dynamics of Materials in the Mesoscopic Domain*, eds. Lal, M., Mashelkar, R. A., Kulkarni, B. D., and Naik, V. M. (London: Imperial College Press and the Royal Society), 340–352.

[206] Frette, V., Christensen, K., Malthe-Sorenssen, A., Feder, J., Jossang, T., and Meakin, P. 1996. Avalanche dynamics in a pile of rice. *Nature* **379** 49–52.

[207] Sander, L. 1991. Growth and aggregation far from equilibrium. In *Solids far from Equilibrium*, ed. Godreche, C. (Cambridge: Cambridge University Press).

[208] Paczuski, M. and Boettcher, S. 1996. Universality in sandpiles, interface depinning, and earthquake models. *Phys. Rev. Lett.* **77** 111–114.

[209] Hwa, T. and Kardar, M. 1992. Avalanches, hydrodynamics, and discharge events in models of sandpiles. *Phys. Rev.* **A 45** 7002–7023.

[210] Carlson, J. M. and Langer, J. S. 1989. Mechanical model of an earthquake fault. *Phys. Rev.* **A 40** 6470–6484.

[211] Nakanishi, H. 1991. Statistical properties of the cellular-automaton model for earthquakes. *Phys. Rev.* **A43** 6613–6621.

[212] Gabrielov, A., Newman, W. I., and Knopoff, L. 1994. Lattice models of failure: sensitivity to the local dynamics. *Phys. Rev.* **E 50** 188–197; Xu, H. J. and Knopoff, L. 1994. Periodicity and chaos in a one-dimensional dynamical model of earthquakes. *Phys. Rev.* **E 50** 3577–3581.

[213] Nagel, S. R., private communication.

[214] Bouchaud, J. P., Cates, M. E., Ravi Prakash, J., and Edwards, S. F. 1994. A model for the dynamics of sandpile surfaces. *Journal de Physique I* **4** 1383–1410; 1995. Hysteresis and metastability in a continuum sandpile model. *Phys. Rev. Lett.* **74** 1982–1985.

[215] Boutreux, Th., Raphaël, E., and de Gennes, P. G. 1998. Surface flows of granular materials: a modified picture for thick avalanches. *Phys. Rev.* **E 58** 4692–4700; Aradian, A., Raphaël, E., and de Gennes, P. G. 1999. Thick surface flows of granular materials: effect of the velocity profile on the avalanche amplitude *Phys. Rev.* **E 60** 2009–2019; de Gennes, P. G. 2000. Tapping of granular packs: a model based on local two-level systems. *J. Coll. Int. Sci.* **226** 1–4.

[216] Hoyle, R. B. and Mehta, Anita 1999. Two-species continuum model for Aeolian sand ripples. *Phys. Rev. Lett.* **83** 5170–5173.

[217] Kroy, K., Sauermann, G., and Herrmann, H. J. 2002. Minimal model for sand dunes, *Phys. Rev. Lett.* **88** 054301.

[218] Franklin, F. C. and Johanson, L. N. 1955. Flow of granular material through a circular orifice. *Chem. Eng. Sci.* **4** 119–129.

[219] Mehta, Anita Bhattacharjee, J. K., and Luck, J. M. 1996. Noisy nonlinear coupled equations – some new insights. *Pramana*, **48** 749–758.

[220] Luck, J. M. and Mehta, Anita 2001. Anomalous aging phenomena caused by drift velocities. *Europhys Lett.* **54** 573–580.

[221] Halpin-Healy, T. and Zhang, Y. C. 1995. Kinetic roughening, stochastic growth, directed polymers and all that. *Phys. Rep.* **254** 215–415.

[222] Edwards, S. F. 1964. The statistical dynamics of homogeneous turbulence. *J. Fluid Mech.* **18** 239–273.

[223] Moore, M. A., Blum, T., Doherty, J. P., Marsili, M., Bouchaud, J. P., and Claudin, P. 1995. Glassy solutions of the Kardar–Parisi–Zhang equation. *Phys. Rev. Lett.* **74** 4257–4260.

[224] Bagnold, R. A. 1941. *The Physics of Blown Sand and Desert Dunes* (London: Methuen and Co.).

[225] Greeley, R., Kraft, M., Sullivan, R., *et al.* 1999. Aeolian features and processes at the Mars Pathfinder landing site. *J. Geophys. Res.* **104** 8573–8584.

[226] Anderson, R. S. and Bunas, K. L. 1993. Grain size segregation and stratigraphy in Aeolian ripples modelled with a cellular automaton. *Nature* **365** 740–743.

[227] Sharp, R. P. 1963. Wind ripples. *J. Geol.* **71** 617–636.

[228] Pye, K. and Tsoar, H. 1990. *Aeolian sand and sand dunes.* (London Unwin Hyman)

[229] Hunter, R. E. 1985. A kinematic model for the structure of lee-side deposits. *Sedimentology* **32** 409–422.

[230] Anderson, R. S. 1988. The pattern of grainfall deposition in the lee of aeolian dunes. *Sedimentology* **35** 175–188.

[231] Anderson, R. S. 1990. Aeolian ripples as examples of self-organization in geomorphological systems. *Earth-Sci.Rev.* **29** 77–96; Forrest, S. B. and Haff, P. K. 1992. Mechanics of wind ripple stratigraphy. *Science* **255** 1240–1243; Landry, W. and Werner, B. T. 1994. Computer simulations of self-organized wind ripple patterns. *Physica* **D 77** 238–260; Nishimori, H. and Ouchi. N. 1993. Computational models for sand ripple and sand dune formation. *Int. J. Modern Physics.* **B7** 2025–2034; Nishimori, H. and Ouchi, N. 1993. Formation of ripple patterns and dunes by wind-blown sand. *Phys. Rev. Lett.* **71** 197–200; Ouchi, N. and Nishimori, H. 1995. Modeling of wind-blown sand using cellular-automata. *Phys. Rev.* **E 52** 5877–5880; Nishimori, H., Yamasaki, M., and Andersen, K. H. 1998. A simple model for the various pattern dynamics of dunes. *Int. J. Modern Phys.* **B 12** 257—272.

[232] Anderson, R. S. 1987. A theoretical model for aeolian impact ripples. *Sedimentology* **34** 943–956.

[233] Hoyle, R. B. and Woods, A. W. 1997. Analytical model of propagating sand ripples. *Phys. Rev.* **E 56** 6861–6868.

[234] Mitha, S., Tran, M. Q., Werner, B. T., and Haff, P. K. 1986. The grain-bed impact process in Aeolian saltation. *Acta Mechanica* **63** 267–278.

[235] Ungar, J. and Haff, P. K. 1987. Steady state saltation in air. *Sedimentology* **34** 289–299.

[236] Lele, S. K. 1992. Compact finite difference schemes with spectral-like resolution. *J. Comput. Phys.* **103** 16–43.

[237] Falcon, E., Wunenburger, R., Evesque, P., Fauve, S., Chabot, C., Garrabos, Y., and Beysens, D. 1999. Cluster formation in a granular medium fluidized by vibrations in low gravity. *Phys. Rev. Lett.* **83(2)** 440—443.

[238] Painter, B. and Behringer, R. P. 2000. Substrate interactions, effects of symmetry breaking, and convection in a 2D horizontally shaken granular system. *Phys. Rev. Lett.* **85(16)** 3396–3399.

[239] Medved, M., Dawson, D., Jaeger, H. M., and Nagel, S. R. 1999, Convection in horizontally vibrated granular material. *Chaos* **9(3)** 691–648.

[240] Hsiau, S. S., Ou, M. Y., and Tai, C. H. 2002. The flow behavior of granular material due to horizontal shaking. *Advanced Powder Technology* **13(2)** 167–180.

[241] Aranson, I. S., Meerson, B., Sasorov, P. V., and Vinokur, V. M. 2002. Phase separation and coarsening in electrostatically driven granular media. *Phys. Rev. Lett.* **88(20)** 204301.

[242] Goldhirsch, I. and Zanetti, G. 1993. Clustering instability in dissipative gases. *Phys. Rev. Lett.* **70** 1619–1622.

[243] Chapman, S. and Cowling, T. G. 1970. *The Mathematical Theory of Nonuniform Gases.* (Cambridge: Cambridge University Press).

[244] Kogan, M. K. 1969 *Rarefied Gas Dynamics* (New York: Plenum Press).

[245] Harris, S. 1971. *Introduction to the Theory of the Boltzmann Equation* New York: Holt, Reinhart and Winston

[246] Cercignani, C. 1975. *Theory and Application of the Boltzmann Equation.* (Edinburgh and London: Scottish Academic Press).

[247] Rericha, E. C., Bizon, C., Shattuck, M. D., and Swinney, H. L. 2002. Shocks in supersonic sand. *Phys. Rev. Lett.* **88(1)** 014302.

[248] Goldhirsch, I. 2003. Rapid granular flows. *Ann. Rev. Fluid Mech.* **35** 267–293.

[249] Ronis, D. 1979. Statistical mechanics of systems nonlinearly displaced from equilibrium I. *Physica* **A99** 403–434, and refs. therein.

[250] Goldhirsch, I. and van Noije, T. P. C. 2000. Green–Kubo relations for granular fluids. *Phys. Rev.* **E61(3)** 3241–3244.

[251] Dufty, J. W. and Brey, J. J. 2002. Green–Kubo expressions for a granular gas. *J. Stat. Phys.* **109(3-4)** 433–448.

[252] Forterre, Y. and Pouliquen, O. 2001. Longitudinal vortices in granular flows. *Phys. Rev. Lett.* **86(26)** 5886–5889.

[253] Goldhirsch, I. 1991. Clustering instability in granular gases, *Proceedings of the DOE/NSF meeting on 'The Flow of Particulates and Fluids'* 211–235 (Worcester Mass.: WPI).

[254] Goldhirsch, I., Tan M.-L., and Zanetti, G. 1993. A molecular dynamical study of granular fluids: the unforced granular gas, *J. Sci. Comp.* **8(1)** 1–40.

[255] McNamara, S. and Young, W. R. 1991. Inelastic collapse and clumping in a one-dimensional granular medium. *Phys. Fluids.* **A4(3)** 496–504.

[256] Luding, S. and Herrmann, H. J. 1999. Cluster-growth in cooling granular media. *Chaos* **9(3)** 673–681.

[257] Livne, E., Meerson, B., and Sasorov, P. V. 2002. Symmetry breaking and coarsening of clusters in a prototypical driven granular gas. *Phys. Rev.* **E66(5)** 05030eds 1.

[258] Goldhirsch, I., Noskowicz, S. H., and Bar-Lev, O. 2003. The homogeneous cooling state revisited. *Granular Gas Dynamics*, Pöschel, T. and Brilliantov, N. eds., Lecture Notes in Physics, 37–64 (Berlin and New York: Springer).

[259] Deltour, P. and Barrat, J.-L. 1997. Quantitative study of a freely cooling granular medium. *J. Phys. I. France* **7** 137–151.

[260] Brey, J. J., Ruiz-Monterro, M. J., and Cubero, D. 1999. Origin of density clustering in freely evolving granular gas. *Phys. Rev.* **E60(3)** 3150–3157.

[261] Hopkins, M. A. and Louge, M. Y. 1991. Inelastic microstructure in rapid granular flows of smooth disks. *Phys. Fluids* **A3(1)** 47–57.

[262] Tan, M.-L. and Goldhirsch, I. 1997. Intercluster interactions in rapid granular shear, flows. *Phys. Fluids* **9(4)** 856–869.

[263] Savage, S. B. 1992. Instability of an unbounded uniform granular shear flow. *J. Fluid Mech* **241** 109–123.

[264] Babic, M. 1993. On the stability of rapid granular flows. *J. Fluid Mech* **254** 127–150.

[265] Nott, P. R., Alam, M., Agrawal, K., Jackson, R., and Sundaresan, S. 1999. The effect of boundaries on the plane Couette flow of granular materials: a bifurcation analysis. *J. Fluid Mech.* **397** 202–229.

[266] Soto, R., Mareschal, M., and Mansour M. M. 2000. Nonlinear analysis of the shearing instability in granular gases. *Phys. Rev.* **E62(3)** 3836–3842.

[267] Kudrolli, A., Wolpert, M., and Gollub, J. P. 1997. Cluster formation due to collisions in granular material. *Phys. Rev. Lett.* **78(7)** 1383–1386.

[268] Mueth, D. M., Debregeas, G. F., Karczmar, G. S., Eng, P. J., Nagel, S. R., and Jaeger, H. M. 2000. Signatures of granular microstructure in dense shear flows. *Nature* **406** 385–389.

[269] Olafsen, J. S. and Urbach, J. S. 1998. Clustering, order and collapse in a driven granular monolayer. *Phys. Rev. Lett.* **81(20)** 4639–4372.

[270] Tsai, J. C., Losert, W., Voth, G. A., and Gollub, J. P. 2002. Two dimensional granular Poiseuille flow on an incline: multiple dynamical regimes. *Phys. Rev.* **E65(1)** 011306.

[271] Blair, D. L., Aranson, I. S., Crabtree, G. W., Vinokour, V., Tsimring, L. S., and Joserand, C. 2000. Patterns of thin vibrated granular layers: interfaces, hexagons and superoscillons. *Phys. Rev.* **E61(5)** 5600–5610.

[272] De Bruin, J. R., Lewis, B. C., Shattuck, M. D., and Swinney, H. L. 2001. Spiral patterns in oscillated granular layers. *Phys. Rev.* **E63(4)** 0413050.

[273] Metcalf, T. H., Knight, J. B., and Jaeger, H. M. 1997. Standing wave patterns in shallow beds of vibrated granular material. *Physica* **A236(3–4)** 202–210.

[274] Eggers, J. 1999. Sand as Maxwell's demon. *Phys. Rev. Lett.* **83(25)** 5322–5325.

[275] Van der Weele, K., van der Meer, D., Versluis, M., and Lohse, D. 2001. Hysteretic clustering in granular gas. *Europhys. Lett.* **53(3)** 328–334.

[276] Liss, E. D. and Glasser, B. J. 2001. The influence of clusters on the stress in a sheared granular material. *Powder Technology* **116(2–3)** 116–132.

[277] Agrawal, K., Loezos, P. N., Syamlal, M., and Sundaresan, S. 2001. The role of meso-scale structures in rapid gas-solid flows. *J. Fluid Mech.* **445** 151–185.

[278] Kadanoff, L. P. 1999. Built upon sand: theoretical ideas inspired by granular flows. *Rev. Mod. Phys.* **71(1)** 435–444.

[279] Brilliantov, N. V. and Pöschel, T. 2000. Granular gases with impact-velocity-dependent restitution coefficient. In *Granular Gases*, Pöschel, T. and Luding, S. eds., 100–124. (Berlin and new York:) Springer-Verlog.

[280] Falcon, E., Laroche, C., Fauve, S., and Coste, C. 1998. Behavior of one inelastic ball bouncing repeatedly off the ground. 1998. *Euro. Phys. J.* **B3(1)** 45–57.

[281] Alam, M. and Hrenya, C. M. 2001. Inelastic collapse in simple shear flow of a granular medium. *Phys. Rev.* **E63(6)** 061308.

[282] Luding, S. and McNamara, S. 1998. How to handle the inelastic collapse of a dissipative hard-sphere gas with the TC model. *Granular Matter* **1(3)** 3416–3425.

[283] Sela, N. and Goldhirsch, I. 1995. Hydrodynamics of a one-dimensional granular medium. *Phys. Fluids* **7(3)** 507–525.

[284] Goldhirsch, I. and Sela, N. 1996. Origin of normal stress differences in rapid granular flows *Phys. Rev.* **E 54(4)** 4458–4461.

[285] Sela, N., Goldhirsch, I., and Noskowicz, S. H. 1996. Kinetic theoretical study of a simply sheared granular gas – to Burnett order. *Phys. Fluids.* **8(9)** 2337–2353.

[286] Sela, N. and Goldhirsch, I. 1998. Hydrodynamic equations for rapid flows of smooth inelastic spheres. *J. Fluid Mech.* **361** 41–74.

[287] Goldshtein, A. and Shapiro, M. 1995. Mechanics of collisional motion of granular materials, Part I: General hydrodynamic equations. *J. Fluid Mech.* **282** 75–114.

[288] Brey, J. J., Dufty, J. W., Kim, C. S., and Santos, A. 1998. Hydrodynamics for granular flow at low density. *Phys. Rev.* **E58(4)** 4638–4653.

[289] Kamenetsky, V., Goldstein, A., Shapiro, M., and Degani, D. 2000. Evolution of a shock wave in a granular gas. *Phys. Fluids.* **12(11)** 3036–3049.

[290] Hørlück, S. and Dimon, P. 2001. Grain dynamics in a two dimensional granular flow. *Phys. Rev.* **E63(3)** 031301.

[291] Jin, S. and Slemrod, M. 2001. Regularization of the Burnett equations for rapid granular flows via relaxation. *Physica* **D150(3-4)** 207–218.

[292] Standish, R. K. 1999. Numerical evidence for divergent Burnett coefficients. *Phys. Rev.* **E60(5)** 5175–5178.

[293] Ernst, M. H. and Dorfman, J. R. 1975. Nonanalytic dispersion relations for classical fluids. *J. Stat. Phys.* **12(4)** 311–361.

[294] Kumaran, V. 2001. Hydrodynamic modes of a sheared granular flow from the Boltzmann and Navier-Stokes equations. *Phys. Fluids.* **13(8)** 2258–2268.

[295] Glasser, B. J. and Goldhirsch, I. 2001. Scale dependence, correlations and fluctuations of stresses in rapid granular flows. *Phys. Fluids* **13(2)** 407–420.

[296] Croteau, T. and Ronis, D. 2002. Nonequilibrium velocity distributions in liquids under shear. *Phys. Rev.* **E66(6)** 066109, and refs. therein.

[297] Soto, R. and Mareschal, M. 2001. Statistical Mechanics of fluidized granular media: short-range velocity correlations. *Phys. Rev.* **E63(4)** 041303.

[298] Luding, S. 2000. On the relevance of 'molecular chaos' for granular flows. *ZAMM* **80** S9–S12.

[299] Grad, H. 1949. On the kinetic theory of rarefied gases, *Comm. Pure Appl. Math.* **2** 331–407.

[300] Jenkins, J. T. and Richman, M. W. 1985. Grad's 13-moment system for a dense gas of inelastic particles, *Arch. Ration. Mech. Anal.* **87** 355–377.

[301] Ramirez, R., Risso, D., Soto, R., and Cordero, P. 2000. Hydrodynamic theory for granular gases. *Phys. Rev.* **E62(2)** 2521– 2530.

[302] Tij, M., Tahiri, E. E., Montanero, J. M., Garzo, V., Santos, A., and Dufty J. W. 2001. Nonlinear Couette flow in a low density granular gas. *J. Stat. Phys.* **103(5–6)** 1035–1068.

[303] Soto, R., Mareschal, M., and Risso, D. 1999. Departure from Fourier's law for fluidized granular media. *Phys. Rev. Lett.* **83(24)** 5003–5006.

[304] Lun, C. K. K., Savage, S. B., Jeffrey, D. J., and Chepurniy, N. 1984 Kinetic theories of granular flow: inelastic particles in a Couette flow and slightly inelastic particles in a general flow field, *J. Fluid Mech.* **140** 223–256.

[305] Jenkins, J. T. and Richman, M. W. 1988. Plane simple shear of smooth inelastic circular disks: the anisotropy of the second moment in the dilute and dense limit, *J. Fluid Mech.* **192** 313–328.

[306] Lun, C. K. K. 1991. Kinetic theory for granular flow of dense, slightly inelastic, slightly rough spheres. *J. Fluid Mech.* **223** 539–559.

[307] Boyle, E. J. and Massoudi, M. 1990. A theory for granular materials exhibiting normal stress effects based on Enskog's dense gas theory. *Int. J. Eng. Sci.* **28** 1261–1275.

[308] Brey, J. J., Ruiz-Montero, M. J., and Moreno, F. 2001. Hydrodynamics of an open vibrated granular system. *Phys. Rev.* **E63** 061305.

[309] Jenkins, J. T. 1992. Boundary conditions for rapid granular flow: flat, frictional walls. *J. Appl. Mech.* **114** 120–127.

[310] Brey, J. J., Ruiz-Montero, M. J., and Moreno, F. 2000. Boundary conditions and normal state for a vibrated granular fluid. *Phys. Rev.* **E62(4)** 5339–5346.

[311] Campbell, C. S. 1993. Boundary interactions for two-dimensional granular flows. Part 1. Flat boundaries, asymmetric stresses and couple stresses. 1993. *J. Fluid Mech.* **247** 111–136; 1993. Boundary interactions for two-dimensional granular flows. Part 2. Roughened boundaries. *J. Fluid Mech.* **247** 137–156.

[312] Loyalka, S. K. 1971. Approximate method in kinetic theory. *Phys. Rev.* **E 14(11)** 2291–2294.

[313] Sone, Y. and Aoki, K. 1977. Slightly rarefied gas flow over a specularly reflecting body. *Phys. Fluids* **20(4)** 571–576.

[314] Goldhirsch, I. 1999. Scales and kinetics of granular flows. *CHAOS* **9(3)** 659–672.

[315] Pekeris, C. L. 1955 Solution of the Boltzmann–Hilbert integral equation, *Proc. N. A. S.* **41** 661–664.

[316] Johnson, P. C. and Jackson, R. 1987. Frictional-collisional constitutive relations for granular materials with applications to plane shearing. *J. Fluid Mech.* **176** 67–93.

[317] Abu-Zaid, S. and Ahmadi, G. 1993. Analysis of rapid shear flows of granular materials by a kinetic model including frictional losses. *Powder Tech.* **77(1)** 7–17.

[318] Cafiero, R. and Luding, S. 2000. Mean field theory for a driven granular gas of frictional particles. *Physica* **A280(1–2)** 142–147.

[319] Moon, S. J., Swift, J. B., and Swinney, H. L. 2004. Role of friction in pattern formation in oscillated granular layers. *Phys. Rev. E* **69(3)** 031301.

[320] Nasuno, S., Kudrolli, A., and Gollub, J. P. 1997. Friction in granular layers: hysteresis and precursors. *Phys. Rev. Lett.* **79** 949–952.

[321] Luding, S., Huthmann, M., McNamara, S., and Zippelius, A. 1998. Homogeneous cooling of rough, dissipative particles: theory and simulations. *Phys. Rev.* **E58** 3416–3425.

[322] Huthmann, M. and Zippelius, A. 1998. Dynamics of inelastically colliding rough spheres: relaxation of translational and rotational energy. *Phys. Rev.* **E56(6)** R6275–R6278.

[323] Herbst, O., Huthmann, M., and Zippelius, A. 2000. Dynamics of inelastically colliding spheres with Coulomb friction: relaxation of translational and rotational energy. *Granular Matter* **2(4)** 211–219.

[324] Luding, S. 1995. Granular materials under vibration: simulations of rotating spheres. *Phys. Rev.* **E52(4)** 4442–4457.

[325] McNamara, S. and Luding, S. 1998. Energy nonequipartition in systems of inelastic rough spheres. *Phys. Rev.* **E58** 2247–2250.

[326] Jenkins, J. T. and Richman, M. W. 1985. Kinetic theory for plane flows of a dense gas of identical, rough, inelastic circular disks. *Phys. Fluids* **28** 3485–3494.

[327] Lun, C. K. K. and Savage, S. B. 1987. A simple kinetic theory for the flow of rough, inelastic spherical particles. *J. Appl. Mech.* **54(1)** 47–53.

[328] Jenkins, J. T. and Zhang, C. 2002. Kinetic theory for identical, frictional, nearly elastic spheres. *Phys. Fluids* **14(3)** 1228–1235.

[329] Dahler, J. S. and Theodosopulu, M. 1975. The kinetic theory of dense polyatomic fluids. *Adv. Chem. Phys.* **31** 155–229.

[330] Shukhman, I. G. 1984. The collisional dynamics of particles in the Saturn rings. *Astronomicheskii Journal (Russian)* **61(5)** 985–1004.

[331] Araki, S. 1998. The dynamics of particle disks 2. Effects of spin degrees of freedom. *Icarus* **76(1)** 182–198.
[332] Hayakawa, H., Mitarai, N., and Nakanishi, H. 2002. Collisional granular flow as a micropolar fluid. *Phys. Rev. Lett.* **88(17)** 174301.
[333] Orza, J. A. G., Brito, R., van, Noije, T. P. C., and Ernst, M. H. 1997. Patterns and long range correlations in idealized granular flows. *Int. J. Mod. Phys. C.* **8(4)** 953–965.
[334] Reif, F. 1965. *Fundamentals of Statistical and Thermal Physics.* (New York: McGraw Hill).
[335] Van Noije, T. P. C. and Ernst, M. H. 2000. Cahn–Hilliard theory for unstable granular fluids. *Phys. Rev.* **E61(2)** 1765–1782.
[336] Venkataramani, S. C. and Ott, E. 2001. Pattern selection in extended periodically forced systems: a contiuum coupled map approach. *Phys. Rev.* **E63(4)** 046202.
[337] Bocquet, L., Losert, W., Schalk, D., Lubensky, T. C., and Gollub, J. P. 2002. Granular shear flow dynamics and forces: experiment and continuum theory. *Phys. Rev.* **E65(1)** 011307.
[338] Hill, S. A. and Mazenko, G. F. 2001. Nonlinear hydrodynamical approach to granular materials. *Phys. Rev.* **E63** 031303.
[339] Grossman, E. L., Zhou, T., and Ben Naim, E. 1997. Towards granular hydrodynamics in two dimensions. *Phys. Rev.* **E55(4)** 4200–4206.
[340] Van Noije, T. P. C., Ernst, M. H., and Brito, R. 1998. Ring kinetic theory for an idealized granular gas. *Physica* **A251(1–2)** 266–283.
[341] Lutsko, J. F. 2001. Model for the atomic-scale structure of the homogeneous cooling state of granular fluids. *Phys. Rev.* **E63(6)** 061211.
[342] Goldhirsch, I. 1999. granular gases: probing the boundaries of hydrodynamics. In *Granular Gases*, T. Pöschel, and S. Luding, eds., Lecture Notes in Physics (Berlin: Springer-Verlag).
[343] Huthmann, M. and Zippelius, A. 1997. Dynamics of inelastically colliding rough spheres: relaxation of translational and rotational energy. *Phys. Rev.* **E56(6)** R6275–R6278.
[344] Knight, T. A. and Woodcock, L. V. 1996. Test of the equipartition principle for granular spheres in a saw-tooth shaker. *J. Phys.* **A 29(15)** 4365–4386.
[345] Rouyer, F. and Menon, N. Velocity fluctuations in a homogeneous 2D granular gas in steady state. *Phys. Rev. Lett.* **85(17)** 3676–3679.
[346] Losert, W., Cooper, D. G. W., Delour, J., Kudrolli, A., and Gollub, J. P. 1999. Velocity statistics in granular media. *Chaos* **9(3)** 682–690.
[347] Goldhirsch, I. and Tan, M.-L. 1996. The single particle distribution function for rapid granular shear flows of smooth inelastic disks. *Phys. Fluids* **8(7)** 1752–1763.
[348] Van Noije, T. P. C. and Ernst, M. H. 1998. Velocity distribution in homogeneous granular fluids: the free and the heated case. *Granular Matter* **1** 57–64.
[349] Garzo, V. and Dufty, J. W. 2002. Hydrodynamics for a granular binary mixture at low density. *Phys. Fluids.* **14(4)** 1476–1490.
[350] Buchholtz, V., Pöschel, T., and Tillemans, H. J. 1995. Simulation of rotating drum experiments using noncircular particles. *Physica* **A216(3)** 199–212.
[351] Pöschel, T. and Buchholtz, V. 1995. Molecular dynamics of arbitrarily shaped granular particles. *J. de Phys. I* **5(11)** 1431–1455.
[352] Jenkins, J. T. and Askari, E. 1991. Boundary conditions for rapid granular flows: phase interfaces. *J. Fluid Mech.* **223** 497–508.
[353] Liu, A. J. and Nagel, S. R. 2001 *Jamming and Rheology: Constrained Dynamics on Microscopic and Macroscopic Scales* (London: Taylor and Francis); Coniglio, A.,

Fierro, A., Herrmann, H. J., and Nicodemi, M. (2004) *Unifying concepts in granular media and glasses* (Amsterdam: Elsevier).

[354] Blumenfeld, R. 2005. Stress transmission and isostatic states of non-rigid particulate systems. In *Modeling of Soft Matter*, IMA Volume in Mathematics and its Applications, **141** eds. Maria-Carme, T. Calderer and Eugene, M. Terentjev (New York: Springer-Verlag).

[355] Hummel, F. H. and Finnan, E. J. 1921. The distribution of pressure on surface supporting a mass of granular material, *Proc. Inst. Civil Eng.* **212** 369–392; Jotaki, T. and Moriyama, R. 1979. On the bottom pressure distribution of the bulk material piled with the angle of repose. *J. Soc. Powder Technol. Jpn.* **16** 184–191; Smid, J. and Novosad, J. 1981. Pressure distribution under heaped bulk solids. In *Proc. of 1981 Powtech Conf., Ind. Chem. Eng. Symp.* **63** D3/V/1.

[356] Knight, J. B., Fandrich, C. G., Lau, C. N., Jaeger, H. M. and Nagel, S. R. 1995. Density relaxation in a vibrated granular material *Phys. Rev. E* **51** 3957–3963.

[357] Chakravarty, A., Edwards, S. F., Grinev, D. V., Mann, M., Phillipson, T. E., and Walton, A. J. 2003. Statistical mechanics and reversible states in quasi-static powders, *Proceedings of the Workshop on the Quasi-static Deformations of Particulate Materials* (Budapest, Budapest: University of Technology and Economics Press); Phillipson, T. E., Blumenfeld, S., Proud, W. G. and Blumenfeld, R., Dynamics of contacts in granular systems from conductance measurements, private communication.

[358] Kirkpatrick, S. 1971. Classical transport in disordered media: scaling and effective-medium theories. *Phys. Rev. Lett.* **27** 1722–1725; Kirkpatrick, S. 1973. Percolation and conduction. *Rev. Mod. Phys.* **45** 574–588; Phillipson, T. E., Blumenfeld, S., Proud, W. G. and Blumenfeld, R., Dynamics of contacts in granular systems from conductance measurements, private communication.

[359] Edwards, S. F. and Oakeshott, R. B. 1989. The transmission of stress in an aggregate. *Physica* **D 38** 88–92.

[360] Edwards, S. F. 1989. The mathematics of powders. *IMA Bulletin* **25** 94–96; Edwards, S. F. and Oakeshott, R. B. 1989. Theory of powders *Physica* **A 157** 1080–1090; Edwards, S. F. and Grinev, D. V. 2001. The tensorial formulation of volume function for packings of particles. *Chem. Eng. Sci.* **56** 5451–5455.

[361] Makse, H. A. and Kurchan, J. 2002. Testing the thermodynamic approach to granular matter with a numerical model of a decisive experiment. *Nature* **415** 614–616.

[362] Ono, I. K., O'Hern, C. S., Durian, D. J., Langer, S. A., Liu, A. J., and Nagel, S. R. 2002. Effective temperatures of a driven system near jamming *Phys. Rev. Lett.* **89** 095703–095706; Fierro, A., Nicodemi, M. and Coniglio, A. 2002. Equilibrium distribution of the inherent states and their dynamics in glassy systems and granular media. *Europhys. Lett.* **59** 642–647; Coniglio, A., Fierro, A., Nicodemi, M., Pica Ciamarra, M. P., and Tarzia, M. 2005. Statistical mechanics of dense granular media. *J. Phys.: Condens. Matter* **17** S2557–S2572; Barrat, A., Kurchan, J., Loreto, V. and Sellitto, M., (2000). Edwards' Measures for Powders and Glasses, *Phys. Rev. Lett.* **85** 5034–5037.

[363] Ball, R. C. and Blumenfeld, R. 2002. Stress field in granular systems: loop forces and potential formulation. *Phys. Rev. Lett.* **88** 115505–115508.

[364] Blumenfeld, R. and Edwards, S. F. 2003. Granular entropy: explicit calculations for planar assemblies. *Phys. Rev. Lett.* **90** 114303–114306.

[364a] Blumenfeld, R. and Edwards, S. F. (2006). Geometric partition functions of cellular systems: Explicit calculation of the entropy in two and three dimensions, *Eur. Phys. J. E* **19** 23–30.

[365] Euler's relation was established circa 1750. For a review, see e.g. Coxeter, H. M. S. 1973. *Regular Polytopes* (New York: Dover); Weaire, D. and Rivier, N. 1984. Soap, cells and statistics – random patterns in 2 dimensions. *Contemp. Phys.* **25** 59–99.

[366] Maxwell, J. C. 1864. On reciprocal figures and diagrams of forces, *Phil. Mag.*, Ser. 4, **27** 250–261; Maxwell, J. C. 1869. On reciprocal diagrams, frames and diagrams of forces, *Trans. Roy. Soc. Edinb.* **26** 1–40.

[367] Levy, M. 1874. *La Statique Graphique et ses Applications aux Constructiones* (Paris: Academie des Sciences).

[368] Edwards, S. F. and Grinev, D. V. 1999. Statistical mechanics of stress transmission in disordered granular arrays *Phys. Rev. Lett.* **82** 5397–5400; Edwards, S. F. and Grinev, D. V. (1998). Statistical mechanics of vibration-induced compaction of powders, *Phys. Rev.* **E 58**, 4758–4762; Edwards, S. F. and Grinev, D. V. (2001). The tensorial formulation of volume function for packings of particles, *Chem. Eng. Sci.*, **56**, 5451–5455.

[369] Edwards, S. F. and Mounfield, C. C. 1996. A theoretical model for the stress distribution in granular matter: I, II, III *Physica* **A 226** 1–33

[370] Bouchaud, J.-P., Cates, M. E., and Claudin, P. J. 1995. Stress distribution in granular media and nonlinear wave equation *J. Phys. I* **5** 639–656.

[371] Ball, R. C. and Blumenfeld, R. 2003. From plasticity to a renormalisation group. *Phil. Trans. R. Soc. Lond.* **360** 731–740.

[372] Blumenfeld, R. 2004. Stress in planar cellular solids and isostatic granular assemblies: coarse-graining the constitutive equation. *Physica* **A 336** 361–368.

[373] Blumenfeld, R. 2004. Stresses in granular systems and emergence of force chains. *Phys. Rev. Lett.* **93** 108301–108304.

[374] Wakabayashi, T. 1957. Photoelastic method for determination of stress in powdered mass. *Proc. 7th Jpn. Nat. Cong. Appl. Mech.*, 153; Dantu, P. 1957. Contribution à l'étude mécanique et géométrique des milieux pulvérulents. *Proc. 4th Int. Conf. Soil Mech. and Found. Eng.*, 144; Howell, D. and Behringer, R. P. 1997. Fluctuations and dynamics for a two-dimensional sheared granular material. In *Powders and Grains 97*, eds. Behringer, R. P. and Jenkins, J. T. (Rotterdam: Balkema), 337; Bagster, D. F. and Kirk, R. 1985. Computer generation of a model to simulate granular material behaviour, *J. Powder Bulk Solids Technol.* **1** 19; Melrose, J. R. and Ball, R. C. 1995. The pathological behavior of sheared hard spheres with hydrodynamic interactions. *Europhys. Lett.* **32** 535; Thornton, C. 1997. Force transmission in granular media. *Kona Powder and Particle* **15** 81.

[375] Vanel, L., Howell, D., Clark, D., Behringer, R. P. and Clement, E. 1999. Memories in sand: experimental tests of construction history on stress distributions under sandpiles. *Phys. Rev.* **E 60** R5040.

[376] Liffman, K. Chan, D. Y. and Hughes, B. D. 1992. Force distribution in a two dimensional sandpile. *Powder Technol.* **72** 225.

[377] Brujic, J. 2004. *Experimental Study of Stress Transmission Through Particulate Matter*, PhD Thesis, Cambridge University.

[378] Blumenfeld, R., Edwards, S. F. and Ball, R. C. 2005. Granular matter and the marginally rigid state. *J. Phys.: Cond. Mat.* **17** S2481–S2487.

[379] Ouaguenoui, S. and Roux, J.-N. 1997. Force distribution in frictionless granular packings at rigidity threshold, *Europhys. Lett.* **39** 117–122.

[380] Antony, S. J. 2001. Evolution of force distribution in three-dimensional granular media. *Phys. Rev.* **E 63** 011302.

[381] Atman, A. P. F., Brunet, P., Geng, J., *et al.* 2005. Sensitivity of the stress response function to packing preparation. *J. Phys. Condens. Matter* **17** S2391.

[382] Atman, A. P. F., Brunet, P., Geng, J., *et al.* 2005. From the stress response function (back) to the sandpile 'dip'. *Eur. Phys. J.* **E 17** 93.

[383] Bagi, K. 1996. Stress and strain in granular assemblies. *Mechanics of Materials* **22** 165.

[384] Bagi, K. 2003. Statistical analysis of contact force components in random granular assemblies. *Granular Matter* **5** 45.

[385] Bathurst, R. J. and Rothenburg, L. 1988. Micromechanical aspects of isotropic granular assemblies with linear contact interactions. *J. Appl. Mech.* **55** 17.

[386] Blair, D. L., Mueggenburg, N. W., Marshall, A. H., Jaeger, H. M., and Nagel, S. R. 2001. Force distributions in three-dimensional granular assemblies: effects of packing order and interparticle friction. *Phys. Rev.* **E 63** 041304.

[387] Breton, L., Claudin, P., Clément, E., and Zucker, J. D. 2002. Stress response function of a two-dimensional ordered packing of frictional beads. *Europhys. Lett.* **60** 813.

[388] Brockbank, R., Huntley, J. M., and Ball, R. C. 1997. Contact force distribution beneath a three dimensional granular pile. *J. Phys. (France) II* **7** 1521.

[389] Calvetti, F., Combe, G., and Lanier, J. 1997. Experimental micromechanical analysis of a 2D granular material: relation between structure evolution and loading path. *Mech. Coh. Frict. Materials* **2** 121.

[390] Cambou, B., Dubujet, P., Emeriault, F., and Sidiroff, F. 1995. Homogenization for granular materials. *Eur. J. Mech.* **A/Solids 14** 255.

[391] Cambou, B., Chaze, M., and Dedecker, F. 2000. Change of scale in granular materials. *Eur. J. Mech.* **A/Solids 19** 999.

[392] Cates, M. E., Wittmer, J. P., Bouchaud, J. P., and Claudin, P. 1998. Jamming, force chains and fragile matter. *Phys. Rev. Lett.* **81** 1841.

[393] Chang, C. S. and Liao, C. L. 1994. Estimates of elastic modulus for media of randomly packed granules. *Appl. Mech. Rev.* **1** S197.

[394] Christoffersen, J., Mehrabadi, M. M., and Nemat-Nasser, S. 1981. A micromechanical description of granular material behaviour. *J. Appl. Mech.* **48** 339.

[395] Claudin, P. and Bouchaud, J. P. 1997. Static avalanches and giant stress fluctuations in silos. *Phys. Rev. Lett.* **78** 231.

[396] Claudin, P., Bouchaud, J. P., Cates, M. E., and Wittmer, J. P. 1998. Models of stress fluctuations in granular media. *Phys. Rev.* **E 57** 4441.

[397] Combe, C. and Roux, J. N. 2000. Strain versus stress in a model granular material: a devil's staircase. *Phys. Rev. Lett.* **85** 3628.

[398] Coppersmith, S. N., Liu, C.-h., Majumdar, S., Narayan, O., and Witten, T. A. 1996. Model for force fluctuations in bead packs. *Phys. Rev.* **E 53** 4673.

[399] Desrues, J. and Viggiani, G. 2004. Strain localization in sand: an overview of the experimental results obtained in Grenoble using stereophotogrammetry. *Int. J. Numer. Anal. Meth. Geomech.* **28** 279.

[400] Didwania, A. K., Cantelaube, F., and Goddard, J. D. 2000. Static multiplicity of stress states in granular heaps. *Proc. Roy. Soc. Lond.* **A 456** 2569.

[401] Drescher, A. and de Josselin de Jong, G., 1972. Photoelastic verification of a mechanical model for the flow of a granular material. *J. Mech. Phys. Solids* **20** 337.

[402] Edwards, S. F. and Grinev, D. V. 2001. Transmission of stress in granular materials as a problem in statistical mechanics. *Physica* **A 302** 162.

[403] Eloy, C. and Clément, E. 1997. Stochastic aspects of the force network in a regular granular piling. *J. Phys. I (France)* **7** 1541.

[404] Falk, M. L. and Langer, J. S. 1998. Dynamics of viscoplastic deformations in amorphous solids. *Phys. Rev.* **E 57** 7192.

[405] Fung, Y. C. 1965 *Foundations of solid mechanics* (Englewood Cliffs, New Jersey: Prentice-Hall, Inc.).

[406] Gay, C. and da Silveira, R. A. 2004. Anisotropic elastic theory of preloaded granular media. *Europhys. Lett.* **68** 51.

[407] Geng, J., Howell, D. W., Longhi, E., *et al.* 2001. Footprints in sand: the response of a granular material to local perturbations. *Phys. Rev. Lett.* **87** 035506.

[408] Geng, J., Longhi, E., Behringer, R. P., and Howell, D. W. 2001. Memory in two-dimensional heap experiments. *Phys. Rev.* **E64** 060301(R).

[409] Geng, J., Reydellet, G., Clément, E., and Behringer, R. P. 2003. Green's function measurements of force transmission in 2D granular materials. *Physica* **D 182** 274.

[410] Goldenberg, C. and Goldhirsch, I. 2002. Force chains, microelasticity and macroelasticity. *Phys. Rev. Lett.* **89** 084302.

[411] Goldenberg, C. and Goldhirsch, I. 2002. On the microscopic foundations of elasticity. *Eur. Phys. J.* **E9** 245.

[412] Goldenberg, C. and Goldhirsch, I. 2004. Small and large scale granular statics. *Granular Matter* **6** 87.

[413] Goldenberg, C. and Goldhirsch, I. 2005. Friction enhances elasticity in granular solids. *Nature* **485** 188.

[414] Head, D. A., Tkachenko, A. V., and Witten, T. A. 2001. Robust propagation direction of stresses in a minimal granular packing. *Eur. Phys. J.* **E6** 99.

[415] Hicher, P. Y. 1996. Elastic properties of soils. *J. Geotech. Engrg.* **122** 641.

[416] Howell, D. W., Behringer, R. P., and Veje, C. 1999. Stress fluctuations in a 2D granular Couette experiment: a continuous transition. *Phys. Rev. Lett.* **82** 5241.

[417] Howell, D. W., Behringer, R. P., and Veje, C. 1999. Fluctuations in granular media. *Chaos* **9** 559.

[418] Johnson, K. L. 1985. *Contact Mechanics* (Cambridge: Cambridge University Press).

[419] Kruyt, N. P. and Rothenburg, L. 2002. Probability density functions of contact forces for cohesionless frictional granular materials. *Int. J. Solids and Structures* **39** 571.

[420] Kruyt, N. P. and Rothenburg, L. 2002. Micromechanical bounds for the effective elastic moduli of granular materials. *Int. J. Solids and Structures* **39** 311.

[421] Kruyt, N. P. 2003. Statics and kinematics of discrete Cosserat-type granular materials. *Int. J. Solids and Structures* **40** 511.

[422] Kruyt, N. P. 2003. Contact forces in anisotropic frictional granular materials. *Int. J. Solids and Structures* **40** 3537.

[423] Kolb, E., Cviklinski, J., Lanuza, J., Claudin, P., and Clément, E. 2004. Reorganization of a dense granular assembly: the 'unjamming response function'. *Phys. Rev.* **E 69** 031306.

[424] Landau, L. D. and Lifshitz, E. M. 1986 *Theory of Elasticity*. (Oxford: Pergamon Press).

[425] Landry, J. W., Grest, G. S., Silbert, L. E., and Plimpton, S. J. 2003. Confined granular packings: structure, stress and forces. *Phys. Rev.* **E 67** 041303.

[426] Lema ître, A. 2002. Origin of a repose angle: kinetics of rearrangement for granular materials. *Phys. Rev. Lett.* **89** 064303.

[427] Lenczner, D. 1963. An investigation into the behaviour of sand in a model silo. *Structural Engineer* **41** 389.

[428] Leomforte, F., Tanguy, A., Wittmer, J. P., and Barrat, J. L. 2004. Continuum limit of amorphous elastic bodies II: linear response to a point source force. *Phys. Rev.* **B 70** 014203.

[429] Liffman, K., Chan, D. Y. C., and Hughes, B. D. 1994. On the stress depression under a sandpile. *Powder Technology* **78** 263.

[430] Løvoll, G., Mløy, K. J., and Flekkøy, E. G. 1999. Force measurements on static granular materials. *Phys. Rev.* **E 60** 5872.

[431] Luding, S. 1997. Stress distribution in static two-dimensional granular model media in the absence of friction. *Phys. Rev.* **E 55** 4720.

[432] Luding, S. 2004. Micro-macro transition for anisotropic, frictional granular packings. *Int. J. Sol. Struct.* **41** 5821.

[433] Makse, H. A., Gland, N., Johnson, D. L., and Schwartz, L. M. 1999. Why effective medium theory fails in granular materials. *Phys. Rev. Lett.* **83** 5070.

[434] Makse, H. A., Gland, N., Johnson, D. L., and Schwartz, L. M. 2004. Granular packings: nonlinear elasticity, sound propagation, and collective relaxation dynamics. *Phys. Rev.* **E 70** 061302.

[435] Makse, H. A., Johnson, D. L., and Schwartz, L. M. 2000. Packing of compressible granular materials. *Phys. Rev. Lett.* **84** 4160.

[436] Matuttis, H. G. 1998. Simulation of the pressure distribution under a two-dimensional heap of polygonal particles. *Granular Matter* **1** 83.

[437] Matuttis, H. G. and Schinner, A. 1999. Influence of the geometry on the pressure distribution of granular heaps. *Granular Matter* **1** 195.

[438] Moukarzel, C. F. 1998. Isostatic phase transition and instability in stiff granular materials. *Phys. Rev. Lett.* **81** 1634.

[439] Moukarzel, C. F. 2002. Random multiplicative processes and the response functions of granular packings. *J. Phys. Condens. Matter* **14** 2379.

[440] Moukarzel, C. F., Pacheco-Martinaez, H., Ruiz-Suarez, J. C., and Vidales, A. M. 2004. Static response in disk packings. *Granular Matter* **6** 61.

[441] Mueggenburg, N. W., Jaeger, H. M., and Nagel, S. R. 2002. Stress transmission through three-dimensional ordered granular arrays. *Phys. Rev.* **E 66** 031304.

[442] Narayan, O. and Nagel, S. R. 1999. Incipient failure in sandpile models. *Physica* **A 264** 75.

[443] Nedderman, R. M. 1992. *Statics and kinematics of granular materials*, (Cambridge: Cambridge University Press).

[444] Ngan, A. H. W. 2003. Mechanical analog of temperature for the description of force distribution in static granular packings. *Phys. Rev.* **E 68** 011301; 2003. Mechanical analog of temperature for the description of force distribution in static granular packings. *Phys. Rev.* **E 68** 069902; see also the comment of Metzger, P. T., 2004. *Phys. Rev.* **E 69** 053301.

[445] Nguyen, M. L. and Coppersmith, S. N. 1999. Properties of layer-by-layer vector stochastic models of force fluctuations in granular materials. *Phys. Rev.* **E 59** 5870.

[446] Nguyen, M. L. and Coppersmith, S. N. 2000. Scalar model of inhomogeneous elastic and granular media. *Phys. Rev.* **E 62** 5248.

[447] O'Hern, C. S., Langer, S. A., Liu, A. J., and Nagel, S. R. 2001. Force distributions near jamming and glass transitions. *Phys. Rev. Lett.* **86** 111.

[448] Oron, G. and Herrmann, H. J. 1998. Exact calculation of force networks in granular piles. *Phys. Rev.* **E 58** 2079.

[449] Ostojic, S. and Panja, D. 2005. Response of a hexagonal granular packing under a localized external force. *Europhys. Lett.* **71** 70.

[450] Otto, M., Bouchaud, J. P., Claudin, P., and Socolar, J. E. S. 2003. Anisotropy in granular media: classical elasticity and directed force chain networks. *Phys. Rev.* **E 67** 031302.

[451] Ovarlez, G. and Clément, E. 2005. Elastic medium confined in a column versus the Janssen experiment. *Eur. Phys. J.* **E 16** 421.

[452] Ovarlez, G., Fond, C., and Clément, E. 2003. Overshoot effect in the Janssen granular column: a crucial test for granular mechanics. *Phys. Rev.* E **67** 060302(R).

[453] Radjai, F., Jean, M., Moreau, J. J., and Roux, S. 1996. Force distributions in dense two-dimensional granular systems. *Phys. Rev. Lett.* **77** 274.

[454] Radjai, F., Roux, S., and Moreau, J. J. 1999. Contact forces in a granular packing. *Chaos* **9** 544.

[455] Radjai, F., Troadec, H., and Roux, S. 2003. Micro-statistical features of cohesionless granular media. *Italian Geotechnical Journal* **3** 39.

[456] Radjai, F., Wolf, D., Jean, M., and Moreau, J. J. 1998. Bimodal character of stress transmission in granular packings. *Phys. Rev. Lett.* **80** 225.

[457] Reydellet, G. and Clément, E. 2001. Green's function probe of a static granular piling. *Phys. Rev. Lett.* **86** 3308.

[458] Rothenburg, L. and Selvadurai, A. P. S. 1981. A micromechanical definition of the Cauchy stress tensor for particulate media. *Proc. Int. Symp. Mechanical Behaviour of Structured Media*, 469.

[459] Roux, J. N. 2000. Geometric origin of mechanical properties of granular materials. *Phys. Rev.* E **61** 6802.

[460] Serero, D., Reydellet, G., Claudin, P., Clément, E., and Levine, D. 2001. Stress response function of a granular layer: quantitative comparison between experiments and isotropic elasticity. *Eur. Phys. J.* E **6** 169.

[461] Silbert, L. E., Grest, G. S., and Landry, J. W. 2002. Statistics of the contact network in frictional and frictionless granular packings. *Phys. Rev.* E **66** 061303.

[462] Šmíd, J. and Novosad, J. 1981. Pressure distribution under heaped bulk solids. *Ind. Chem. Eng. Symp.* **63** D3V 1.

[463] Snoeijer, J. H. and van Leeuwen, J. M. J. 2002. Force correlations in the q-model for general q-distributions. *Phys. Rev.* E **65** 051306.

[464] Snoeijer, J. H., van Hecke, M., Somfai, E., and van Saarloos, W. 2003. Force and weight distributions in granular media: effects of contact geometry. *Phys. Rev.* E **67** 030302(R).

[465] Snoeijer, J. H., van Hecke, M., Somfai, E., and van Saarloos, W. 2004. Packing geometry and statistics of force networks in granular media. *Phys. Rev.* E **70** 011301.

[466] Snoeijer, J. H., Vlugt, T. J. H., Ellenbroek, W. G., van Hecke, M., and van Leeuwen, J. M. J. 2004. Ensemble theory for force networks in hyperstatic granular matter. *Phys. Rev.* E **70** 061306.

[467] Socolar, J. E. S. 1998. Average stresses and force fluctuations in noncohesive granular materials. *Phys. Rev.* E **57** 3204.

[468] Tanguy, A., Wittmer, J. P., Leonforte, F., and Barrat, J. L. 2002. Continuum limit of amorphous elastic bodies: a finite-size study of low-frequency harmonic vibrations. *Phys. Rev.* B **66** 174205.

[469] Tatsuoka, F., Sato, T., Park, C. S., Kim, Y. S., Mukabi, J. N., and Kohata, Y. 1994. Measurements of elastic properties of geomaterials in laboratory compression tests. *Geotechnical Testing Journal* **17** 80.

[470] Thornton, C. and Antony, S. J. 1998. Quasi-static deformation of particulate media. *Phil. Trans. R. Soc. Lond.* A **356** 2763.

[471] Tixier, M., Pitois, O., and Mills, P. 2004. Experimental impact of the history of packing on the mean pressure in silos. *Eur. Phys. J.* E **14** 241.

[472] Tkachenko, A. V. and Witten, T. A. 1999. Stress propagation through frictionless granular material. *Phys. Rev.* E **60** 687.

[473] Tkachenko, A. V. and Witten, T. A. 2000. Stress in frictionless granular material: adaptative network simulations. *Phys. Rev.* E **62** 2510.

[474] Tsoungui, O., Vallet, D., and Charmet, J. C. 1998. Use of contact area trace to study the force distributions inside 2D granular systems. *Granular Matter* **1** 65.

[475] Vanel, L., Claudin, P., Bouchaud, J. P., Cates, M. E., Clément, E., and Wittmer, J. P. 2000. Stresses in silos: comparison between theoretical models and new experiments. *Phys. Rev. Lett.* **84** 1439.

[476] Walsh, S. D. C. and Tordesillas, A. 2004. The stress response of a semi-infinite micropolar granular material subject to a concentrated force normal to the boundary. *Granular Matter* **6** 27.

[477] Walton, K. 1987. The effective elastic moduli of a random packing of spheres. *J. Mech. Phys. Solids* **35** 213.

[478] Williams, J. C., Al-Salman, D., and Birks, A. 1987. Measurement of static stresses on the wall of a cylindrical container for particulate solids. *Powder Tech.* **50** 163.

[479] Wittmer, J. P., Cates, M. E., and Claudin, P. 1997. Stress propagation and arching in static sandpiles. *J. Phys. (France) I* **7** 39.

[480] Wittmer, J. P., Claudin, P., Cates, M. E., and Bouchaud, J. P. 1996. An explanation for the central stress minimum in sandpiles. *Nature* **382** 336.

[481] Wittmer, J. P., Tanguy, A., Barrat, J. L., and Lewis, L. 2002. Vibrations of amorphous, nanometric structures: when does continuum theory apply? *Europhysics Lett.* **57** 423.

[482] Wood, D. M. 1990. *Soil Behaviour and Critical State Soil Mechanics*, (Cambridge: Cambridge University Press).

[483] Goldhirsch, I., Noskowicz, S. H., and Bar-Lev, O. 2005. Nearly smooth granular gases. *Phys. Rev. Lett.* **95**, Art. no. 068002.

[484] Goldhirsch, I., Noskowicz, S. H., and Bar-Lev, O. 2005. Hydrodynamics of nearly smooth granular gases. *J. Phys. Chem.* **B 109**, 21449–21470.

Index

Printed in the United States
by Baker & Taylor Publisher Services